Creativity for Scientists and Engineers
A practical guide

Online at: https://doi.org/10.1088/978-0-7503-4967-3

Creativity for Scientists and Engineers
A practical guide

Dennis Sherwood
The Silver Bullet Machine Manufacturing Company Limited, Rutland, UK

IOP Publishing, Bristol, UK

© IOP Publishing Ltd 2022

All rights reserved. No part of this publication may be reproduced, stored in a retrieval system or transmitted in any form or by any means, electronic, mechanical, photocopying, recording or otherwise, without the prior permission of the publisher, or as expressly permitted by law or under terms agreed with the appropriate rights organization. Multiple copying is permitted in accordance with the terms of licences issued by the Copyright Licensing Agency, the Copyright Clearance Centre and other reproduction rights organizations.

Certain images in this publication have been obtained by the authors from the Wikipedia/Wikimedia website, where they were made available under a Creative Commons licence or stated to be in the public domain. Please see individual figure captions in this publication for details. To the extent that the law allows, IOP Publishing disclaim any liability that any person may suffer as a result of accessing, using or forwarding the image(s). Any reuse rights should be checked and permission should be sought if necessary from Wikipedia/Wikimedia and/or the copyright owner (as appropriate) before using or forwarding the image(s).

Permission to make use of IOP Publishing content other than as set out above may be sought at permissions@ioppublishing.org.

Dennis Sherwood has asserted his right to be identified as the author of this work in accordance with sections 77 and 78 of the Copyright, Designs and Patents Act 1988.

ISBN   978-0-7503-4967-3 (ebook)
ISBN   978-0-7503-4965-9 (print)
ISBN   978-0-7503-4968-0 (myPrint)
ISBN   978-0-7503-4966-6 (mobi)

DOI   10.1088/978-0-7503-4967-3

Version: 20221001

IOP ebooks

British Library Cataloguing-in-Publication Data: A catalogue record for this book is available from the British Library.

Published by IOP Publishing, wholly owned by The Institute of Physics, London

IOP Publishing, No.2 The Distillery, Glassfields, Avon Street, Bristol, BS2 0GR, UK

US Office: IOP Publishing, Inc., 190 North Independence Mall West, Suite 601, Philadelphia, PA 19106, USA

To all those who wish to 'make the world a better place', and who might find this book helpful in achieving that.

Contents

Preface	xvi
Acknowledgements	xix
Author biography	xxi
Foreword	xxii
Introduction	xxiv
Prologue	**0-1**
The right answer	0-2
Kepler's *BIG IDEA*	0-2
The egg-shaped orbit	0-4
0.00429 and 5° 18′	0-6
What happened next	0-9
Kepler re-visited	0-10
Rudolf II	0-21
Tycho Brahe	0-22
Johannes Kepler	0-23
Kepler's ellipse	0-24
References	0-25

Part I   Koestler's Law of Creativity

1   What, precisely, is creativity?   **1-1**

1.1	Some dictionary definitions	1-1
1.2	My 'sound-bite' definition – just five words	1-2
1.3	Ideas as outcomes, ideas as questions	1-3
1.4	Invention and discovery	1-5
1.5	What's missing from the sound-bite?	1-6
1.6	What is 'new'?	1-6
1.7	It's difference that's important, not novelty…	1-8
1.8	…and the best way to discover differences is to be observant	1-9
1.9	Value	1-12
	References	1-14

2   Creativity in context   **2-1**

2.1	Creativity alone is not enough	2-1

2.2	A richer picture	2-4
2.3	Process 1 – Creativity	2-7
2.4	Process 2 – Evaluation	2-9
2.5	Processes 3 and 4 – Development and Implementation	2-11
2.6	The Target Diagram and skills	2-11
	References	2-14

3   The six domains of creativity   3-1

3.1	Creativity is not just about 'the better mousetrap'	3-1
3.2	Content	3-2
3.3	Process	3-2
3.4	Strategy	3-3
3.5	Structures	3-3
3.6	Relationships	3-4
3.7	You!	3-5
3.8	The importance of the organisational culture	3-6

4   Koestler's Law   4-1

4.1	Arthur Koestler's definition of creativity	4-1
4.2	The 'Eureka moment' myth	4-2
4.3	'But I'm not a creative person'	4-3
4.4	Creativity is all about patterns	4-3
4.5	'Bisociation' and 'thinking aside'	4-7
4.6	The more familiar the parts, the more striking the new whole	4-8
4.7	What Koestler's Law does, and doesn't, do	4-10
4.8	The Koestler Challenge	4-12
	Arthur Koestler	4-13
	References	4-14

5   Some more examples of Koestler's Law   5-1

5.1	Literature	5-1
5.2	Art	5-2
5.3	Chemistry	5-4
5.4	How chemistry made impressionist art happen…	5-4
5.5	…and how physics has facilitated contemporary art	5-7
5.6	History, politics, philosophy, and economics	5-9
5.7	Newton's Laws of Motion and Gravitation	5-11

5.8	A brief digression – coincidence, co-invention and the zeitgeist	5-12
5.9	The light bulb	5-15
5.10	Casa Batlló	5-18
5.11	The DC electric motor	5-20
5.12	The impossible building	5-22
5.13	Special relativity	5-26
5.14	The structure of DNA	5-30
	5.14.1 DNA and X-ray crystallography	5-30
	5.14.2 Photo 51	5-31
	5.14.3 Florence Bell	5-32
	5.14.4 Sven Furberg	5-33
	5.14.5 The diffraction pattern of a helix	5-36
	5.14.6 James Watson	5-36
	5.14.7 The triple helix	5-37
	5.14.8 The missing Koestler's Law fragment	5-37
	5.14.9 Jerry Donohue	5-39
	5.14.10 The Nobel Prize	5-40
	5.14.11 A moment to reflect	5-41
5.15	DNA – a final word	5-44
	Thomas Edison	5-46
	Erwin Chargaff	5-47
	References	5-48

Part II How to have great ideas, deliberately

6 The 'da Vinci problem' 6-1

6.1	Building on Koestler's Law	6-1
6.2	The helicopter that couldn't fly	6-3
6.3	The problem of the missing component	6-5
6.4	You might be a 'victim', now	6-6
6.5	Identify the missing component(s) as precisely as you can	6-7
6.6	Keep your eyes – and ears – open	6-10
6.7	Be patient	6-12
6.8	In conclusion	6-13
	Leonardo da Vinci	6-14
	References	6-15

7 Emergence – why some patterns are better than others 7-1
7.1 Emergence 7-1
7.2 Same components, different patterns 7-2
7.3 Not too little, not too much 7-2
7.4 Patterns within patterns 7-3
7.5 Emergence is often subjective 7-4
7.6 An enriched definition of creativity 7-7
References 7-8

8 Knowledge, experience, learning, and unlearning 8-1
8.1 Where are the Koestler's Law 'components'? 8-1
8.2 Donald Hebb's Theory of Learning 8-3
8.3 The learning trap 8-7
8.4 Unlearning 8-8
8.5 Why is unlearning so difficult? 8-11
   8.5.1 Love 8-11
   8.5.2 Laziness 8-11
   8.5.3 Fear 8-11
   8.5.4 Arrogance 8-12
   8.5.5 Unlearning is indeed difficult... 8-12
8.6 Hegel, and genetics 8-13
8.7 A brief pause… 8-14
Donald Hebb 8-16
References 8-17

9 How to have great ideas 'on demand' 9-1
9.1 *InnovAction!* 9-1
9.2 Step 1: Define the 'focus of attention' 9-2
9.3 Step 2: Individually and in silence, write down everything you know about the agreed focus of attention 9-4
9.4 Step 3: Share 9-7
   9.4.1 What the 'share' is about 9-7
   9.4.2 Managing the group dynamics 9-7
   9.4.3 Self-facilitate 9-8
   9.4.4 One point at a time 9-9
   9.4.5 Writing on the flip-chart 9-9
   9.4.6 Be succinct 9-9

		9.4.7 Don't worry about structure	9-10
		9.4.8 Don't argue…	9-10
		9.4.9 …but do ask questions	9-11
		9.4.10 Don't duplicate points already captured	9-11
		9.4.11 Don't start generating ideas (yet!)	9-11
		9.4.12 And eventually…	9-11
	9.5	Step 4: Then choose one feature, and ask 'How might this be different?'	9-12
	9.6	Step 5: Let it be…	9-13
	9.7	Step 6: …and then, when that discussion runs out of steam, choose another feature and repeat steps 4 and 5	9-14
	9.8	The nine dots puzzle revisited	9-15
		References	9-15
10	***InnovAction!* in action**		**10-1**
	10.1	Ideas for games based on chess	10-1
	10.2	Some things we know about chess	10-1
	10.3	Ideas, ideas, ideas…	10-4
		10.3.1 Self-capture	10-4
		10.3.2 'Hobbled' bishops	10-4
		10.3.3 Taking turns	10-4
		10.3.4 Square occupancy	10-4
		10.3.5 The initial layout	10-5
		10.3.6 Combinations	10-5
	10.4	It really is as simple as that!	10-5
	10.5	The central step – Step 4: 'How might this be different?'	10-6
		10.5.1 Three key rules	10-6
		10.5.2 Don't be negative	10-6
		10.5.3 Don't lose the key questions…	10-7
		10.5.4 …and keep a record of all the ideas	10-7
	10.6	Different ways of being different	10-7
		10.6.1 Size and scale	10-7
		10.6.2 Sequence, flow and configuration	10-8
		10.6.3 Function and scope	10-8
		10.6.4 Roles and responsibilities	10-9
	10.7	Some examples	10-9
		Reference	10-13

11 Springboards and retro-fits — 11-1

11.1 *InnovAction!* is not the only way to have idea 'on demand' — 11-1
    11.1.1 The 'mountains and valleys' metaphor — 11-1
    11.1.2 *InnovAction!* – a springboard — 11-2
11.2 Some other springboards — 11-2
    11.2.1 Challenge assumptions — 11-2
    11.2.2 Decomposing and recombining — 11-3
    11.2.3 Edward de Bono's 'PO' — 11-3
11.3 Random words – a retrofit — 11-5
11.4 Some other retro-fits — 11-8
    11.4.1 PO-2 — 11-8
    11.4.2 Simile, metaphor and analogy — 11-8
    11.4.3 Other people's shoes — 11-9
    11.4.4 Journeying — 11-9
    11.4.5 Visioning — 11-10
    11.4.6 Working backwards — 11-10
11.5 Springboards and retrofits – which to use? — 11-11
    References — 11-12

12 Creativity workshops — 12-1

12.1 Observation, curiosity and permission made real — 12-1
12.2 The workshop themes — 12-2
12.3 Who should participate? — 12-3
12.4 How workshops are structured — 12-5
    12.4.1 Why off-site, and why two days? — 12-5
    12.4.2 The workshop agenda — 12-5
    12.4.3 Longer, and shorter, durations — 12-7
12.5 The idea generation group briefs — 12-7
12.6 Don't impose constraints on cost and resources — 12-20
12.7 Creativity, not evaluation — 12-21
12.8 Quantity, quantity, quantity — 12-21
12.9 After the workshop — 12-22
    12.9.1 The workshop report — 12-22
    12.9.2 People keep thinking — 12-23
    12.9.3 The next step – wise evaluation — 12-23
    References — 12-24

13 Creativity in science and engineering   13-1

13.1 What this chapter is about   13-1
13.2 Detecting gravitational waves   13-1
13.3 Building Nemo   13-8
13.4 Synthetic synapses   13-12
13.5 Biomimetic adhesives   13-15
13.6 The magic colouring sheet   13-18
13.7 Quantum entanglement, single-pixel cameras, and novel endoscopes   13-22
13.8 Keeping the UK's railways safe   13-26
13.9 The 'Medusa Effect'   13-29
13.10 Mixing things up: ellipsometry and strong coupling   13-32
13.11 Reducing noise   13-34
13.12 How nanopatterns made it from a semiconductor facility to an artist's print room   13-37
13.13 Newton's rings and flat screens   13-41
13.14 Blue Plan-it® and Water ARC®   13-44
References   13-47

Part III How to evaluate ideas, wisely

14 Evaluation in context   14-1

14.1 Why wise evaluation is important   14-1
14.2 A very bad idea indeed   14-2
14.3 Not all ideas are good ones…   14-3
14.4 …and even good ideas can be fiercely opposed   14-5
14.5 How do you, and your organisation, evaluate ideas now?   14-6
     14.5.1 Evaluating ideas wisely is really important   14-6
     14.5.2 Some frequently-used methods   14-6
     14.5.3 The boss knows best   14-7
     14.5.4 (Apparently) no process   14-7
     14.5.5 'Trial by ordeal'   14-7
     14.5.6 Likes and dislikes   14-7
     14.5.7 'Pros' and 'cons'   14-8
     14.5.8 Evaluation by numbers   14-9
References   14-10

15 How to evaluate ideas wisely — 15-1

15.1 Features of a wise evaluation process — 15-1
    15.1.1 What does a well-designed evaluation process look like? — 15-1
    15.1.2 Consistency — 15-1
    15.1.3 Fairness — 15-2
    15.1.4 Balance — 15-2
    15.1.5 Completeness — 15-2
    15.1.6 Speed — 15-2
    15.1.7 Pragmatism — 15-3
    15.1.8 Transparency — 15-3
    15.1.9 Openness — 15-3
    15.1.10 How does your organisation's evaluation process rate? — 15-3

15.2 An ideal process for wise evaluation — 15-3
    15.2.1 Overview — 15-3
    15.2.2 Step 1: Imagine the future — 15-4
    15.2.3 Step 2: What are the consequences? — 15-5
    15.2.4 Step 3: The journey — 15-6
    15.2.5 Step 4: How might all the problems be solved? — 15-7
    15.2.6 Step 5: The numbers — 15-7
    15.2.7 Step 6: The decision — 15-9

15.3 The half-way house — 15-11
15.4 Wise evaluation, Edward de Bono's 'hats', and the importance of language — 15-12
15.5 'Evaluation Lite' — 15-14
15.6 And so to development and implementation — 15-17
    References — 15-17

Part IV Building an innovative culture

16 What is 'culture'? — 16-1

16.1 The Covid-19 vaccine miracle — 16-1
16.2 Language — 16-3
16.3 Observation, curiosity and permission revisited — 16-4
16.4 The wider picture – 'enablers' and 'motivators' — 16-5
    References — 16-9

17 Enablers — 17-1

17.1 Budgets — 17-2
17.2 Funding — 17-4
17.3 Managing development and implementation — 17-6
17.4 The idea archive — 17-7
17.5 Physical environment — 17-8
17.6 Behaviours — 17-11
References — 17-12

18 Motivators — 18-1

18.1 Reward and recognition — 18-2
18.2 Performance measures — 18-3
18.3 Training — 18-4
18.4 The role of senior management — 18-6
18.5 Embedding innovation in the day-job — 18-7
18.6 So, what next? — 18-8
References — 18-9

19 Epilogue — 19-1

20 Further reading — 20-1

Index — I-1

Preface

Throughout my entire education, and in my work career too, I have never been taught how to be creative – or, more fundamentally, how to think for myself.

At school and at university, I was pumped full of knowledge, and shown how to solve problems that had already been solved by others. I learnt what happens when acids and alkalis react, how to find the roots of a quadratic equation, how to determine the energy levels of a particle in a square potential well. And being a diligent fellow, I was able to replicate solutions-that-others-had-found-before in my exams, perhaps with some gentle tweaks or stretches.

But when I was doing my PhD, I was tackling a problem that, to my knowledge, no-one had explicitly written about in the books. And I was stuck. I had no idea how to go about *discovering how* to solve the problem. And when my supervisor, in a rare conversation, asked, 'well, what ideas have you got?', I felt a combination of threat and shame. *I didn't have any! That's why I was stuck!* My supervisor then shrugged his shoulders and went away, and I stumbled on. Eventually, I stumbled over something that worked, and wrote my thesis accordingly. Phew!

And in my career as a partner in a large consulting firm, the same thing happened. All the work was about doing things others had done before elsewhere, but not for this particular client. That helped the client. But frustrated me. I wanted to work with my clients – as appropriate of course – to do something new, so that they could obtain a real competitive advantage, not just copy something from somewhere else to catch up.

But how could I do that? How could I stand in front of a client and say, 'let's do something new'? For that required that I, and my team, could come up with novel ideas; and not just novel ones, but ones that would be of genuine benefit to the client too.

At that time – around the early 1990s – I believed that ideas 'just happened', and, for the fortunate few, those ideas would be good and 'happen' more frequently. But I was not among those fortunate few, and so there was no way in which I could say to a client 'trust me, an idea will pop up soon...'.

That then posed a question.

Is it possible to have ideas – and good ones – 'on demand', now?

To which, if you believe – as I did then – that ideas 'just happen' (or not), the answer must be 'no'.

But I was still puzzled. So I read shelves of books, I thought a lot, and I had lively conversations with many superbly talented people. That all led me to Arthur Koestler's *Act of Creation*, which provided an important clue that the answer might not be 'no'. As I describe in chapter 4, Koestler's insight was that having ideas – creativity – is not about the out-of-the-blue discovery of something new but about the formation of a new pattern from things that already exist. Ah! That's important. If I want to have an idea, I simply combine together things that already exist, things that are out there, somewhere, now.

That was intriguing, for I had never considered creativity in that way before, and it made a lot of sense – especially with hindsight, for this is exactly what Newton said about 'standing on the shoulders of giants'; it is also consistent with chemistry, for all molecules are just different patterns of the same elements. And music too: Mozart used the same notes as Paul McCartney, but to form very different patterns.

Intriguing, yes; but also tantalising. There were so many loose ends. What are the 'existing things' that I might combine together for the idea I need right now? Where are they? Given that there are so many 'existing things' out there, how do I choose the 'right' ones and ignore the 'wrong' ones? How do I combine them together in the 'right' way?

My head was reeling. I felt Koestler's concept was powerful, but abstract. There was something missing – a way to use 'Koestler's Law' as the guiding principle for a process, a process that I could actually use, for real, with a client, to generate ideas. All that was in my mind for some time...

...and then, in late 1997, I was looking for a birthday present for my elder son, Torben: a two-player board game, *Tri-Tactics*, that I had much enjoyed when I was around his age. Briefly, the board is a 12 × 12 grid, and each player has around 50 cardboard pieces, each with a picture of, say, a battleship on one side, and a plain colour on the other. Before the game starts, the players place their pieces wherever they wish within the five nearest rows, the picture towards the player, and the plain colour towards the opponent. The play of the game is to move the pieces towards one another, and when two opposing pieces are adjacent back-to-back, there is an opportunity to 'challenge', resulting in the removal of the weaker piece. Everything therefore depends on the choice made by each player as to where to position each piece at the outset.

That all seems like a digression, but it is a vital link in the chain. For one afternoon, I found myself in front of a game shop, looking – in vain, as it turned out – for *Tri Tactics*. In the shop window, I saw a chess board, with the pieces neatly laid out in the familiar two rows on each side.

I was a looking at the game of chess, with all the pieces positioned on their specific, pre-determined, squares for the start of every game. I was looking for the game of *Tri Tactics*, in which the key feature is that the pieces are positioned wherever the player wishes. And I was thinking about Koestler.

Chess. *Tri Tactics*. Koestler. All in my head at the same time.

And then... suppose, like in *Tri Tactics*, the chess pieces could be placed in *different starting positions, as the player might wish?* If the key concept of *Tri Tactics* were combined with the game of chess? This would give a new initial pattern of chess pieces, and the subsequent play of the game would be different.

That new pattern of pieces is exactly as Koestler describes – a new pattern of existing components. And the outcome is a new idea, a different evolution of the game, even if all the other rules of chess – such as how each piece can move – remains the same.

That also defines a process of discovery. Firstly, to notice a feature of chess-as-we-know-it – for example, that all the pieces have pre-determined starting positions – and then to ask 'how might this be different?' to 'force' a new pattern, one being that the positions of the knights and bishops might be swapped. And once that possibility is in your mind, it is very easy to think of many others, each of which would result in a different game. Like what might happen if the feature of chess that 'pawns capture diagonally forwards' is made different by adopting the rule 'pawns capture immediately in front'… Yes…

Over the following weeks, I developed the concept further, and crafted it into the *InnovAction!* process for idea generation as described in chapter 9. I also had the benefit of working with some clients who were willing to give it a try. And it worked! At last, ideas on demand!

Since then, with the help of many good people, and many invigorating clients, I have refined the process, and have also developed some other associated ones too, such as a method for evaluating ideas wisely, as described in chapter 15.

Which is what this book is all about. Deliberate processes for the discovery of ideas, and also to evaluate them wisely, all operating within an organisation's culture. But not just rules-of-thumb, or arbitrary-seeming cook-book recipes, for I also describe *why* these processes work, giving 'first-principles' explanations so that they can be applied in any context. Along the way, I tell many stories about real ideas from the past, from Kepler's discovery of the elliptical orbit of Mars in 1605 to the unlocking of the secret of DNA's double helix by Watson and Crick in 1953. To bring things up-to-date, in chapter 13, some current-day, highly distinguished, scientists and engineers tell their stories too: my thanks to them all.

And my thanks to you too for picking this book up. I do hope you will find it engaging, interesting, informative. And, most importantly, useful too. If only, when I was doing my PhD, I had known that when I was stuck, I should have observed, very carefully, what I was doing, and identified, very precisely, all the assumptions I was making. Having done that, had I asked 'how might this be different?', I would have avoided all that stumbling around.

Dennis Sherwood
Exton, Rutland
July 2022

Acknowledgements

Creativity – which to me is fundamentally the capability and confidence to think cogently for oneself – is an immensely valuable attribute. And it has been my privilege, over the last 30 years or so, to work with many individuals, teams and organisations to think, hard, about just what 'creativity' is, to develop reliable, effective, methods to generate ideas 'on demand' and to evaluate them wisely, and to put those methods into practice for real. All have contributed to my own thinking and understanding, and I acknowledge and thank everyone with whom I have interacted. But let me mention in particular those whose knowledge, help and challenge have been especially influential: Alison Bott, Professor Bill Barnes, David Waller, Judith Hackett, Professor Nicholas Higham FRS, Mark Selway, Professor Miles Padgett FRS, Dame Muffy Calder, Rob Eastaway, Tony Cohen, to name just a few. Thanks too to the UK's Engineering and Physical Sciences Research Council (EPSRC), whose *Creativity@home* programme stimulated an interest in creativity across the scientific research community, and provided funding for many projects I have undertaken with academic groups across the country.

A distinctive feature of this book is the collection of 'guest stories' in chapter 13, in which the authors describe their own experience of creativity in their respective, and very varied, fields. It was most gracious of all these people to take the time and trouble to contribute in this way, so particular thanks to:

- Adam Middleton, Siemens BV
- Professor Anatoly Zayats, King's College London
- Professor Bill Barnes, University of Exeter
- Charles Williams, GGB LLC
- Colin Haynes, Alstom
- Professor Harish Bhaskaran, University of Oxford
- Dr Hermione Cockburn, Dynamic Earth, Edinburgh
- Professor Sir James Hough FRS, University of Glasgow
- Dr Jess Brown, Charlie He and Dr Justin Sutherland, Carollo Engineers Inc.
- Mike Semens-Flanagan, IMI Critical Engineering
- Professor Miles Padgett FRS, University of Glasgow
- Professor Nikolaj Gadegaard, University of Glasgow
- Professor Rob Jenkins, University of York
- Professor Sheila Rowan FRS, University of Glasgow

I also thank those who granted permission for me to reproduce materials for which they hold the copyright: all are explicitly identified and acknowledged. Thanks too to all in the production team at the Institute of Physics.

And profound thanks – of course – to my wife, Anny, and two sons, Torben and Torsten.

Throughout the book, I make many statements that are not based on my own experience, but are from information I have learnt or read. I have therefore included many references to my sources, including links to websites – all of which were accessed successfully on 6 December 2021.

Author biography

Dennis Sherwood

For the last twenty years or so, I have been running my own consulting firm, The Silver Bullet Machine Manufacturing Company Limited, which – despite the somewhat unconventional name – does 'what it says on the tin'. 'Silver Bullet' is a metaphor for a great idea, but for an organisation to rely on a single 'silver bullet' is fragile. Far better to have a 'machine' that 'builds' them, so that great ideas can be generated again and again, whenever and wherever they might be helpful. A 'Silver Bullet Machine', of course, is not a physical piece of kit like a lathe or a spectrometer; rather, it's an organisational capability. So my firm helps enterprises 'manufacture' that capability, hence the name. And it has been great to work with organisations of all scales in all sectors, from retailers to universities, from media companies to engineers, to do just that. Our role is not to 'have the ideas' – that's what the clients do, for the ideas must be their own. What we do is to orchestrate the process. Which works.

Some background… over the years, I have been the Managing Director of the UK Operations of the SRI Consulting, an Executive Director at Goldman Sachs, and a consulting partner firstly in Deloitte Haskins + Sells, and then Coopers & Lybrand. And before that, I studied for a PhD in Biology at the University of California, San Diego, and an MPhil in Molecular Biophysics and Biochemistry at Yale, having read Natural Sciences (Part II Physics) at Clare College, Cambridge.

I am a great 'evangelist' about creativity and innovation, so I enjoy giving presentations and participating in conferences; I write quite a lot too – I've published many articles, and this is book number 15 (not all of these are about creativity – some of the others are on subjects such as thermodynamics and X-ray crystallography!).

Foreword

It is rare that our education includes any course on creativity, and even rarer to be taught ways of having ideas, yet this is one of the most important aspects of life. Having such a creative capability is surely one of the most effective activities to enhance our world. This is revolutionary! Ideas change the world.

The creative ability to have significant ideas on-demand seems unrealistic, and a healthy scepticism would seem to be a reasonable response to such a suggestion. There is something almost magical about a great idea suddenly occurring to us without conscious effort. Perhaps there is a feeling that this is a delicate process that should not be interfered with in case any subsequent creative ability will be damaged.

I had had the feeling that I might lose the dynamic of my own creativity if the process could somehow be codified and written down. Ideas can come to us in a flash, seemingly outside our control.

Where do ideas come from? From our subconscious? The origins of creativity have been thought about for millennia. The Ancient Greeks ascribed creative ideas not to humans but to Muses, some unknowable distant divinity – this is not so far from the general perception of where our creative ideas come from until recently… In this book, a practical technique for generating and assessing ideas is given along with many examples. The method is distilled from modern studies of the nature of creativity, and finally synthesised and extended by Dennis Sherwood, the founder of The Silver Bullet Machine Manufacturing Company.

I first encountered this method of creativity when asked by the EPSRC, the UK's main funding agency for engineering and physical sciences, to trial Dennis's technique at a workshop for my research group. I was dubious that creativity could be taught but prepared to try what was being offered. I have never been so happy to be proven wrong.

As a group activity, the process was endlessly rewarding and so much fun. It was indeed possible to have useful and exciting ideas to order and the pleasure of arriving at them was as good as ever. And it didn't just stay within the workshop. My group went on to use Dennis's method for both research ideas and also for better logistical planning of group resources. Personally, even after years of experience in atomic force microscopy (AFM), and learning and writing about ideas and new developments for grants, publications and even spin-out companies, this gave me a new toolkit and allowed me to discover how things could be different. Indeed, it led to a new, much simplified, patentable type of AFM.

This book is in a physics series but is equally relevant to other sciences and to engineering. Many of the example cases work in a wider context. This begs the question of whether creativity outside science and engineering, for example, in the humanities, is fundamentally different and whether this method of asking 'how could this be different?' would work. After reading this, I believe it could.

The quality of this book is outstanding. It's a delight to read and gives a set of exciting priceless tools with excitement and so interesting, but the real buzz is in doing – experiencing the process and the birth of ideas for yourself. In reading this book, it feels like Dennis is right there with you. All of his expertise is in this book – it's universal but it's personal.

Mervyn Miles FRS
Emeritus Professor of Physics
University of Bristol

Introduction

Successful scientists and engineers are highly creative – creativity goes with the job.

Creativity, though, is often believed to be 'magical', 'serendipitous', 'intuitive'; something that just 'happens' to 'creative people'.

That's all great, provided that the ideas flow. But sometimes they don't, at which point it's harshly evident that relying on being 'intuitive' is fragile. Quite often, we need a useful idea *now*. For that to happen, hoping for 'magic' just isn't good enough – we need a process, a set of well-defined steps we can follow to find creative solutions to problems; a process we can consciously apply when intuition seems to have dried up; a process we can teach to those who are less experienced or less confident in creative thinking.

The good news is that such a process exists. Here is a thumbnail sketch of how it works, and how you, too, can have great ideas. It can be carried out by an individual, but it works particularly well in small groups:

How to have great ideas – now!

1. **Define the topic, or area of interest, for which an idea would be helpful. This might be a problem to be solved, a method to be improved, a process to be made more efficient, a product to be improved...**

2. **Individually and in silence, write down everything you know about the topic, including even the most trivial details.**

3. **If working in a group, share the results of step 2, so compiling a comprehensive list of features.**

4. **Choose a feature, and ask 'How might this be different?' or 'What might happen if this assumption is removed or changed?', and follow through the consequences.**

5. **Repeat step 4 for another feature identified in steps 2 and 3.**

That's just the barest outline. As can be appreciated, this process is simple, and very easy to use. But despite its simplicity, it's underpinned by rigorous principles, principles that get to the heart of what creativity actually *is*, principles that explain how it is possible to make creativity happen, deliberately, now. These principles, and the details of the process, are described fully in chapters 1–12, and I sincerely hope that reading these will build your confidence in using the process to generate your own ideas.

I fully appreciate that it's very natural – especially for scientists and engineers – to be sceptical not only about the process I have just sketched, but even more so about the possibility that a 'process' can actually be used for something as apparently ephemeral as 'creativity'. But the success of the many creativity workshops I've run over the last twenty years provides abundant evidence of the process's effectiveness, productivity and power. Throughout the book, I give examples that illustrate the practical application of both the principles and the process, and in chapter 13, there are some further examples kindly contributed by some of today's leading scientists and engineers, drawing on their own experience. I trust that reading these examples will be both interesting and inspiring, and will encourage you to try the process out yourself, and (better) with colleagues. That will give you your own evidence that the process works. And once you've done that, your scepticism will be diminished and you'll have the confidence to use the process again and again, whenever and wherever a good idea would be useful.

But how do you know that any particular idea is 'good'? You might like the idea, but that's rather different... As I'll describe, being creative – having an idea – is not the whole story, for once an idea has been discovered, the next step is to determine whether it's 'good' or 'not so good'. I call this 'wise evaluation', and this too has its own process, as detailed in chapters 14 and 15.

Being creative, and evaluating ideas wisely, can be done by a lone individual working in isolation, but in my view, that is not only sub-optimal (the processes work better in small groups) but also very rare, for most of us work in organisations. Creativity and evaluation are therefore necessarily embedded within an organisational context, and that context – as made real by the organisational culture – has a significant influence. That influence can encourage creativity. Or kill it. Stone dead.

'Culture' though, is a slippery concept, hard to pin down, and perhaps regarded as something that the Vice Principal, Chief Executive and HR people have to fix, not the scientists and engineers, not 'me'. But as shown throughout the book, and especially in chapters 16, 17 and 18, that is not the case: much can be achieved locally to put in place the conditions in which creativity can flourish.

For as the subtitle says, this book is 'A Practical Guide', presenting many pragmatic actions that individuals can take, and that groups and teams can implement, to enhance individual and collective creativity. This does not imply that, right now, individuals and the team are 'not creative' or 'not creative enough'. On the contrary – as I stated right at the start of this Introduction – everyone reading this will already be highly creative. The (high) 'base level', however, is not the point; rather, the point is 'what can I individually, and we collectively, do differently to become even more creative than we are now?' And, as we shall see, the generic question 'How might [this] be different?' – whatever [this] might be – is *the* fundamental question underpinning creativity, the key to discovering how to make the world a better place. Which, for me, is a good thing to do. And if you agree, then I trust you will enjoy reading this book, and will benefit from putting its messages into practice.

IOP Publishing

Creativity for Scientists and Engineers
A practical guide
Dennis Sherwood

Prologue

Prague, 4th February 1605

He sat hunched over his desk, furs wrapped tightly against the bitter winter cold. The candle flame, buffeted by the draft, cast flickering shadows around a room strewn with scraps of crumpled paper, walls lined with books.

He sighed, shook his head. For five years, he had been working, calculating, thinking, drawing, struggling. He had been up countless blind alleys, had proposed and disproved any number of hypotheses, had rejoiced in euphoric glimpses of possible success, only to fall into the abyss of abject gloom.

Many others would have long since given up the quest.
But not this man. He was dogged. Committed. Relentless. Single-minded.

Days, weeks, passed. The he spotted it. Totally by accident.

He'd been looking through some trigonometric tables for something else, and there it was.

1.00429. The secant of 5° 18'.

'Surely that can't be a coincidence,' he thought. 'No. It must have some significance… but what?'

More days. More weeks. More despair.

But then, after 1.00429 and 5° 18' had been churning in his mind, disturbing his sleep, intruding on other thoughts, he spotted it.

Yes. That *must* be the answer. It isn't a circle. It isn't an egg-shape.

It must be an ellipse.

Yes. That *must* be the answer…

The right answer

And it was. An answer that was not only right, but an answer that transformed our understanding of the cosmos, and triggered the scientifically-based world that we continue to live in right now.

In the history of not just mathematics and science, this story is a true landmark of creativity. So what better way to begin a book about how scientists and engineers can enhance their own creativity, and that of their teams? But before we get into all that, let's go behind this story and explore in more detail what actually happened during that chilly winter in Prague in 1605...

Kepler's *BIG IDEA*

The man in the furs was Johannes Kepler, and in 1605, he held the position of 'Imperial Mathematicus' to the Holy Roman Emperor, Rudolf II of the House of Habsburg, who ruled his subjects from Hradčany Castle, high above the river Vltava and the Charles Bridge in Prague, the capital of Bohemia [1].

Rudolf II was (to say the least) somewhat 'strange', and not at all the archetypal renaissance monarch. More interested in arts and science than religion and warfare, Rudolf II liked to collect both 'interesting' things (his 'Cabinet of Curiosities') and 'interesting' people, including the artist Arcimboldo (whose speciality was to paint portraits in which the faces are assembled from an intriguing variety of fruits and vegetables, as shown at the end of this chapter!), the Danish nobleman-astronomer Tycho Brahe, and the central character in our current story, Johannes Kepler.

Brahe, Europe's premier astronomer, arrived in Prague in 1599 to continue the celestial observations he had been conducting over the previous 20 years from the island of Hven, positioned in the narrow waters between Denmark and what is now Sweden. At that time, Kepler was a teacher of mathematics and astronomy at a school in the Austrian city of Graz, but he and Brahe had been in correspondence for several years, for Kepler had received considerable renown for his book *Mysterium Cosmographicum,* published in 1596, when Kepler was 25.

The core of the *Mysterium* was Kepler's *BIG IDEA*.

Kepler had accepted the concept – or rather the rediscovery by Copernicus of a concept first known to have been proposed by the Greek philosopher Aristarchus around 250 BCE – that it is the Sun, not the Earth, that lies at the centre of the Universe.

At that time, there were six known planets: Mercury, Venus, Earth, Mars, Jupiter and Saturn. According to Copernicus, these six planets each rotate about the Sun, suggesting that there are six 'planetary spheres', with each planet following a circular path on the surface of the appropriate sphere.

If there are six planetary spheres, there must be five 'voids' between them. The number 'five' crops up in another context: there are five and only five 'Platonic solids', these being the only three-dimensional closed surfaces composed from identical regular two-dimensional shapes: a tetrahedron has four faces, each an equilateral triangle; a cube has six square faces; an octahedron, eight equilateral triangles; a dodecahedron, 12 pentagons; and an icosahedron, 20 equilateral triangles.

Since these five shapes are regular, each can be circumscribed by a sphere, such that the sphere touches each vertex, whilst simultaneously being inscribed by a sphere of smaller radius, such that the inscribed sphere has each face as a tangential plane.

Put all that together, and you get Kepler's *BIG IDEA* – that the structure of the Universe comprises six spheres, one for each planet, with the five intervening voids containing each of the five Platonic solids, such that the outer planetary sphere is the circumscribed sphere to the appropriate solid whilst the inscribed sphere is that of the next inner planet, as shown in figure P.1.

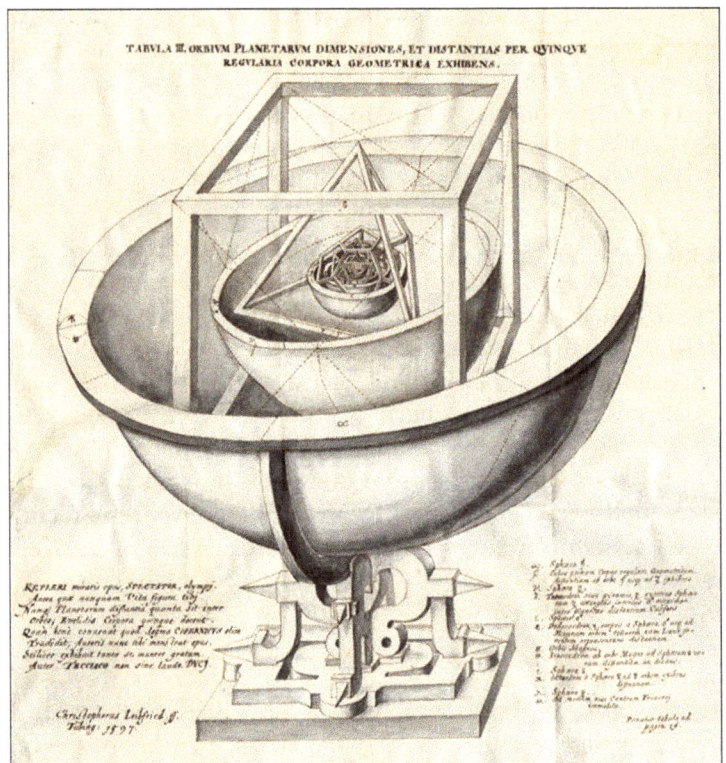

Figure P.1. Kepler's *BIG IDEA*, showing how the five Platonic solids fit neatly within the six planetary spheres.
Source: Wikimedia 'Kepler's Platonic solid model of the Solar System from Mysterium Cosmographicum' https://upload.wikimedia.org/wikipedia/commons/thumb/f/f1/Kepler_Platonic_Solids.tif/lossy-page1-6526px--Kepler_Platonic_Solids.tif.jpg.

To prove his *BIG IDEA*, Kepler needed astronomical data. Which he didn't have, but Brahe did: for the last 20 years, Brahe had amassed a gold-mine of highly accurate astronomical observations of the positions of the planets and the stars, all measured without the use of telescopes (which had not yet been invented), but by using sighting sticks and huge protractors.

On his arrival in Prague, Brahe needed an assistant, and so Brahe invited Kepler to join him. Kepler duly came to Prague in 1600, and was given the task of analysing Brahe's experimental data on his observations of the orbit of Mars.

In October 1601, Brahe died in rather strange circumstances – it has even been conjectured that he was murdered by Kepler [2]. But whatever the circumstances of Brahe's death, Kepler remained in Prague, taking over Brahe's appointment as 'Imperial Mathematicus', and – more importantly – gaining access to many more of Brahe's astronomical observations, beyond those relating to the orbit of Mars.

At last he could analyse accurate data on the movement of all the planets, trace their paths through space, and so prove his *BIG IDEA* that each of the five Platonic solids fits snugly into the five voids between the six planetary spheres.

The egg-shaped orbit

We know now, of course, that Kepler's task was hopeless. The planets do not have circular orbits which define a sphere; rather, the orbits are elliptical – so no regular Platonic solid could ever fit neatly in the ellipsoid-of-revolution voids. And the existence of Uranus and Neptune is something of a problem too, for there are only five Platonic solids, but we now know seven voids.

But in 1601, Kepler had no such knowledge. He sincerely believed that the planets had circular orbits, for the commonly accepted view (and everything that Kepler had been taught) was that the Universe is a perfect structure, and only the circle is a perfect closed curve. Furthermore, the two additional planets would not be discovered until many years later: Uranus, in 1781 by William Herschel [3], and Neptune in 1846, visually by Johann Galle, after the mathematical prediction of its existence, independently, by both John Couch Adams and Urbain le Verrier [4].

So Kepler spent years, from before Brahe's death until 1605, going round in his own circular orbit, getting nowhere, and nowhere again. To test his Platonic solid idea, he firstly had to determine the orbits of each planet around the Sun. He had to start somewhere, so, after Brahe's death, he decided to continue the work he had already begun but not yet completed, the study of the orbit of Mars.

The data he had available were tables compiled by Brahe and his assistants, over many years, giving the positions of many stars, the planets and the Moon, as seen in the night sky from his observatory on the Danish island of Hven. For each observation, there were three data items: the time at which that measurement was taken, and also two angular measurements defining either the celestial 'latitude' and 'longitude' of the astronomical object being studied, or the north/south and east/west angular distances between, say, a planet and a nearby fixed star. Given that the observatory on Hven was itself moving through space (by virtue of both the daily rotation of the Earth about its axis, and also the rotation of the Earth about the Sun), and that Mars is moving too (along its orbit around the Sun), the task of computing the orbit of Mars in relation to the 'fixed stars' was prodigious indeed.

All the calculations were by hand – of course there were no calculating or computing machines, and no logarithms either, for it wasn't until 1614 that they were invented [5] (or is it discovered? I'll return to that question later in the book!). But even if Kepler was able to benefit from the long-forgotten calculating method rejoicing in the name 'prosthaphaeresis' [6] – as is quite likely, for this is known to have been used by Brahe, and one of its originators, Joost Bürgi moved to Prague in 1604 and worked closely with Kepler – the number of numbers he had to crunch was huge, and the process of crunching was laborious, time-consuming, and highly error-prone.

Most people, without any doubt, would have given up after a few failures. But not Kepler. He struggled on and on and on…

From our current-day perspective, it is unremarkable that Kepler failed to prove his *BIG IDEA*, for we now know that his task was impossible. What is truly amazing, however, is what Kepler actually did discover, despite the fact he wasn't looking for those results! Even more amazing is that in discovering those results, he had to *throw away* his most cherished beliefs. Not just the belief in his *BIG IDEA*. But his belief in circular orbits too. And his belief that the Sun was at the centre of those orbits. As we shall see, 'being in love with one's own ideas' is one of the major barriers to creativity, and Kepler had every reason to be 'in love' with circular orbits and a (literally) heliocentric universe. But as his analysis, his mathematics and his reason showed again and again that the measured reality did not fit the data, he changed his mind.

That is truly remarkable. For the tendency of most humans is not to change one's mind, but – whilst avoiding the temptation to falsify the data – to take the more subtle approach of changing the *interpretation* of the data, or to point out that uncertainties in the data imply that the observed results are still 'totally consistent with my idea'.

For which Kepler had every opportunity. Take, for example, his growing realisation that the orbit of Mars was not a circle, but something else. One of his methods was to determine the position of Mars from Brahe's observations at, say, three positions around its orbit. Those three points define a circle, from which he could predict the position of Mars in other positions around its orbit. He could then check his predictions against Brahe's observations, and see whether or not his predictions were valid. In fact, he found discrepancies. And some of those discrepancies were very small, just a few *minutes* of arc.

But instead of saying 'those observations must be ever-so-slightly in error – after all, it was probably a bit misty that night when Brahe made them, perhaps he didn't read his protractor quite right, or perhaps the sighting stick wasn't quite straight', he firstly said 'Ah! I'll try again, for some other positions' (and again and again...), until he ultimately said 'This isn't working. Maybe my belief that the orbit is circular is wrong... perhaps the orbit is something else'.

But what might that 'something else' be?

He had three clues. Since ancient times, it had been known that Mars did not wander arbitrarily across the sky, but maintained an orbit in a well-defined celestial plane, returning to the same position every few years. The orbit must therefore be a closed curve. Secondly, he noticed that the discrepancies between his predictions of where Mars 'should' be if the orbit were circular, and the corresponding actual positions of Mars as determined from Brahe's observations, consistently indicated that the actual position was within the circle, not outside it. The orbit must therefore be some form of 'squashed circle'.

The third clue came from Kepler's measurements of the planet's speed, for he had established that this varied at different positions of Mars around its orbit, being fastest when Mars was closest to the Sun (the position known as 'perihelion'), and slowest when furthest from the Sun ('aphelion'). The speed is therefore 'lopsided'.

Putting these inferences together, Kepler asked the question 'what closed curve is a lopsided squashed circle?' His answer was 'egg-shaped'.

That was wrong too.

0.00429 and 5° 18′

Having come to terms with the likelihood that the orbit of Mars was a 'squashed' circle, Kepler asked 'how 'squashed' is it?' To answer this question, Kepler could identify the locations of Mars at the perihelion, and diametrically opposite at the aphelion, and draw the circle that Mars would follow between those two known points, if the orbit were in fact circular. He then used Brahe's observations to determine the position of Mars when exactly half-way between the perihelion and aphelion, and measured how far the actual position was from the corresponding position on the circle.

Kepler had also known for some time that the position of the Sun was not, as might have been expected, at the centre of the orbit – whether circular or otherwise – but rather displaced along the perihelion–aphelion axis. Indeed, the realisation that there is a difference between the perihelion and the aphelion is itself evidence that the Sun is not at the geometric centre, even of a circle: if the orbit of Mars were a circle with the Sun at the centre, the distance from the Sun to Mars would remain constant at all times.

The measurement of the 'squashing', and the known location of the Sun, enabled Kepler to draw a diagram, a simplified version of which is shown in figure P.2.

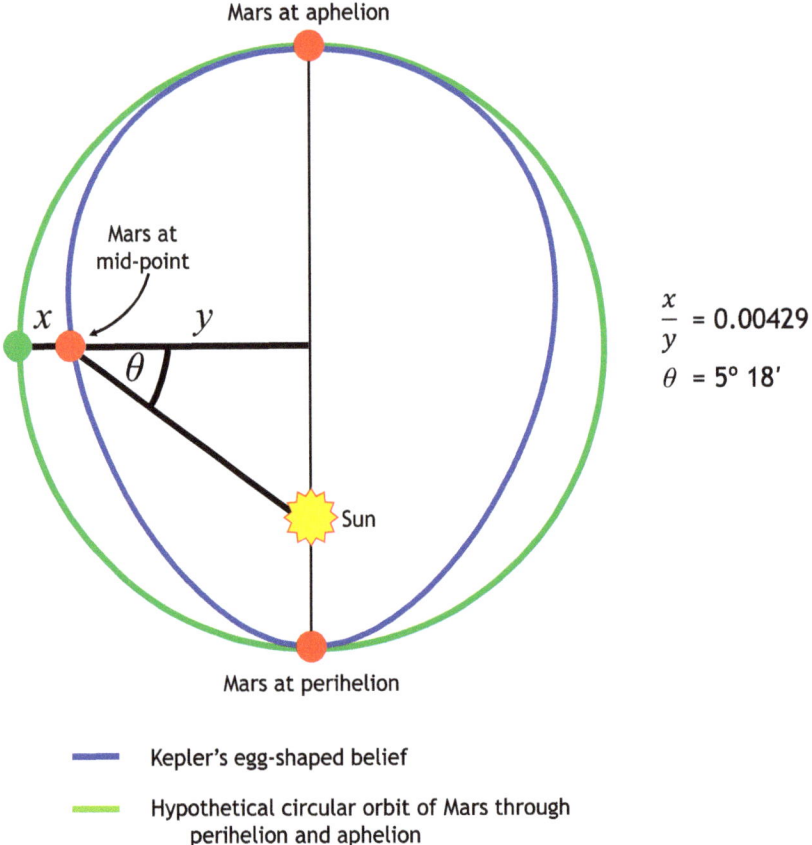

Figure P.2. Kepler's three key measurements. Note that this diagram is NOT to scale: for the orbit of Mars, both the distance x and the angle θ are very much smaller than as shown.

Kepler measured the degree of 'squashing' as the distance x between the actual position of Mars at the mid-point of its orbit, and the position at which Mars would have been, had the orbit been circular – that's the green 'blob' in figure P.2. And when he expressed that distance as a ratio as compared to the distance y between the actual position of Mars and the centre of the orbit, the result was $x/y = 0.00429$.

Think about that for a moment. If the radius of the circle, as perhaps drawn on a normally-sized sheet of paper, is 10 cm = $x + y$, and if $x/y = 0.00429$, then the distance y is 9.9573 cm, and the distance x just 0.0427 cm. Not 0.0428 cm. Not 0.0426 cm. 0.0427 cm. That's about the width of a pencil stroke. It would have been so easy, so understandable, for Kepler to say, 'That position of Mars at mid-orbit must be wrong. Surely it should lie on the circle. Maybe I made a mistake in one of my calculations? After all, there are hundreds, thousands, of them, all done by hand. Or Maybe Brahe's measurement wasn't quite right – those sight-line observations are really difficult to do accurately.'

But he didn't. He trusted his work; he trusted Brahe's work too. And he believed that the tiny ratio $x/y = 0.00429$ was right.

The other measurement he could make was of the angle θ, the direction of Mars to the Sun. That turned out to be 5° 18′.

And then nothing much happened. Kepler continued with his calculations, his drawings, his grinding labours. But in the back of his mind, and sometimes in the front, were those two numbers, 0.00429 and 5° 18′.

And then something did happen. Many of Kepler's computations were about the relationships between distances and angles, so he often used trigonometric tables, looking up sines and cosines, tangents and cotangents. The first comprehensive tables had been published in 1551 by Georg Rheticus, the only student of Copernicus (ah! a cosmological connection!) [7], and after he died, his student, Valentinus Otho, continued his work, producing, in 1596, a volume of about 1,500 pages, containing about 100,000 trigonometrical items, each to 10 decimal places.

So, one day, Kepler was looking through the tables. And he happened to see the number 1.00429. 'That rings a bell', he thought, 'or at least the .00429 bit...'. He then glanced across the page, and saw the angle to which 1.00429 referred.

5° 18′.

Yes, the secant of 5° 18′ is 1.00429!

Here is a translation of a passage from Kepler's book *Astronomia Nova*, published in 1609, in which (referring to the area between the circle and the egg, as shown in figure P.2, as a 'sickle'), he describes that moment:

> ... *I was wondering why and how a sickle of just this thickness, 0.00429, came into being. Whilst this thought was driving me around, while I was considering again and again that ... my apparent triumph over Mars had been in vain, I stumbled entirely by chance on the secant of the angle 5° 18′ ... When I realised that this secant equals 1.00429, I felt as if I had been wakened from a sleep...*

At that instant, Kepler realised something more than a coincidence had happened, but, to use his own metaphor, he was still 'waking up'. His thoughts were still muzzy from slumber, a piece of the puzzle remained missing. Then, after some further work... YES! For he suddenly realised that the fact that the secant of 5° 18′ is 1.00429 implied that the orbit of Mars was not an egg-shape. It was an ellipse. And not just any ellipse. An ellipse with the Sun as focus.

And the history of science, and of our world, was transformed at that moment.

> **You too can be Kepler**
>
> Imagine it's 1605, and that you are Kepler.
>
> You've just noticed that the secant of 5° 18' is 1.00429.
>
> What is the proof that the mid-point of the orbit lies on an ellipse, with the sun at one focus?
>
> How would you prove that the total orbit is that ellipse?

What happened next

Picking up that last question, Kepler's insight applied only to the position of Mars half-way along its orbit. Yes, that particular point might lie on the ellipse with the Sun at one focus, but that might have been a coincidence. The orbit itself might in fact still be egg-shaped – an egg-shape that just happens to pass through that point, as illustrated in figure P.3.

But Kepler was convinced the orbit was an ellipse, and so he did some more computations to prove it. Assuming that the orbit is an ellipse, he calculated where Mars 'should' be at a number of other points around the orbit, and then checked those predictions against Brahe's observations; he also took some of Brahe's readings and determined where they would lie on his chart. It all worked. Yes, an ellipse. With the Sun at one focus.

Gone were the circles and the spheres. Gone were the Platonic solids, fitting neatly in the voids between the spheres. The *BIG IDEA* was wrong. And it wasn't just the orbit of Mars that was elliptical: further computations showed that the other planets had elliptical orbits too – hence what later became known as Kepler's First Law: all planets move in elliptical orbits, with the Sun at one focus. Kepler's labours resulted in two other statements that also achieved the status of 'Laws of Physics' too: 'planets each sweep out equal areas in equal times' and (the rather more cumbersome) 'for each planet, the square of the time taken to complete an orbit is proportional to the cube of the ellipse's longer axis'.

Kepler died in 1630, aged 58, in Regensburg, the Bavarian city astride the river Danube. Tycho Brahe's tomb in Prague's Church of Our Lady before Týn, with its life-size richly-sculptured memorial stone, is visited by hosts of tourists every year. No such veneration for Kepler. The church in which Kepler's funeral took place, and his gravestone in the neighbouring churchyard, were destroyed in 1633 during the ravages of the Thirty Years War.

But Kepler's work and Laws lived on, and his legacy passed to Newton, whose story we will explore in section 5.6. And we all know what happened after that.

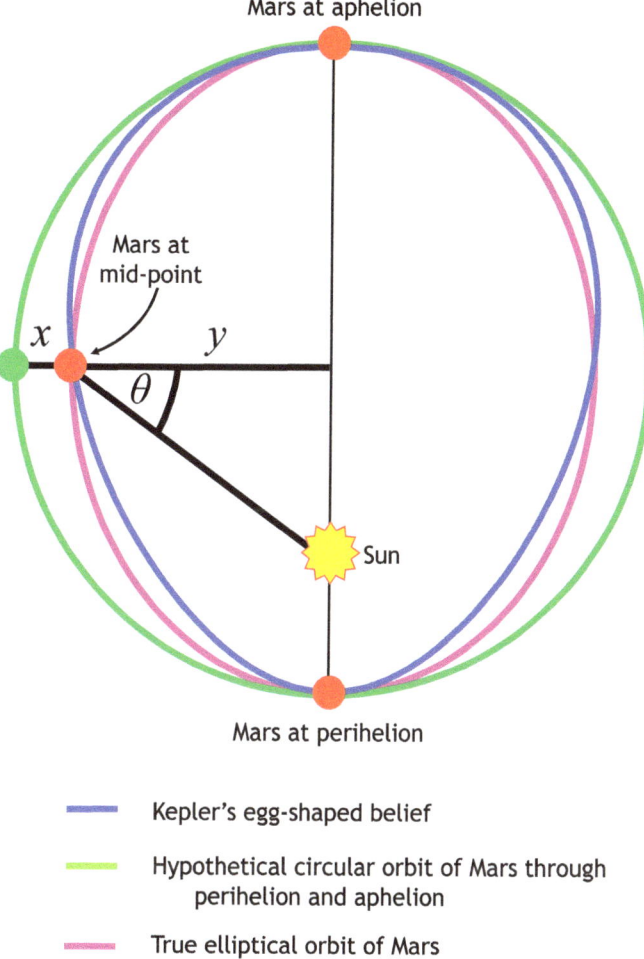

Figure P.3. The position of Mars at its mid-point might lie on both an egg-shape and an ellipse. To prove that the orbit is an ellipse, Kepler had to verify the position of Mars at other positions around the orbit.

Kepler re-visited

There can be no doubt that Kepler's contribution to science, and to us all, is that of a true giant. Had the Nobel Prize been available at the time, he surely would have won the award for Physics outright, and probably a Fields Medal too – he was 34 years old when he discovered that the orbit of Mars was elliptical (and 38 when his findings were published in his book *Astronomia Nova*), so he safely satisfied the Fields Medal's under-40 rule twice over.

But his story also vividly illustrates all the features of creativity that we will explore in this book – which is why I thought it would be a good way of starting.

Kepler's Laws are the tangible manifestations of creative genius. But as the story shows (and the story as related here is far less complete and rich than the story as told by Arthur Koestler is his masterful book *The Sleepwalkers*), Kepler did not make his momentous discoveries 'all-of-a-sudden'. There was no 'Eureka' moment, no instantaneous 'strike of lightning'.

So how did it happen? What can we learn about the creative process? And can we use that learning to enhance our own creative ability?

We'll explore these questions, and their answers, in detail as this book progresses. But drawing on Kepler's story, here are the key features:

- Observation
- Curiosity
- Permission
- Luck
- Bisociation
- Knowledge
- Hard work
- Tenacity
- Failure
- Learning from mistakes
- Risk
- Don't fall in love with your own ideas
- No problem to solve

Some thoughts on each...

Observation

Everything that Kepler did was based on observation: in particular, the astronomical observations recorded, over decades, by Brahe and his assistants. And when his predictions were not in agreement with the observations, Kepler trusted the observations and concluded that the problem was with his predictions.

In the early 1500s, this reliance on hard evidence was 'new'. The tradition derived from the Greek philosophers, and perpetuated through the middle ages, was about the purity of thought, regardless of the reality, all in the context of religious teaching. One aspect of the renaissance, and of the emergence of the modern age, was the recognition that evidence counts, and that if an 'elegant' idea did not match reality, it was the idea that was wrong, not the reality.

Curiosity

Kepler asked very good questions, the most fundamental being 'what is the simplest closed curve that matches Brahe's observations of the orbit of Mars?' There were already many descriptions of the orbit of Mars, based on increasingly complex combinations of circles within circles. As more, and more accurate, observational information became available, the advocates of celestial epicycles, deferents and equants drew ever more baroque arabesques, which looked charming, but had no underlying explanation or logic whatsoever. Kepler's curiosity challenged all that: his quest was to find a simple solution, and one that had meaning – meaning as expressed in his ideas about there being some type of force acting between the Sun and a planet, causing the planet to sweep around its orbit (Kepler even used the metaphor of a broom to describe it). Kepler didn't get the right answer as regards gravity – that was for Newton. But he did take the first steps down the right path.

Kepler was also very willing to ask 'lower level' questions too. So when something he was trying didn't work, he'd try again and again, as most of us do. And when the repeated attempts still failed, he realised that, to unblock the impasse, he had to do something *differently*. So if, after years of trying, a circular orbit just wouldn't work, he asked that most important question 'how might [this] be different?' – where, in this case [this] is 'my belief that the orbit of Mars is circular'.

And to answer that question, he didn't guess, or take a (necessarily subjective) view as to what might be 'perfect' or 'beautiful'. He trusted Brahe's observations and his own calculations, which, together, led him to conclude that the orbit is egg-shaped.

He then noticed the relationship between 0.00429 and 5° 18', triggering the idea of the ellipse. Problem solved.

Permission

Sometimes, it can be very difficult to ask questions. For asking a question can be interpreted by others as an admission of ignorance, or as a challenge to the tried-and-tested, to cherished beliefs, to authority. Those who ask 'difficult' questions can often be at the wrong end of the disapproval of those who have power. So we all learn that sometimes asking questions can be a very dangerous thing to do. With the result that we don't ask. Especially if we are junior.

Creativity, however, demands that questions *are* asked: nothing will ever be discovered or invented if the *status quo* is accepted as the only possible way of doing things, if no one dares ask that oh-so-important question 'how might [this] be different?'

Fortunately for Kepler, after Brahe's death, he needed no 'permission' from anyone other than himself, for his 'boss', the Holy Roman Emperor Rudolf II, was not a religious dogmatist (as were many other kings, dukes, counts, bishops and general members-of-the-public at that time), nor an oppressive task master: Kepler could more or less do as he pleased. And fortunately for us, Kepler gave *himself* permission to think, to explore, to experiment, to try. He did not say 'I know that the orbit must be a circle, for that is what everyone has ever stated, from Plato and Aristotle to Copernicus and Brahe', or 'I'd better not ask that question – Galileo might not like it'. Rather, he allowed himself to think the then-unthinkable. 'Suppose that the orbit of Mars is *not* a circle? What else might it be? Well... perhaps a really complex shape, like the diagrams of all those people who believe in epicycles. Or maybe something simpler... like an egg-shape... or maybe an ellipse...'

It is quite possible that someone before Kepler had (bravely!) asked the question, 'Suppose planetary orbits weren't circles...?' But we'll never know. For whoever that person might have been, their boss said 'Don't be so stupid. Just get on with the day-job, will you? And stop asking dumb questions!' And even someone fortunate enough to be allowed to ask the question would not have been able to discover the answer, for only Brahe's heirs and Kepler had access to Brahe's observations. They were the only sufficiently accurate observations available at the time, and Brahe's heirs didn't have the required mathematical skills and tenacity...

Luck

...which makes Kepler lucky that he was, initially, Brahe's assistant, with access to Brahe's data after Brahe's death (which is a key feature of the conspiracy theory surrounding the circumstances of that event...), working for a benign non-interfering boss, Rudolf II, and having the 1596 book of the trigonometric tables. Right place; right time. If Rudolf II had said 'Stop doing all that planetary stuff, it's been getting nowhere for years. I need a horoscope – now!' If Brahe's observations had been less accurate, or lost, or destroyed in a fire. If he didn't have tables of secants. If he hadn't noticed, among all those numbers on the page, that particular number 1.00429. If any of these had not happened, then we would not be talking about Kepler in this book. Perhaps we would never have heard of him, just as we have never heard of all the other teachers of mathematics at schools across Europe in the early 1600s. And sooner or later, someone else would have come to the same conclusions, and it would be their name, not Kepler's, that would be remembered. Yes, Kepler was indeed in the right place at the right time.

Kepler was lucky too in studying Mars. All planetary orbits are elliptical, but each has a different eccentricity. The current estimate of the eccentricity of Mars is 0.09341 [8], corresponding to $x/y = 0.00439$ and $\theta = 5° 22'$ [9], very close to Kepler's values of $x/y = 0.00429$ and $\theta = 5° 18'$. The orbit of Venus, by contrast, is much closer to a circle: the eccentricity is now measured as 0.006773 [10], corresponding to $x/y = 0.00002$ and $\theta = 0° 23'$. These values are so small that, had Kepler been studying data for Venus rather than Mars, Kepler would almost certainly have concluded that the orbit of Venus is circular. He would doubtless have been pleased with that result, and encouraged to continue his quest for the Platonic solids. It's possible that he might subsequently have discovered that the orbit of Mars was elliptical, but perhaps not...

So luck does play a role. But sometimes luck can, at least in part, be engineered.

That Kepler and Brahe happened to meet in Prague was 'accidental', in that neither man had planned for that specific event to happen. But it was no accident that the two men met. Since about 1595, Kepler had wanted to prove his *BIG IDEA* about the Platonic solids, and he knew that the only person who had the right data was Brahe. So Kepler had contacted Brahe, and was in correspondence with him, before Brahe extended the invitation to Kepler to join him in Prague as his assistant. Kepler's approach was deliberate, with the intent that by getting to know Brahe, he might then be able to verify his hypothesis. Kepler did not know that he and Brahe would actually get together, he did not know that the Platonic solid idea was flawed, he did not know he would discover the elliptical orbit. But he did some very deliberate things to help make their meeting happen.

Yes, there was some luck. But Kepler was not merely a passive victim of fate: he did everything he could to put himself in the best possible place, and with the best possible person, to help get things going.

Bisociation

Just before Kepler 'emerged from his sleepwalk', there were three things, simultaneously, in his mind, albeit in different mental 'compartments':

- the ratio 0.00429
- the angle 5° 18′
- closed, egg-shaped, curves.

Then he notices that $1.00429 = \text{secant } 5° 18'$.

At that instant, these previously unrelated concepts converged, and in so doing formed a meaningful story in Kepler's mind: since 1 + the ratio 0.00429 = secant 5° 18′, then rather than being egg-shaped, the closed curve might be an ellipse, with the Sun at one focus. And that 'might be' became 'is' when Kepler tested his elliptical hypothesis for other positions around the orbit.

This 'convergence', this coming-together of hitherto unrelated fragments of information so that the result is 'interesting', is fundamental to all creativity. As I shall explore in more detail in chapter 4, Arthur Koestler (yes, the same Arthur Koestler as the author of *The Sleepwalkers*) calls this 'bisociation' [11], and one of the key skills in making creativity more deliberate is to find ways to increase the likelihood that 'bisociations' will happen – and to spot the 'interest' in the result.

A further feature of Kepler's story is that each 'fragment' was noticed at different times. He had been thinking about closed curves for many years, and when he measured $x/y = 0.00429$ and $\theta = 5° 18'$, he still did not have sufficient information to identify the orbit as an ellipse. And by the time he spotted, accidentally, that $1.00429 = \text{secant } 5° 18'$, he might have forgotten his measurements of x/y and θ, or not recognised the relationship between the number 1.00429 and the measurement $x/y = 0.00429$. And let's not forget that Kepler's mind was crammed full of numbers and angular measurements, and there would have been umpteen numbers on that page of the trigonometric table that happened to contain secant 5° 18′. Yet he made the right connections.

There's another important 'bisociation' – or rather 'trisociation' – in Kepler's story too. I've mentioned that Kepler suggested that there was some type of force acting between the Sun and each planet, 'sweeping' the planet around its orbit. Kepler's concept of a force, acting at a distance, combined three 'fragments'. The first was the work of William Gilbert on magnetism, and the magnetic properties of the Earth, published in 1600 in his book *De Magnete* [12]; the second was Kepler's own discovery, in 1603, that the intensity of light diminishes as the square of the distance from the source. And the third was his observation that Mars travelled faster when it was positioned closer to the Sun, and more slowly when further away, implying some kind of inverse relationship between speed and distance.

Putting these three 'fragments' together, Kepler formed the idea that the force acting between the Sun and Mars was proportional to the inverse of the intervening distance. Having already discovered the inverse square law for light, it is perhaps surprising that he suggested a simple inverse law for what we now know to be gravity, but that's what he did, and that, as we now know, was wrong. But for Kepler, there was a 'happy ending': although he had made the wrong assumption about the law for the force, he made a second, compensating, error in some calculations, yet still discovered the truth that planets sweep out equal areas in equal times – what became known as his Second Law. Luck strikes again.

As we shall see, a fundamental principle of all creativity is the 'assembly' of apparently unrelated 'fragments' into 'interesting patterns', and those 'fragments' will be noticed at different times and in different places. If any of those 'fragments' are forgotten then they can never be 'assembled'; if the set of 'fragments' is incomplete, so that a necessary fragment is 'missing', then the 'partial assembly' just won't make sense.

Creativity therefore requires an observant, and an alert, mind: observant, so that all 'fragments' are 'harvested'; alert so that you can identify that moment at which those hitherto random-looking 'fragments' come together to form a meaningful pattern. As we have seen, Kepler was (extremely) observant – he noticed, and cared about, details, and remembered what he had noticed. He was also alert, so when he saw, in those trigonometrical tables, that $1.00429 = \text{secant } 5°\ 18'$, he knew he had noticed something very important.

Knowledge

Suppose that Kepler had noticed the 'coincidence' that $1.00429 = \text{secant } 5°\ 18'$, but had no knowledge of ellipses. What would have happened? Probably nothing – other, perhaps, than Kepler thinking to himself 'that's a funny coincidence', and then going on to think about something else.

Or suppose that Kepler did in fact know something about ellipses, but not enough to recognise that the relationship that $1.00429 = \text{secant } 5°\ 18'$ implies that the mid-point of the orbit of Mars must lie on one. Once again, nothing would have happened.

This is just one, specific, example of a general, and very important, principle: all creativity is knowledge-based, and the more knowledge you have, the more likely it is that you will be creative.

Many people find that principle a total paradox. 'Surely,' they say, 'isn't creativity all about discovering something new? So what has knowledge got to do with it? Doesn't knowledge actually get in the way? Isn't that why younger people – who inevitably have less knowledge than older ones – are more creative than older people?'

Explaining (I hope clearly!) how this paradox is resolved is one of the key objectives of this book – as you will be able to judge as you read on. But for the moment, think about the principle: the more knowledge you have, the larger your 'library of fragments' and the greater the likelihood that you will be able to use those 'fragments' to discover 'interesting patterns'. And, very fundamentally, that's what creativity is all about.

Hard work

There is a widespread myth that creativity happens in a flash, out-of-the-blue, like a strike of lightning. This myth is partly true in that the event of 'bisociation' – the moment at which those hitherto different, separate, fragments of knowledge all come together to form an 'interesting pattern' – can happen quite suddenly. But to identify, gather and assemble the right 'fragments' can take many years, and much, very hard, work. Especially since you rarely know, in advance, what those 'fragments' actually are – so in collecting the 'right fragments', you will inevitably collect a huge number of 'wrong fragments' along the way. But you won't know which fragments are 'right' and which 'wrong'. And to complicate matters even further, not only might the same fragment be 'right' in one context but 'wrong' in another, but also some of the 'wrong fragments', such as Kepler's egg, might be seductively attractive.

So be in no doubt. Creativity requires hard, hard, work. It's no accident that one of the most quoted of the sayings attributed to Thomas Edison – who must be almost everyone's 'Mr Creative' – is 'Genius is 1% inspiration, 99% perspiration'. At best.

Tenacity

Allied to 'hard work' is tenacity – the determination to keep going, and going, and going, and still going, even though it's tough. Really tough. But there is a fundamental truth. If you give up, you will never find what you're looking for. But if you continue, you just might.

But it's only 'might'. There are no guarantees.

So those who wish to enhance their personal creativity, and the creativity of their team, are well advised to enrich their skills at judging, wisely, when you should keep going, and when it makes good sense to stop and do something else.

Everyone will agree that, alongside Thomas Edison, another towering creative genius was Leonardo da Vinci. As will be discussed in chapter 6, in the 1490s he imagined, and drew, some pictures of 'flying machines' that were truly fanciful at that time, and that look intriguing now. Even for a genius of the calibre of da Vinci, there was no way in which any of the designs, so beautifully drawn in his diagrams, could have been made to work. Yes, great ideas. But, in the 1490s, best discarded, and left as treasures of art.

Failure

All creativity is strewn with failure. And failure hurts. 'Why am I so stupid?', 'Why did I waste so much time on that!?!' are bad enough. And 'Why are *you* so stupid?' and 'Why did *you* waste so much time on that!?!' hurt even more.

Creativity is inevitably the exploration of the as-yet unknown. And no explorers know where they're going because they haven't been there before, and anyone who might have been there before didn't leave a map for others to follow. So expect to fail. Again and again. But don't feel it's 'my fault'. It's not a question of 'fault'.

How many times did Kepler fail? He spent about five years, sitting at his desk, in the draught, doing countless calculations by hand. But he couldn't fit the orbit of Mars to a circle, however hard he tried. And so he could never fit a dodecahedron between the spheres of Mars and the Earth. What a failure!

So failure is part of the process. That's tough. But true.

As encapsulated in another quotation attributed to Edison, in relation to his labours in developing his electric light-bulb: 'I haven't failed – I've just found 10 000 ways that didn't work' [13].

Learning from mistakes

But maybe 'failure' is the wrong word.

For there is a significant difference between 'the results of my calculations did not agree with the known observations' (as experienced by Kepler time and again), and 'I failed to get to the shops because I was thinking so hard as to why the fact that $1.00429 = \text{secant } 5° \ 18'$ implies that the orbit of Mars is an ellipse that I didn't look where I was going and fell off my bike'.

The English language can use the single word 'failure' in both cases, but whereas the latter is thoughtlessness-verging-on-negligence, the former is a discrepancy between a plausible hypothesis and actuality.

What's important about 'failure' in the sense of the discrepancy, rather than negligence, is that negligence is about carelessness, thoughtlessness, the intentional (or indeed unintentional) flouting of rules which are there for good reason; a discrepancy is something that happens that is not 'my fault', something we notice and learn from, something we don't repeat (or, perhaps more realistically, don't repeat three times, for in science we often repeat a 'failed' experiment to verify that it really didn't work – but not too often!).

Furthermore, what might be interpreted as a 'failure' might be a clue to something really important. So when Kepler 'failed' to validate a circular orbit, rather than giving up, or blaming 'bad observations', or fudging the results, he learnt: learnt that his hypothesis that the orbit is a circle might be wrong, and that it might be something else.

There are many examples of creative ideas resulting from 'mistakes'. For instance... in 1856, an 18-year old chemist had been told by his boss to mix some chemicals together with the intention of discovering how to synthesise quinine, a valuable product for treating malaria. He did as he was instructed, and saw, at the bottom of his test tube, a dark sludge. It certainly wasn't quinine. So his experiment had failed.

Most of us would have thrown the sludge away, and started again. But the chemist noticed it and thought 'that's interesting – I think I'll take a deeper look...'.

He didn't have permission from his boss to spend time on this, so, out of interest, he devoted his own time, working in a shed in his garden, and keeping it a secret from his boss.

He then discovered that he had synthesised a beautifully-coloured dye – the dye we now know as mauveine [14]. And as a result, William Perkin founded the entire synthetic dyestuff industry, which itself was the origin of the modern pharmaceutical industry.

All from that dark sludge. That failure.

Risk

Failure and learning both happen after-the-event: you can only fail after you've tried, and you can only learn from that failure after you've experienced it – and also only if your mind is open enough to recognise the failure and to be willing to learn from it, rather than to explain it away as an 'unintended consequence', or blame others.

Risk is different. Risk is about what might happen in the future, and is considered and assessed before-the-event.

The message here is that *all* creativity is risky. By definition, in being creative, in thinking about an idea, and in attempting to make that idea real, you are journeying where you have never been before, into the unknown. And, as they say on the old maps, 'Here be dragons'.

There's a risk that others will laugh at your idea. Or be jealous of it. Or of you. There's a risk that the idea will take far longer than you had hoped to make real, and cost a lot more too. There's a risk that someone else in the field might be working on a similar idea, and get the answer first. There's a risk that the idea might not actually work, and that all that time, effort, money, and emotional energy, will be wasted.

Kepler took all these risks. Maybe others were jealous of the Platonic solid idea. Working on it certainly took a very long time, and consumed much energy. And what was Galileo up to?

Anyone who is extremely risk-averse is ill-suited to being creative, and even less so to working on an idea to make it real. Far better to keep to the tried-and-tested, to follow the instruction manual, to comply with the rules.

To be creative, you must be willing to take some degree of risk. Of course, it makes good sense to do everything you can to minimise those risks, and to mitigate their effects should they happen. But risk is still there.

Yes, it is profoundly true that creativity is associated with risk. But sometimes – quite often, in fact – following the instruction manual, complying with the rules, maintaining the *status quo*, might not be as risk-free as might appear. As, for example, Tsar Nicholas II of Russia tragically discovered when his persistent refusal to adopt, let alone think of, new political ideas over the previous two decades proved to be a major cause of the revolution that swept him aside in 1917. Sometimes the risk of being creative is less than the risk of not being so.

Don't fall in love with your own ideas

It's a very human trait to be proud of one's own ideas. But that can be taken to a dysfunctional extreme, especially in someone who wishes to be creative. For if you are too much 'in love' with your own ideas, you will be resistant to the ideas of others, you will be blind to those aspects of your idea that don't quite work, and you risk 'bending' the reality to fit your idea, rather than being willing to 'bend' your idea to fit reality – or discard it altogether.

As we saw in the story of Kepler, if he had been too 'in love' with circles, he would have attributed that oh-so-small ratio 0.00429 to a consequence of an error in Brahe's data. But he wasn't, and he didn't. He changed his idea.

No problem to solve

One final feature of the story of Kepler.

Without any doubt, Kepler solved the problem of discovering the true orbit of Mars. But in truth, the orbit of Mars was not a 'problem' – at least not in the mind of almost everyone else in Europe (and everywhere else in the world, too) at that time. In so far as anyone thought about it, everyone was perfectly happy with the belief that the orbit of Mars was a circle. Which of course Kepler, at the time, believed too.

And it is certainly true that no living person – at that time, and probably before then and ever since then too – considered that proving that the five Platonic solids could fit between the planetary spheres was a 'problem' that needed any 'solution' at all.

But by thinking creatively about a total non-problem, well...

So I mention 'no problem to solve' as the final feature of Kepler's story, for it too is a deep principle.

Many people think that creativity is primarily a skill that is only useful in solving problems. Which indeed it is – if there is a problem to be solved, if something doesn't work properly, then yes, creativity is very valuable in helping solve the problem, or fixing whatever it is so that it does indeed work properly. There are many examples of this, one of the most notable being the work of John Harrison, who laboured for some 30 years to perfect a reliable, accurate, maritime chronometer, culminating with his successful 'H4' of 1759, so solving the problem of measuring longitude [15] – a problem of great importance for safe marine navigation, and to which a solution was vitally needed.

There are also many beliefs associated with proverbs such as 'leave well alone', or 'if it ain't broke, don't fix it', which are incentives not to tinker with things that work.

But some of the most wonderful examples of creativity are associated with contexts in which there is no obvious problem-to-solve, and everything appears to be working just fine. And there an enormous number of instances of this – examples that are held in the highest regard as the very pinnacles of creativity.

One such is Beethoven's 5th Symphony. Beethoven wrote that symphony from his heart, because he wanted to express his innermost feelings. Not because someone had said, 'Beethoven, we have a problem. We don't have enough symphonies starting 'Ba-ba-ba-boom'. Fix it!'. Likewise, Shakespeare didn't write *Hamlet* to solve the problem of a lack of plays about introspective royalty, Monet did not paint his sequence of water-lily paintings to solve the problem of a dearth of pictures of flowers.

Those examples relate to the arts – the archetypal 'creative' activities – but the principle that 'you don't need to have a problem to solve to be creative' is fundamental, and applies to all fields. Including Kepler. Very few people cared whether the orbit of Mars was a circle, an ellipse, or an elaborate confection of epicycles. No one was saying, 'Kepler, we have a problem with Mars. Fix it!'. But by thinking about that orbit creatively, by matching theory to real observations, by not being overwhelmed by failure, by not being in love with his ideas, and by drawing on his deep knowledge, Kepler opened a new era.

So I end this Prologue with a final thought.

Creativity is certainly valuable when solving a problem, when wishing to make something work better.

But creativity can be even more valuable when applied when there is no problem-to-solve, when everything is working well. For by thinking, by exploring, and by asking that oh-so-important question 'how might [this] be different?', you can discover ways to make things work even better, and so help build a better world.

Rudolf II

Rudolf II was born in 1552 in Vienna, the eldest son of the Austrian half of the Habsburg dynasty [16]. His father, Maximilian II, was, amongst many other things, Holy Roman Emperor, King of Bohemia and King of Hungary. The head of the other half of the Habsburg family was Philip II, ruler of Spain, the Netherlands and the recently-conquered territories in the Americas, and also, simultaneously, Rudolf's mother's brother, and father's cousin (the genealogy of the Habsburgs is singularly convoluted!) [17].

Rudolf grew up in Vienna and Madrid, and succeeded his father as Holy Roman Emperor in 1576, moving his court from Vienna to Prague in 1583.

Giuseppe Arcimboldo's 'portrait' of Rudolf II as the Roman god Vertumnus (yes, that nose is a pear!).
Source: Wikimedia 'Vertumnus and Portrait of Rudolf II von Habsburg' https://en.wikipedia.org/wiki/File:Porträtt,_Rudolf_II_som_Vertumnus._Guiseppe_Arcimboldo_-_Skoklosters_slott_-_87582.jpg.

He lived through turbulent times. The Counter-Reformation was in full swing, driving the tension between the Catholic and Protestant communities that was to erupt in the tragedy of Europe's Thirty Years War in 1618. And on his empire's Eastern border, there was a protracted war with the Ottoman Empire from 1593 to 1606. Unlike many of the contemporary rulers in Europe (of which Elizabeth I of England was one), Rudolf's response was to withdraw into the seclusion of his castle, with his curiosities and his artworks.

Fortunately for us, Rudolf's interests in astronomy, astrology and the occult led him to invite Tycho Brahe to Prague, and subsequently, to allow Brahe to approach Kepler to join him.

Rudolf's withdrawal, however, was, for him, politically unwise. In the 1600s, rulers ruled, they did not hide. His younger brothers declared him insane in 1606, initiating a sequence of events in which he was progressively stripped of his powers, culminating in his being held a prisoner in his own castle, where he died in 1612, three days after the death of his pet lion [18]. It is said that Brahe had identified that both Rudolf and the lion shared similar horoscopes, and would suffer similar fates. So maybe Brahe was not just a great astronomer, but a pretty good astrologer too…

Tycho Brahe

Denmark is famous for its egalitarian, liberal and fair society, and so the idea of Danish 'lords' and 'nobles' might be somewhat unexpected. But for most of Denmark's history, just as elsewhere across Europe, there were serfs and nobles, peasants and Kings. In fact, Denmark has maintained its monarchy (the current incumbent, Margrethe II, ascended to the throne in 1972, and is a most intelligent and artistic lady [19]), and it still has some 'of noble blood', although they keep that quite quiet.

In 1546, however, when Tycho (the Danes call him Tyge) was born, the Brahes were amongst the most noble of them all: many of Tycho's forbears had been royal counsellors and high officials at the Danish Court, and the Bille–Brahe family exists to this day.

Brahe's tombstone in Prague.
Source: Photograph of Tycho Brahes's tombstone reproduced courtesy of Dr Carlos Dorce i Polo, https://themathematicaltourist.wordpress.com/2014/08/09/tycho-brahes-tomb/.

Whilst studying at Copenhagen University, Tycho developed an interest in astronomy, which he pursued in Leipzig and Rostock – where he lost part of his nose in a duel, resulting in his wearing a nasal prosthesis for the rest of his life [20]. And from about 1567, with the support from his family to forego politics and the law for star-gazing, he became a full-time astronomer with a particular talent for designing instruments, and an obsession for ever more accurate, and precise, measurement.

The talent and the obsession were both fulfilled following King Frederick II's gift of the island of Hven in 1576. His castle-cum-observatory, Uraniborg, and underground complex Stjerneborg (which protected Brahe's instruments from the wind) soon became a 16th Century version of a modern-day research hub, with some 100 students, craftsmen and assistants working there over a 20-year period.

After the death of Frederick II, Tycho became increasingly at loggerheads with Frederick's successor Christian IV, and also with the local inhabitants on Hven, to the extent that he dismantled all his instruments, taking them first to Copenhagen, then Hamburg, and ultimately, at Rudolf II's invitation, to Prague.

Ceded to Sweden in 1658, Hven, now Ven, can be visited by boat from Copenhagen, Helsingborg and Landskrona; there's a Tycho Brahe museum and some remnants of Stjerneborg, but all that's left of Uraniborg is a restored garden [21]. What a project it would be to rebuild it – and the instruments too!

Johannes Kepler

Kepler, perhaps, looking intense.
Source: Wikimedia 'Portrait of an unknown man, often mistaken for Kepler' https://commons.wikimedia.org/wiki/File:Johannes_Kepler_1610.jpg.

Like so many of the young men who had the good fortune to receive some education in the late 1500s, Kepler's original intention was to enter the Church. At the University of Tübingen, he studied theology, with some mathematics and astronomy too, and throughout his life, religion would play a major role, both in his fundamental beliefs and also as a backdrop to his day-to-day existence: he was a convinced Lutheran during a time of much religious strife, his mother was accused of witchcraft, and the Thirty Years War raged around him.

His first job, in 1594, was as a teacher of mathematics and astronomy at a school in Graz, in the Austrian province of Styria. And it was at Graz that he had his *BIG IDEA* about the Platonic solids, gaining considerable fame across the community of European astronomers with its publication in his book *Mysterium Cosmographicum*.

As described in the text, Kepler discovered the elliptical orbit of Mars in 1605, but his results were not published, in his book *Astronomia Nova*, until 1609. One of the reasons for the delay was a long-running legal dispute with Brahe's heirs about the ownership of Brahe's data: Kepler's relationship with Brahe, when he was alive, and with Brahe's family and colleagues, after his death, was fractious. But Kepler had problems with relationships in general: his 'autobiographical notes', written when he was about 25, make a disturbing read.

After Rudolf II's death in 1612, Kepler moved to Linz, and then, when Linz was besieged in 1626, to Ulm – where, some 250 years later, Einstein would be born. Throughout this time, Kepler had continued to compile astronomical data, resulting in the Rudolphine Tables (after Rudolf II), the most accurate and comprehensive catalogue of the stars and the planets for many years to come. But even the publication of that was fraught with legal wrangling.

In 1608, Kepler compiled a horoscope for Albrecht von Wallenstein, who was to become the commander of the Catholic forces in the Thirty Years War. Some authors allege that in a revised version of the horoscope dated 1625, Kepler predicted 'horrible disorders' in March 1634. If he did so, he was wrong. Wallenstein was assassinated on 25th February 1634 [22]. But perhaps the discrepancy might be within 'experimental error'. Or would Kepler have fretted about that difference of 4 days – just as he fretted about differences of a few minutes (of arc) – leading him to discover a new paradigm for horoscopes? We'll never know, for Kepler was not aware of his 'mistake', having died, seriously ill, in 1630.

Kepler's ellipse

Here is one solution to the puzzle posed about Kepler's solution. The information that Kepler had is represented in figure P.4:

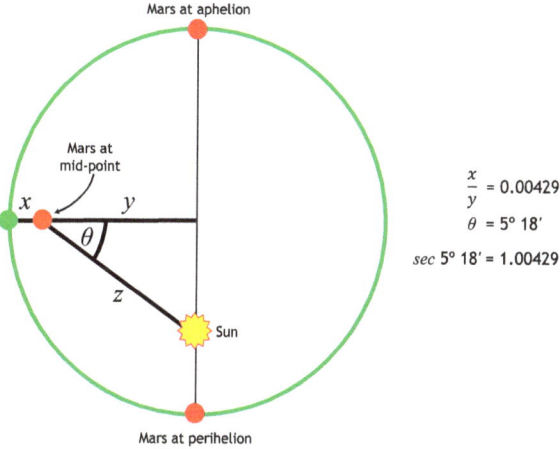

Figure P.4. Kepler's ellipse.

The green curve is the circular orbit that Mars 'should' have, but at the mid-point, the actual position of Mars is displaced a distance x inwards, such that, according to Brahe's observations, $x/y = 0.00429$, and the angle θ between Mars and the Sun is 5° 18′. Kepler than noticed that $sec\ 5° 18′ = 1.00429 = 1 + x/y = (x + y)/y$.

The magnitudes 0.00429 and 5° 18′ don't matter, but the secant relationship does. By definition, $sec\ \theta = z/y$, where z is the distance between Mars and the Sun. But since, for that particular position of Mars, $sec\ \theta = (x + y)/y$, then $z = x + y$.

That's the key result. The sum $x + y$, in this context, is a very special quantity. As the diagram shows, it is the radius of the circle of the orbit Mars 'should' have. So Kepler showed that the distance of Mars to the Sun at the point at which Mars is half-way around its orbit is the radius of that circle. Also, by symmetry, there is a 'mirror image' position of the Sun closer to the aphelion, which is also at a distance z from Mars. The total distance from this 'mirror image' position, to Mars, and then to the actual position of the Sun is therefore $2z$, a distance equal to the diameter of the circle. And, importantly, for any orbit, a constant.

One definition of an ellipse is the locus of a point such that the sum of the distances from that point to each focus is a constant. That's what Kepler showed for the mid-point, giving him a hypothesis that he could test for other points.

References

[1] The primary source for the material in this chapter is Arthur Koestler's book *The Sleepwalkers*, originally published by Hutchinson in 1959, and most recently reprinted by Penguin in 2014. Other sources are Ferguson K 2002 *The Nobleman and his Housedog: Tycho Brahe and Johannes Kepler – The Strange Relationship that Revolutionised Science* (London: Review) and Love D 2015 *Kepler and the Universe: How One Man Revolutionised Astronomy* (Amherst, NY: Prometheus Books). The date of 4th February 1605, as in the introductory story, is my invention, but plausible: various sources indicate that Kepler's discovery of the elliptical orbit of Mars happened some time before Easter 1605.

[2] See, for example, Gilder J and Gilder A-L 2004 *Heavenly Intrigue* (New York: Doubleday), and also www.theguardian.com/science/2012/nov/15/astronomer-tycho-brahe-death-scientists

[3] www.sciencemuseum.org.uk/objects-and-stories/Uranus-first-planet-discovered-telescope

[4] https://astronomy.com/news/2020/11/who-discovered-Neptune

[5] https://digital.nls.uk/scientists/biographies/john-napier/discoveries.html

[6] http://euler.nmt.edu/~brian/prost.pdf

[7] https://mathshistory.st-andrews.ac.uk/Biographies/Rheticus/

[8] https://nssdc.gsfc.nasa.gov/planetary/factsheet/Marsfact.html

[9] Referring to figure P.2, Kepler measured the ratio $x/y = 0.00429$. For an ellipse of semi-major axis $a = x + y$, and semi-minor axis $b = y$, the eccentricity $e = \sqrt{(1 - b^2/a^2)}$, which, in terms of the ratio x/y becomes $e = \sqrt{(1 - (1/(1 + x/y)^2))}$. For a circle, $a = b = y$, and $e = 0 = x$.

[10] https://nssdc.gsfc.nasa.gov/planetary/factsheet/Venusfact.html

[11] Koestler A 1949 *Insight and Outlook* (London: Macmillan) 36ff p
Koestler A 1964 *The Act of Creation* (London: Hutchinson) 35ff p

[12] Gilbert W 1991 *De Magnete* (New York: Dover)

[13] www.forbes.com/sites/nathanfurr/2011/06/09/how-failure-taught-edison-to-repeatedly-innovate/?sh=16c4516c65e9—and https://edison.rutgers.edu/newsletter9.html too!

[14] https://blog.sciencemuseum.org.uk/mauve-mania/

[15] www.rmg.co.uk/stories/topics/harrisons-clocks-longitude-problem

[16] www.holyromanempireassociation.com/holy-roman-emperor-rudolf-ii.html

[17] www.gutenberg.org/files/42025/42025-h/images/house_of_hapsburg.jpg

[18] Marshall P 2007 *The Mercurial Emperor: The Magic Circle of Rudolf II in Renaissance Prague* (London: Pimlico) p 224

[19] www.meridian.org/wildswans/aboutartist.html

[20] Christianston J R 2000 *On Tycho's Island: Tycho Brahe and his Assistants 1570–601* (Cambridge: Cambridge University Press)

[21] www.visithven.dk/en/tycho-brahe-museum

[22] Kaspar M 2003 *Kepler* (New York: Dover) p 324

Part I

Koestler's Law of Creativity

IOP Publishing

Creativity for Scientists and Engineers
A practical guide
Dennis Sherwood

Chapter 1

What, precisely, is creativity?

1.1 Some dictionary definitions

This book is all about creativity, so a good place to start is to ensure that we're all 'on the same page' as regards precisely what 'creativity' means. You might like to think about that for a few moments, and write down your own definition or explanation, using your own words…

…and while you do that, I'm looking at some dictionaries to see what they say, for example…

> *The use of skill and imagination to produce something new or to produce art.*
> Oxford Learner's Dictionary [1]

> *The ability to produce or use original and unusual ideas.*
> Cambridge Dictionary [2]

> *The ability to create.*
> Merriam-Webster [3]

> *Creative ability; artistic or intellectual inventiveness.*
> Collins [4]

> *The ability to use your imagination to produce new ideas, make things etc.*
> Longman [5]

Defining creativity as 'the ability to create' doesn't take us very far, and the others are not much better, simply equating creativity with words such as 'new', 'original' and even 'unusual'. Might 'unusual' things be 'odd' or 'strange' too? So does that imply that creativity is about doing something odd or strange? That's rather problematic…

In my view, these dictionary definitions are unsatisfactory. We need something much more informative. So the purpose of this chapter is to deliver a richer definition of 'creativity', to dig deeper, to identify some truly fundamental insights. And to do that, I'll explore the word creativity at three levels: the 'sound bite', the 'contextual', and the 'fundamental'.

1.2 My 'sound-bite' definition – just five words

Creativity is having an idea.

That's my 'sound-bite' definition, just five words.

Even such a simple sentence carries significant meaning, so let's get behind those five words.

Firstly, although not stated explicitly, this definition assumes that 'having an idea' is an activity that takes place within a brain, a single brain. So creativity is an activity that is localised deep within my own head. Or your own head. But not in the 'space' between us. Creativity is therefore associated with the firing of neurons in the brain, and the connectivity between neurons, and so a more fundamental explanation of creativity takes us into cognitive psychology, neurophysiology and brain science, to which we shall return in chapter 8.

If creativity is localised within a single brain, and not in the intervening 'space' between brains, what happens in that 'space'?

The answer is enrichment.

Once I have an idea in my head, I can then, if I wish, talk about it and describe it to you. As I talk about it, I am transferring it from the interior of my brain to the 'space' between us, and, having done so, you can then 'observe' the idea, incorporate it into your mind, consider it, think about it, and make some comments. As a result, the idea is enriched, for your thought processes 'mingle' with mine, contextualising the idea from a different point-of-view, and perhaps modifying or changing it. My thought processes are therefore influenced by yours, leading me to think harder too. As I shall explore in more detail later, this dialogue – or, if more people are involved, 'multilogue' – is a key, and very valuable aspect, of the wider process of creativity, which goes beyond just having a 'raw' idea towards crafting ideas that could be very good indeed.

But at the very outset, all ideas are incomplete, un-thought-through, fragile, full-of-holes. And, knowing that, when I have an idea in my brain, I might be very reluctant, if not downright fearful, of saying anything at all. I might get shot down. Someone might say 'have you thought about…?'. Someone might say 'that won't work because…'. And they're right. The idea might not work because…, and, no, I haven't thought about that. Mmm. I don't think I want to go through all that, so far better to stay silent. I'll say nothing at all.

That, very understandable, human reaction poses a significant problem. If I don't say anything about my idea, I deny you the opportunity to enrich it. And even if my idea is incomplete, un-thought-through and all the rest – as it most certainly is – my failure to talk about it not only denies you the opportunity to enrich it, but also denies us, together, the opportunity to explore it. This therefore blocks the possibility of making the idea a little more complete, a little more thought-through, a little better. The idea will therefore remain incomplete and fragile in my own head. For ever. And absolutely nothing will happen. What opportunities might be missed as a result of my silence, my fear?

That's a very important message about organisational cultures, which I will explore in more detail in chapters 16, 17 and 18.

One other point about creativity being an activity that is localised within a single brain. That does not say 'human' brain. It says brain. Humans are not the only creatures with brains, complex structures of richly interconnected neurons. So it's quite likely that many animals can have ideas too, and there is much evidence that animals can be very good at solving problems – from tortoises that find their way through mazes [6] to orangutans that can use tools [7]. There is also evidence, for example, that whales can communicate using sound [8], and that birdsong is not just random warbling, but can contain messages. Just like human speech. And if I use speech to articulate that idea in my head, perhaps that whale, and that chestnut-crowned babbler [9], are doing the same thing…

1.3 Ideas as outcomes, ideas as questions

The word 'idea' is, of course, integral to my sound-bite, so I note here that ideas can take two different forms. One form is as a partially-formed outcome – so, for example, a musical composer might be walking in a park, and a few notes come together in her mind. Over the following days, perhaps weeks, that phrase continues to play in her head, at different pitches, at different paces, played by different instruments, embedded within a longer phrase, preceding a different phrase. She then sits down at her piano, and creates the next big pop hit.

That's not a specific example; rather an illustration of a creative process in which the idea is a fragment of the end result, in this case, a few notes that form the basis of the final song. Similar examples might be a writer's idea for the plot of a novel, an artist's idea for an abstract painting.

Another possibility is that the idea not formulated as a fragment of an outcome, but as a question. This is particularly the case in science, for the most fundamental and powerful manifestation of scientific creativity is in asking good questions: if you ask a good question, that directs your search for answers in a purposeful way, and so that search more likely to be effective, increasing the likelihood of finding a good outcome.

But not always, for it is quite possible for a question to be very good, and so worth pursuing, only to arrive, after much work, at the wrong end of a blind alley. It's usually very difficult to anticipate that in advance, and so we should expect such journeys to happen; as we saw in the Prologue, this is not failure, but learning. So, for example, I now consider Kepler's question 'Can I prove that the Platonic solids can fit within the celestial spheres?' to be barking-up-the-wrong-tree and perhaps somewhat odd, but in the context of the time, it was a good one: it was building on the then-current beliefs that the celestial spheres were real, and that the Platonic solids were 'perfect' and therefore 'cosmically appropriate'; furthermore, no one had asked that question before, so Kepler was opening up a hitherto unexplored avenue of enquiry. When Kepler discovered that his question wasn't yielding meaningful results, he then asked another: 'what is the simplest closed curve that matches Brahe's observations of the orbit of Mars?' – in my terms, a much deeper, better, question, and one that ultimately led to one of the greatest discoveries in the history of science. In answering this second question, Kepler proved that the orbit was not a circle; he then posed the hypothesis of the egg. And in process of trying to prove that the orbit is an egg, he spotted the evidence that it might be an ellipse, and the end-game was in sight.

Some questions are 'closed' in that they imply a preferred outcome and lead to the testing of the corresponding hypothesis, for example, 'Do the Platonic solids fit within the celestial spheres?' and 'Is the orbit of Mars egg-shaped?' Other questions are 'open', inviting the questioner, without constraints, to think, to explore and to discover, an archetypal example being 'what is the simplest closed curve that might satisfactorily describe the orbit of Mars?' My experience is that the more open the question, the more fundamental the resulting enquiry, and the more likely that the result will be deeper, more profound, more insightful.

The importance of asking good, fundamental, questions has been recognised for a very long time, for example:

The art of proposing a question must be held of higher value than solving it.
Georg Cantor [10]

This is the essence of science: ask an impertinent question, and you are on the way to a pertinent answer.
Jacob Bronowski [11]

If I had an hour to solve a problem and my life depended on the solution, I would spend the first 55 min determining the proper question to ask... for once I know the proper question, I could solve the problem in less than five minutes.
attributed to Albert Einstein [12]

The wise man is not a person who gives the right answers; he is one who asks the right questions.
Claude Lévi-Strauss [13]

and the somewhat more earthy variant

Give me six hours to chop down a tree, I will spend the first four sharpening the axe.
attributed to Abraham Lincoln [14]

There are many more.

1.4 Invention and discovery

One further point. Some people will say, 'Beethoven is the true example of creativity, not Kepler. Beethoven created the 5th Symphony totally out-of-the-blue, for there was nothing like that before. Total creativity, an absolutely new invention. Kepler didn't do that. Mars has been following an elliptical orbit for billions of years. What's creative about discovering something that is already there?'

This drives a wedge between 'invention' (good) and 'discovery' (dull), as well as implying that only the arts can be 'creative', whilst the sciences are merely boring.

Needless to say, I don't like such statements, for two reasons.

Firstly, the assertion is incorrect. The claim that 'there was nothing like Beethoven's 5th Symphony before Beethoven's genius created it' is false. I do not challenge the claim that Beethoven was a creative genius, for that it surely true. But Beethoven was by no means the first person to write symphonies, or to deploy ensembles of musical instruments in a well-coordinated manner. Nor did Beethoven invent the musical notes, which he combined together in such a powerful way. Yes, that specific combination of notes, that exact mixture that we now call his 5th Symphony, had not been heard before, but many musical compositions like it had – such as the symphonies of Haydn and Mozart, and especially Beethoven's own 3rd Symphony, which some might argue is even more stirring, greater, than the 5th.

Secondly, I do not subscribe to 'invention, good; discovery, dull'. I recognise that the words 'invention' and 'discovery' are subtly different. 'Discovery' is primarily used in connection with the revelation of something that is already there, but hidden, not easily visible, covered up – as indeed implied by the word itself, dis-cover. 'Invention', on the other hand, is used in relation to things that were not there before: we speak of the light bulb as being invented, rather than discovered, and although we don't usually apply the term 'invention' to works of art, we could.

But although the result of Kepler's prodigious labour was the discovery of something-that-was-already-there, the *process of discovery* required immense imagination, perseverance, creativity. As did the labours of Newton and Einstein, of Darwin and Bardeen. In my view, all these discoverers, and many others too, have their place in the pantheon of creative geniuses, alongside inventors such as Edison and Marconi, artists such as Rembrandt and Picasso, and writers such as Shakespeare and Zola, to name just a very few.

Nor do I wish to expend too much energy splitting hairs between 'invention' and 'discovery'. Yes, gravity, as a phenomenon, had always been there, but was the inverse-square law, the explicit statement of how the force attributable to gravity acts between two masses, 'there' in the same sense? Was Newton's formulation of that law an act of 'discovery' of the pre-existing, or the 'invention' of an explanation? Likewise, when both Newton and Leibniz published their independently-derived fundamental theorems of calculus, was this a 'discovery' or an 'invention'? Philosophers, linguists – and perhaps pedants too – might wish to debate that at length; but for me, the distinction is not hugely helpful. To me, both invention and discovery are embraced within the broader concept of creativity, and that's the word that I will primarily use from here on.

1.5 What's missing from the sound-bite?

The 'sound-bite' definition is very brief, just five words. And there are some words that some people might consider to be missing – for example, words such as 'new' or 'original', as well as, for example, 'valuable', 'useful' and 'good'.

Yes, those words are missing. Deliberately so. To explain...

1.6 What is 'new'?

> ### My first brainstorming workshop
>
> I'm young, new to the team. I'm keen to make a good impression, and even more keen not to do anything wrong. And this morning, I'm excited; but apprehensive too. I've been invited to participate in a brain-storming workshop – which is great because the whole team will be there, including the boss, and the topic is one I'm interested in. But it's also a bit nerve-racking – I don't want to put my foot in it.
>
> I enter the room, sit down. The facilitator does the usual opening stuff, and says, 'Great. This session is all about generating as many new ideas as we can. To make that safe, let's remember that all ideas are good ones, and that we don't put each other down. OK?'.
>
> People nod.
>
> The discussion gets going, and I listen carefully. Then, I have an idea. Fantastic! But wait a minute... The facilitator said 'new ideas'. She didn't make a fuss about that, it was a very natural statement. But no one wants to waste time on old ideas, do they?
>
> So before I open my mouth, I'd better think about that idea a bit harder, and check that it really is new. Well... I don't think I recognise it... so it's new to me... at least as far as I remember... but now I think about it, I do remember something a bit like it when I was doing some reading a couple of weeks ago... perhaps it isn't new after all...
>
> My temperature is rising a bit, and I realise I'm not paying attention to what is going on in the workshop. Oh... And then a real fear. Even if the idea is new to me, it might not be to others! They've read a lot more than I have, they know a lot more than I do. And I don't, just don't, want the boss to lean forward and say, 'Ah. Yes... thank you for suggesting that. I realise you're new to the team, so you won't know that we discussed that one last year, and rejected it...'. That would crush me!
>
> So better to stay silent...

You might recognise that story; you might even have experienced something very similar yourself. Part of it is about the natural apprehension of being young, and having only just joined the team; part of it is about the organisational culture as a whole. But a large part of the problem is caused by that innocent-looking word 'new'. Since the agreed objective of the workshop was 'all about generating as many new ideas as possible', then the search is, as stated and agreed, for 'new ideas'. And, yes, the word combination 'new idea' might have been just a turn-of-phrase, a familiar expression used without too much thought. But it still influences what people think, what people do. And the safe thing – for most people – is to stay quiet, and only follow the lead of the bosses.

For the truth is that no one – not even the boss – knows what a genuinely 'new' idea is, and no human being on the planet can be sure, when an idea passes through the brain, that the idea is indeed new. No one has total knowledge of all ideas that have ever been published, let alone proposed, and so no one can truly answer the question 'Is this idea new, yes or no?'

And even if the question could be answered, the answer could be highly misleading. The point here is that the manifestation of any idea does not exist by itself; it exists in a context. So it may be that the idea itself is not new at all, but that the current context is different from the context in which the idea was originally suggested. Although an idea might have been examined before, and perhaps rejected, the context might now have changed so that an old idea might be very relevant now. But under the 'rule' that only new ideas are to be discussed, that old idea might be never be tabled, and perhaps a valuable opportunity might be lost.

An example. The idea that man might fly goes back to antiquity (for example, the story of Icarus, who flew too close to the Sun), and around 1490, polymath genius Leonardo da Vinci designed several intriguing flying machines, depicting them in exquisite drawings [15]. Great designs, but they didn't actually work. So, four hundred years later, in the late 1890s, when Orville Wright was having breakfast one morning and said, 'Why don't we try to build a flying machine?', it's good that his brother Wilbur didn't reply, 'What? That old idea? Da Vinci had it centuries ago! No. We need to do something new!'

Around 1900, the idea of a flying machine was not new. But the context was. In particular, the key reason why da Vinci's designs didn't work was because they were driven by the only power source available to him at that time: human muscles. But however strong men might be, it was impossible for human beings to develop sufficient power to lift their own weight, plus that of any of the da Vinci machines. But by 1900, the internal combustion engine had been invented, and much more power could be derived from much less weight. So, after some careful and comprehensive experimentation on the shapes of wings, the Wright machine successfully took to the air for the first time in 1903 [16]. And, to complete the picture, it was not until 1961 that a sophisticated aeronautical design, combined with the use of light-weight materials such as balsa wood, plywood, aluminium and nylon, resulted in SUMPAC, a machine that could take off, and fly a short distance, under muscle power alone [17].

Words such as 'new', 'novel' and 'original' are, in my view, not only unnecessary in the definition of creativity, but also unhelpful, possibly setting up barriers to clear, exciting and relevant thinking. Furthermore, as we shall see shortly, there is a significant philosophical problem as to what the word 'new' actually means, which makes the use of 'new' within a definition of creativity even more troublesome.

So to me, a definition along the lines of 'creativity is having a new idea' is restrictive and misleading. That's why I use just the words 'creativity is having *an* idea'.

1.7 It's difference that's important, not novelty…

There is another, important, aspect of the word 'new'. It distracts attention, and encourages us to look in the wrong place. For although something 'new' is enticing, exciting and attractive, 'newness' or 'novelty' is usually what we neither need nor want.

That might come as a surprise.

So here's something to think about.

What we usually need, what we usually want, is not 'novelty'. It's 'difference'.

In the vast majority of real situations, situations in which a good idea would be really helpful, what we are seeking is *something different from, and ideally better than, what is happening now*. Whether or not it's 'new' might be 'interesting'; what's important is that it is *different* and, hopefully, *better*.

An example: any manufacturer of any product. Currently, the manufacturer makes 'these'. They're selling well, and there are no commercial problems, but the manufacturer knows that the competitors are out there, and that it makes good business sense to refresh the product range from time to time. So the manufacturer needs to sell 'those'.

The key attribute of 'those' is that they are *different* from 'these', for – quite obviously – if they were the same, nothing will have happened. Perhaps the difference between 'these' and 'those' is significant, as, for example, happened when Nokia shifted from manufacturing rubber boots to mobile phones. Much more frequently, the change is modest, incremental, such as the changes from any one iPhone to the next model. Yes, the manufacturers make a huge fuss about 'new', but that's just marketing hype. The degree of 'newness' between any two successive smartphone models can often be *de minimis*. But they are different in some aspect, even if it's only the colour.

Another example: teamwork. The leader of a team senses that the team isn't working quite as well as she feels it should. To change that, the leader, and the team members, need to have an idea about what might be better. But in practice, that's all about discovering something that an individual, or a small subset of the team, or the team as a whole, *needs to do differently*. Perhaps it's an agreement that everyone turns up to meetings on time; perhaps it's a recognition that [person X] is genuinely over-committed and should allow someone else to undertake a particular task; perhaps the leader should show a stronger sense of direction. None of these are 'new' in any sense at all. But they could each, or all, be *different* from what is happening at the moment – and, if implemented, have a hugely beneficial effect on the team's performance, and perhaps on the sense of satisfaction experienced by each team member.

A third example: how to make an algorithm run faster. To make that happen, something about that algorithm, or the environment in which the algorithm operates, must be different. The specific difference that produces the improvement might be hard to discover, and that's what research is all about. But the issue is not the novelty of the answer; it's about being different.

1.8 ...and the best way to discover differences is to be observant

There is also another, immensely significant, attribute of 'different' as compared to 'new'. To search for 'novelty' is open-ended. Where do I look? What am I looking for? Those questions are really hard to answer. In contrast, the search for 'difference' is very well-directed, focused and efficient. For 'different' means *'different from what is happening right now'*.

That's important. Very important. For what is happening right now is indeed happening right now. You, and I, can see it, touch it, feel it, describe it. So the manufacturer knows what today's product is, how it works, what customers like about it... and indeed about competitors' products too; every team member can sense what is happening in the team now; that algorithm is there, doing what it does and taking the time it takes.

So you, and I, can describe 'what is happening right now', in great detail, for it is 'there', before our very eyes. And if the things you and I see aren't the same – as inevitably is the case if you are a user of a product that I designed – then we can compare notes, and enrich our collective understanding of the *status quo*. That then gives a most solid platform for asking the single most important question in the entire canon of creativity – *'How might [this] be different?'*, where [this] is any feature of what-happens-now.

Searching for difference is much more pragmatic than searching for novelty. Searching for difference is grounded in the tangible, visible, verifiable reality of what is actually happening now.

So if you wish to be creative, don't just stare into the blue space, hoping for the lightning to strike. That's looking in exactly the wrong direction. What you should do is look at *what is happening now*, in great detail, and share your observations with other people, who might perceive the same reality rather differently. And then, once those observations have been gathered together, ask that 'killer question' – *'How might [this] be different?'*.

> **Careful observation underpins all creativity**
>
> Many novelists and dramatists have acknowledged that much of their creative output is based on their personal experience, and observation.
>
> Charles Dickens, in particular, based many of his characters on people he had seen on London's streets, in court rooms, in factories.
>
> David Hockney, the great contemporary artist, has made many statements about the fundamental importance of careful observation, including, referencing his own work, 'Teaching people to draw is teaching people how to look', and of Van Gogh, 'Lots of people just scan the ground in front of them so that can walk, but they don't really look at things. Van Gogh really looked.'
>
> Charles Darwin, surely, was one of the greatest, and most insightful, observers of the natural world. But he went much further, questioning, seeking explanations, searching for patterns. So, for example, he might have noticed himself thinking, 'I see all these living forms as just that, individual living forms. How might that be different? Might there be some hidden, underlying patterns?'
>
> Andre Geim, joint winner of the 2010 Nobel Prize in Physics for the discovery of graphene, noticed a thin layer of graphite on some sticky tape. Most people would have thrown it away. Geim didn't do that. He asked 'Suppose I don't throw it away. What else might I do? I might look, very carefully, at that thin deposit…'
>
> And the fictional detective, Sherlock Holmes, so renowned for his powers of observation and deduction, says to his colleague Dr Watson, who has just missed some vital clues, 'You see, but you do not observe. The distinction is clear.' That's a general statement that pervades all the Sherlock Holmes tales; those particular words are to be found in the opening few pages of the very first short story, 'A Scandal in Bohemia'.
>
> **Careful observation underpins all creativity. For it provides the platform for asking 'How might this be different?'.**

For sources, see [18–22].

Yes, careful observation really does underpin all creativity. And to illustrate that, take a moment to solve the problem posed by figure 1.1 on the following page.

You might like to copy the pattern onto a sheet of paper and give it a try…

…and if you do, don't be surprised if, after a few moments and some frustrating trial-and-error, you throw your pencil down and exclaim, 'What a stupid question! It's impossible!'

Most people do.

But not everyone. Like the person who just happens to spot that if you fold the paper, with each of the three rows of dots along three successive 'crests', then the crests can be pushed together, and a single straight line can join them all.

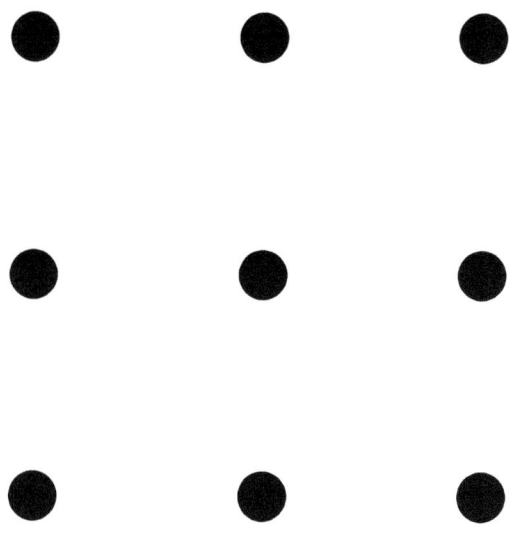

Figure 1.1. How many different ways can you discover of joining all nine dots by drawing a single straight line?

Yes, when the paper is flattened out, there are three straight lines, one through each row. But by folding the paper, only one line was actually drawn... But how did the person spot that? The person doesn't know... so let's call that 'a flash of inspiration'.

Like the person who is bored with the silly puzzle, and looks out of a window, and sees a house decorator painting a nearby wall, using a paint roller. A paint roller... Yes! If I use a paint roller rather than a pencil, then I can join all the nine dots with one – rather fat – straight line! As long as the width of the 'pencil' is greater than the side dimension of the dots, that works – for any number of dots too! Although this solution appears to be 'out of the blue', its discovery is not so mysterious: a combination of luck (looking through the window and seeing a house decorator, which is quite accidental), observation (noticing the paint roller), and 'bisociation' (putting the concept of the paint roller together with the concept of the puzzle, as discussed towards the end of the Prologue).

And then the person who notices things, who is observant. Like the fact that the paper is still, not moving. Like the fact that the paper is a single continuous surface. Like the fact that the paper is flat. Like the fact that the width of the pencil lead is about a millimetre or two.

That person is curious too, and asks questions. What might happen if the paper were no longer still, but could move? That would allow the paper to move underneath the pencil, which is itself moving in a straight line. Mmm... how might the paper move? Well... perhaps it rotates about the central dot. If it rotates whilst the pen moves in a straight line from the edge of the paper towards the central dot, and if the paper is rotating quite fast (relative to the speed of moving the pencil), then all the dots will have moved underneath the pencil by the time the pencil reaches the centre. The pencil has drawn a straight line as required, and all nine dots have been joined, as also required. When the paper stops moving, the line doesn't look straight at all – it will be a spiral, perhaps like the groove on a vinyl record. But the pencil moved in a straight line…

And if the paper were not a continuous surface, what might it look like? Cut up, of course. Yes! I can cut the paper, lay the dots out in a straight line and join them, and then reassemble the paper. And there are many, many ways of cutting the paper – an infinite number of ways…

The paper is flat. How might that be different? Any three-dimensional surface. Like a cube, or a tetrahedron. Or a cylinder. Cylinder! Yes, if I form a cylinder, and make it rotate along the long axis, I can move the pencil in a straight line, so it ends up looking rather like a barber's pole, or a screw thread, with all nine dots joined… And another three-dimensional shape is a wave… so the paper could be folded…

The pencil lead is a millimetre or two wide. How might that be different? Well, it could, in principle, be a nanometre… or a metre… Ah! A 'fat pencil'. That will do it!

How many other solutions can you discover by being observant and then curious? And permission helps too – yes, you can fold the paper, or cut it, or rotate it…

1.9 Value

Back to the definition of creativity… and some further words that some people might consider should be included are, for example, 'valuable', 'useful', 'good' and their various synonyms.

Of course, no one wants ideas that are wasteful, useless or bad.

But when an idea is buzzing around in your mind, you do not know whether the idea is powerful or weak, a potential prize winner or a 'dog', a financial goldmine or a sink-hole. Yes, you might have a strong, even overwhelming, belief that the idea is truly blockbusting, and you will do everything you can to convince others accordingly. But the truth is that you don't *know*. After all, it's only an idea. Reality is different. And until the idea is made real, whether the idea actually *is* good or bad, money-making or money-losing, a great benefit to the public good or riddled with adverse 'unintended consequences', no one can know.

There are many examples of the disconnect between belief and reality, of ideas that were originally believed to be good, but were ultimately shown to be poor, and – the other way around – of ideas that were judged originally as poor but turned out to be very good indeed.

We have already seen that Kepler's Platonic solids, and his egg, were both wrong, but not without value: the (fruitless) quest for the Platonic solids, and the blind alley of the egg, were critical steps along the (highly zigzagged) path to the blockbuster idea, the ellipse. Another, more recent, idea that proved to be a commercial disaster, despite being an intrinsically 'good' idea, was an electrically-powered personal vehicle, the C5, invented and marketed in 1985 by the most famous entrepreneur in Britain at the time, Sir Clive Sinclair. From today's standpoint, an electric car is very sensible, and so perhaps, in 1985, the C5 was an idea ahead of its time [23]. But in the reality of 1985, commercially, it was a total failure.

Some of the best examples of ideas considered to be bad, but ultimately proving to be enormously valuable, are in the publishing industry. The manuscript of JK Rowling's first *Harry Potter* book was turned down by 12 publishers [24]; when William Golding, the winner of the 1983 Nobel Prize in Literature, was looking for a publisher for *The Lord of the Flies*, he was rejected 21 times [25]. In the commercial world, one of the candidates for 'worst-ever judgement of an idea' must be that made by the tech giant Yahoo. In 1998, Yahoo was approached by two 25-year-olds, Larry Page and Sergey Brin, who were seeking to sell a search algorithm, at that time called *PageRank*, that they had just developed. Yahoo said 'no'. With no buyer, Page and Brin decided to make use of the algorithm themselves. The company they formed is called Google [26].

Attributes such as value, utility and goodness are attributes of the reality, not of the idea as first articulated, long before it is realised. Rather, high value, beneficial utility and overall goodness are the prizes of getting the whole process of having an idea, and then doing everything that is needed to transform that idea into reality, right. So in my view it is inappropriate to include these words within the definition of creativity itself.

That said, value, utility and goodness are undoubtedly immensely important. But – once again in my view – they are not associated with creativity. They are, however, central to the process that follows creativity, the process I call 'wise evaluation'. As we shall see in outline shortly, and in detail in chapters 14 and 15, 'wise evaluation' is the rigorous scrutiny of an idea to test, as far as possible, for these attributes, and to exercise wise judgement as to whether the idea is likely to be valuable, useful and good at some time in the future when it will have been implemented – or not. But this scrutiny can only take place after an idea has been well understood and thought-through. So I make what I believe to be an important distinction between creativity – having an idea – and the subsequent process of wise evaluation, judging whether or not that idea is likely to be a good one. Muddling the two is confusing and unhelpful: far better to generate ideas, non-judgementally, first, and then, later, go through them all to judge which are likely to be the good ones, and which the poor.

So creativity is having an idea. Just that. Not a new idea or an original idea. Not a valuable idea or a good one. Just an idea.

But although 'creativity is having an idea' is true, and a helpful sound-bite, it isn't very insightful, and it certainly doesn't give any guidance as to what you can actually *do* to have an idea. We need to go deeper. So, firstly, I'll put creativity in context, and then I'll explore a definition of creativity that is richer, more fundamental, and that helps us actually do it.

References

[1] www.oxfordlearnersdictionaries.com/definition/english/creativity?q=creativity
[2] https://dictionary.cambridge.org/dictionary/english/creativity
[3] www.merriam-webster.com/dictionary/creativity?src=search-dict-box
[4] www.collinsdictionary.com/dictionary/english/creativity
[5] www.ldoceonline.com/dictionary/creativity
[6] www.animalcognition.org/2015/04/03/tortoises-can-master-mazes/
[7] Mulcahy N J 2018 An orangutan hangs up a tool for future use *Nat. Sci. Rep.* **8** 12900
[8] Kaplan M 2008 Marine biologists interpret whale sounds *Nature* https://doi.org/10.1038/news.2008.984
[9] Sengesser S, Holub J L, O'Neill L G, Russell A F and Townsend S W 2019 Chestnut-crowned babbler calls are composed of meaningless shared building blocks *PNAS* **116** 19579–84
[10] https://mathshistory.st-andrews.ac.uk/Biographies/Cantor/quotations/
[11] Bronowski, J 1974 *The Ascent of Man* (Boston, MA: Little Brown) p 153
[12] https://amorebeautifulquestion.com/einstein-questioning/ and https://quoteinvestigator.com/2014/05/22/solve/
[13] Lévi-Strauss C 1964 Le savant n'est pas l'homme qui fournit de vraies réponses; c'est celui qui pose les vraies questions *Mythologiques vol 1, Le Cru et le Cuit* (Paris: Plon)
[14] https://quoteinvestigator.com/2014/03/29/sharp-axe/
[15] https://theconversation.com/leonardo-da-vincis-helicopter-15th-century-flight-of-fancy-led-to-modern-aeronautics-116241
[16] https://airandspace.si.edu/collection-objects/1903-wright-flyer/nasm_A19610048000
[17] www.southampton.ac.uk/engineering/about/making-history/sumpac.page
[18] www.theguardian.com/books/2012/feb/01/charles-dickens-real-character-names
[19] Interview with Jasper Gerard 2 October 2005 *Taking the Fight to the Dreary People* (London: The Sunday Times) and www.vangoghmuseum.nl/en/stories/hockney-van-gogh-two-painters-one-love
[20] https://fs.blog/charles-darwin-thinker/
[21] www.nobelprize.org/uploads/2018/06/geim_lecture.pdf
[22] Doyle A C July 1891 *A scandal in Bohemia* (The Strand Magazine)
[23] www.bbc.com/future/article/20141209-sinclair-c5-30-years-too-soon
[24] https://ew.com/books/2017/10/24/harry-potter-jk-rowling-original-pitch/
[25] www.classicbooks.com/classic-authors/william-golding/
[26] https://blogs.cornell.edu/info2040/2016/10/18/yahoos-greatest-regret/

IOP Publishing

Creativity for Scientists and Engineers
A practical guide
Dennis Sherwood

Chapter 2

Creativity in context

2.1 Creativity alone is not enough

Creativity, having an idea, might give great satisfaction to the originator, but from anyone else's point-of-view, while an idea might be interesting, by itself it's not of much use. Beethoven might have 'heard' his 9th Symphony in his mind (as he was obliged to do since he was quite deaf when he wrote it), but until an orchestra actually played it for us all to hear, no one else was able to benefit from his genius. Likewise, Kepler might, on awakening from his 'sleep', have felt relief now that he had proven that the orbit of Mars was an ellipse, but if he had not documented his proof in *Astronomia Nova*, Newton would have been unable to read it and might never have discovered the laws which now bear his name.

From a practical point of view, creativity alone is not enough; rather, it is the first of a sequence of four distinct processes:

- Creativity
- Evaluation
- Development
- Implementation

as represented schematically by the four coloured zones on the 'Target Diagram' shown in figure 2.1 on the following page.

Collectively, the entire diagram represents 'innovation', which I define as 'making an idea real', encompassing all the activities required from generating the idea in the first place to making the idea available for others to use. And this is a good point at which to define 'entrepreneurship', which is the capability to manage all those activities successfully. Importantly, entrepreneurs do not themselves need to be creative and generate the idea for which they perhaps become famous. They might be, but it's not a necessary condition. What, however, is necessary, 'in spades', is the ability to manage, and to drive through,

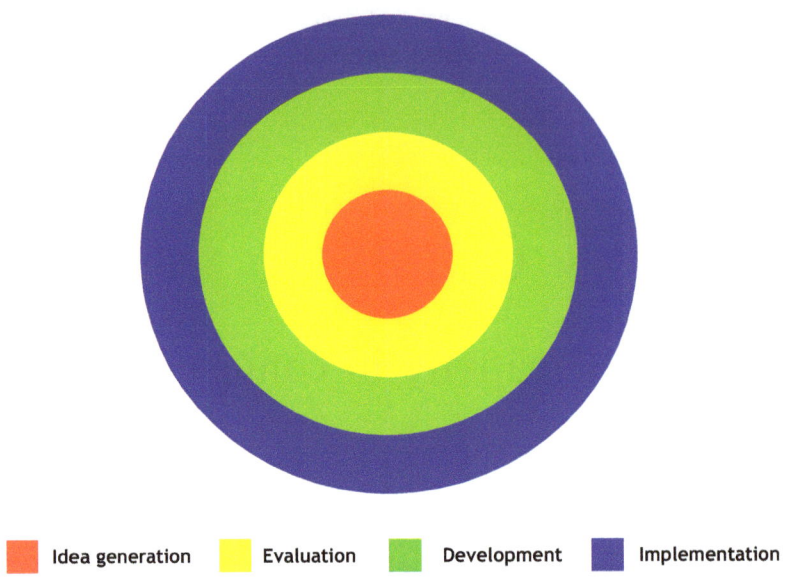

Figure 2.1. The innovation 'Target Diagram'.

a host of activities, and have the tenacity to overcome any number of difficulties, along the complete journey from creativity, through evaluation, development, and finally implementation.

Ray Kroc, for example, the entrepreneur behind McDonalds, neither invented the hamburger, nor opened the original restaurant, nor had the idea for those golden arches, nor was he the first to use the commercial concept known as franchising to grow and run a multi-outlet business. Rather, in 1954, Kroc was a salesman for machines that made milkshakes. On a sales call to a hamburger restaurant in San Bernadino, California, run by two brothers named McDonald, he was impressed by the restaurant's cleanliness and efficiency – so impressed that in 1955 he became a business partner of the brothers, buying them out in 1961 [1]. In so far as Kroc was creative, that was in having the idea that the McDonalds' restaurant might be the basis for building what was to become the world's highest revenue-earning restaurant chain, and his evaluation of that idea as a good one turned out, in practice, to have been hugely valuable (to him) indeed. That said, Kroc can hardly be rated as a 'creative genius' (!): his (most considerable) skills were in the subsequent development and implementation, transforming the McDonalds' original business from just a few restaurants in California and Arizona into today's (that's 2021) chain of some 40,000 outlets in more than 100 countries, serving about 70 million customers *each day*!!! [2]

Another example of the Target Diagram, closer to home: a student has a good idea for a PhD project (creativity, red zone), which is subsequently discussed with, and approved by, the supervisor (evaluation, yellow zone). The research project itself – and all that hard work – is the development zone (green), and implementation (blue) represents the writing and presentation of the thesis, the delivery of conference presentations, and the publication of papers.

The activities associated with each zone of the Target Diagram might cross organisational boundaries, either within the organisation (such as when a research project is collaborative across different departments) or involving a different organisation (for example, when the evaluation is carried out by a funding agency, whose approval is required for the award of a grant). Sometimes, each process is carried out by different organisations altogether, as in the example of penicillin…

Who discovered Penicillin?

Many people would answer 'Sir Alexander Fleming', and some would add 'when he noticed that something had killed some bacteria growing on a Petri dish he'd inadvertently left by an open window at St Mary's Hospital in London whilst he'd been on holiday for a few days'.

Yes, Sir Alexander did indeed do this, and the event, which took place in 1928*, is commemorated by one of London's 'blue plaques' beneath that very window, which can be seen today, quite close to London's Paddington railway station.

In fact, Fleming had blundered. He'd left the Petri dish near the window by accident, and had wanted the bacteria to grow, not to be killed. So many people, seeing the spoiled results, would have said, 'What a mess!!! That's all gone wrong!!!', and then thrown the contents of the dish away.

Fleming did not just see, he observed. And he observed something that was contrary to his expectation: he had expected the bacteria to grow, but some had been killed. He had made a mistake. But he had not failed. So rather than throwing the contents away, he said, 'that's funny… I wonder what has caused the bugs to be killed?'. He had a clue as to where to look, for some fungus was also growing on the Petri dish – fungus that might have blown in through the window – and it was the bacteria close to the fungus that had been killed. He subsequently identified the fungus as being *Penicillium notatum*, and that it had produced some sort of 'mould juice' (his term) that can kill the bacteria, a substance he named 'penicillin'. Fleming published his findings in 1929*… and then nothing much happened. Fleming did a little more research on it, but soon turned to other things.

But did Fleming discover penicillin? He was unable to purify whatever it was that the mould produced, so his discovery was not the active agent, but the rather vaguer observation that this particular *Penicillium* fungus produced some substance that killed bacteria. Well… in 1860, Louis Pasteur noticed that bacteria would not grow in a broth that also contained *Penicillium*, an observation confirmed in 1871 by both John Burdon-Sanderson* (who coincidentally had served as Medical Officer of Health for the London borough of Paddington in the 1850s) and Joseph Lister*. Furthermore, Arab horsemen had long known that smearing mould on the underside of their saddles would help prevent saddle-sores, leading Ernest Duchesne, in his PhD thesis of 1897, to identify the protective mould as a strain of *Penicillium**. And clues can be identified from much earlier times too: the Ancient Egyptians knew the medical benefits of mouldy bread*, and the Bible mentions the use of the herb hyssop for purification and cleansing*; it is now known that hyssop is a good host for the *Penicillium* mould*.

So the question 'Did Fleming discover penicillin?' is 'interesting', and does not have a simple answer. As we shall see shortly, this is but one story about the difficulty often associated with the attempt to attribute the discovery of an idea to a single person. As Newton famously said 'If I have seen further, it is by standing on the shoulders of giants'*.

> Back to the main story… As just noted, nothing much happened after Fleming's 'discovery' in 1928, and subsequent publication 1929. Not, that is, until 1938, when Ernst Chain, working at Oxford University's Sir William Dunn School of Pathology, came across Fleming's paper whilst doing a literature search. 'That's interesting', he thought, 'if we could purify the active agent, that might be very useful…'. Chain then discussed it with his boss, Howard Florey, so launching a research programme to do just that. And that's what they did – but the real hero here is a then-junior member of their team, Norman Heatley. One of the reasons Fleming had been unable to purify penicillin was because the molecule is very unstable, and all the conventional methods of purification destroyed, rather than purified, it. Heatley, a true master of chemical and biochemical creativity, ingenuity and (importantly) pragmatism, discovered some innovative ways to achieve the required purification, to the extent that sufficient, and sufficiently pure, penicillin was manufactured for clinical trials, which ultimately proved successful*.
>
> But the university research team could produce only laboratory-scale quantities. It was now 1941, the Second World War was raging, and it was clear that penicillin would save many lives. Discussions with the UK chemical and pharmaceutical industry proved cumbersome, so Florey and Chain went across the Atlantic to meet representatives of the US Department of Agriculture, and representatives of some major American pharmaceutical companies, including Merck and Pfizer*. The result was that the bulk of industrial production took place in America, under US patents.
>
> This story maps neatly onto the Target Diagram: creativity, the discovery (or perhaps re-discovery) by Fleming; evaluation, Chain's spotting Fleming's paper and Florey's agreement that trying to purify it would be a 'good idea'; development, Heatley's astounding work at Oxford; implementation, in the US at the Department of Agriculture, Merck and Pfizer.
>
> And in December 1945, Fleming, Florey and Chain received the Nobel Prize in Physiology or Medicine. But not Norman Heatley. But then no one has ever claimed that Nobel Prizes are fair.

For sources, see [3–13].

One further important point relating to figure 2.1: this representation is schematic only, emphasising creativity as the originating process, and that it is not possible to leap directly from creativity to implementation. On the contrary: it is obligatory to pass through firstly evaluation and then development. Also, the sizes of the rings have no significance apart from the visual image, and – in particular – imply nothing about, for example, the resources likely to be consumed in accomplishing each process, or the time required: indeed in these regards, the diagram might be quite misleading, for as a general rule (and of course all ideas are different in detail and in scope), the most costly is usually the green zone, development, and the time spent in developing an idea can be many, many times greater than the time required to generate the idea.

2.2 A richer picture

The Target Diagram, as shown in figure 2.1, captures the key message that creativity is the first of the four processes required to make an idea real. A more insightful representation is figure 2.2.

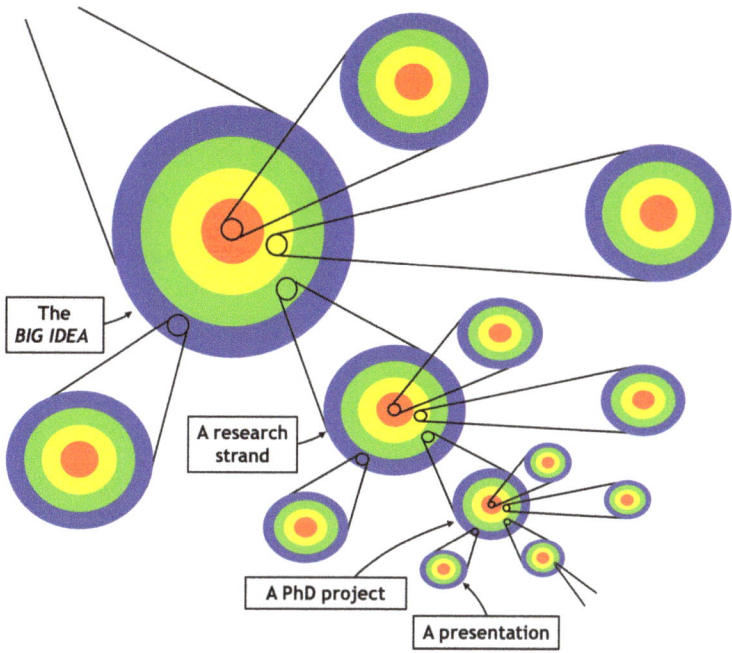

Figure 2.2. A more realistic representation.

This diagram shows the fundamental Target Diagram concept as successively embedded within itself, operating at different scales.

To explain: the large target at the top left represents a *BIG IDEA*, for example, a major research programme funded by a large grant. The idea was generated by a team of top academics, evaluated by the funding agency, and given the go-ahead. The overall research programme is structured as a 'family' of mutually-interdependent, but distinct, 'research strands', some theoretical, some developing new instruments, others experimental. Each different strand has its own host of problems-to-solve, requiring great ideas and much creativity, but within a narrower focus than that of the overall *BIG IDEA*. One such strand is represented by the rather smaller target, the second along the top-left to bottom-right diagonal, which is embedded within the green development zone of the target representing the *BIG IDEA*.

Within the development zone of each strand are all the individual research projects, some of which will be undertaken by post-doctoral research associates, some by PhD students; each of these can be represented by its own target, embedded within the appropriate development zone – as exemplified by the third target along the diagonal.

And if a PhD student, towards the end of the project, is about to give a presentation, but wants to deliver it in a rather more creative way, then the Target Diagram can apply to that too. This is embedded with the (blue) implementation zone of the target representing the project, and itself requires creativity (what ideas can I generate to present this data in a more informative way?), evaluation (is this slide clear, or maybe a bit cluttered?), development (getting the font sizes right, and the colours), and then finally implementation (delivering the presentation).

Florence Nightingale's 'Rose Diagram'

One of the most powerful examples of 'delivering results in a more creative way' is the 1858 'Rose Diagram' designed by Florence Nightingale.

On 4 November 1854, Florence arrived with a team of 38 nurses in the Turkish town of Scutari on the Asian shore of the Bosphorus, opposite Constantinople. Their task was to care for soldiers wounded in the Crimean War*.

The conditions she found were atrocious: overcrowded, insanitary, poorly ventilated, with ten times more soldiers dying from disease than from wounds suffered in battle. She and her colleagues worked tirelessly to improve hospital hygiene, to great effect: the number of deaths attributable to disease plummeted.

Florence recognised, however, that the problems she encountered at Scutari were a particular case of a more systemic problem about how poorly the army was organised to look after its wounded, within the broader context of a lackadaisical governmental attitude to public health in general. But to change 'the system', Florence had to influence the 'top brass' – to gain their attention and to motivate them to take action.

Yes, Florence was a compassionate nurse, and a brilliant organiser/manager. And in addition, she was a talented mathematician and statistician too, maintaining meticulous records of relevant data*. Data that she presented in a novel and powerful way. At that time, data sets were largely presented as tables of numbers. But Florence wanted the numbers to have more impact. So, building on the pioneering work of William Playfair, who developed bar charts and pie charts* and working with the leading epidemiologist and medical statistician William Farr*, Florence created three 'rose diagrams', one of which, from 1858, is reproduced as Figure 2.3:

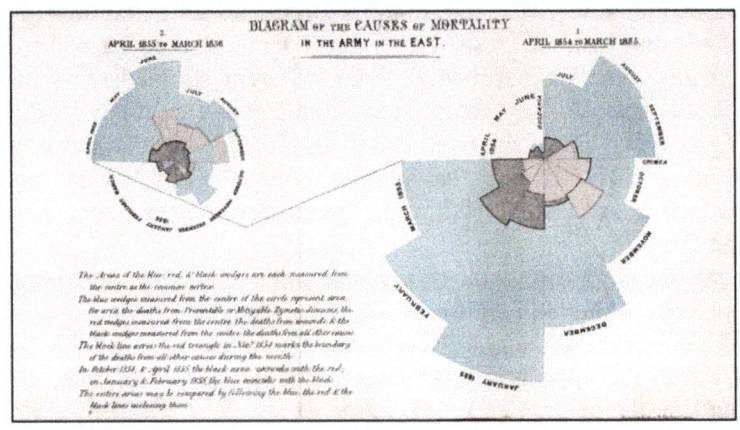

Figure 2.3: Florence Nightingale's 'Rose Diagram' of 1858*

Source: Wikimedia 'Diagram of the causes of mortality in the army in the East'
https://commons.wikimedia.org/wiki/File:Nightingale-mortality.jpg

> The area of each blue wedge, from the centre to the outer circumference, is proportional to the numbers of death from disease; each red wedge, death from wounds; each black wedge, death from other causes. The first landings of troops took place on 14 September 1854, and the large number of deaths from wounds in November 1854 is attributable to the aftermath of the Battle of Balaclava (25 October, and which included 'The Charge of the Light Brigade'), and the subsequent, and brutal, Battle of Inkerman (5 November).
>
> From February 1855 onwards, the effects of Florence's improvements are vividly shown by the progressive shrinkage of the blue wedges (noting that between June and September 1855, there were major battles around Sebastopol – hence the larger red wedges).
>
> And her work had impact. People – and important people too – took notice. And took action. Florence Nightingale's creative data visualisations saved lives*.

For sources, see [14–19].

Figure 2.2 is therefore more realistic: the four processes of creativity-evaluation-development-implementation operate at all scales, but are fundamentally the same; the diagram is fractal. Towards the top left are the *BIG IDEAs,* towards the bottom right are the solutions to today's problems. Towards the top left, the ideas are 'bigger'; towards the bottom right, 'smaller'. In general, generating *BIG IDEAs* would be something that happens relatively rarely, perhaps once a year; in contrast, invoking the Target Diagram for 'little ideas' is something that could happen every day, for there are always today's problems to solve, and opportunities to grasp that make 'today's world a little bit better'. And don't fall into the trap of believing that only *BIG IDEAs* are valuable and that 'little' ideas are trivial. 'Making the world a better place' is always valuable in context: including making the font on that presentation rather bigger so that people at the back can read the words.

The more complex fractal representation of figure 2.2 is more realistic than the simpler diagram of figure 2.1, but even figure 2.2 isn't fully right, for it fails to capture one further feature of reality: that the process is recursive, and that sometimes, having followed a particular trail for a long time, having solved all sorts of problems, and having generated and tested all sorts of perfectly plausible ideas along the way, I get truly stuck, or discover that the original *BIG IDEA* was in fact wrong. At that point, I have to stop and start again, so there is a big feedback loop across the whole diagram too. But I haven't yet discovered a good, intelligible, way of capturing that visually, so my representations have progressed only as far as figure 2.2. I clearly need to invoke the Target Diagram myself to generate some good ideas as to how to improve the Target Diagram … and so if you have any suggestions, please let me know!

2.3 Process 1 – Creativity

Before the 'world can be made a better place', you have to have the idea first. So the very first step is creativity, having that idea. Accordingly, the purpose of this chapter is to start exploring just what 'creativity' is, so helping set the scene for the next few chapters, especially chapter 9, in which I will describe in detail precisely how you can generate great ideas, deliberately, 'on demand', whenever and wherever an idea might be helpful.

A lot more on creativity is to come; the point I would like to make here is that, in general, creativity can be done very effectively in a short time, and by a small number of people.

We have already seen that, fundamentally, creativity happens in the brain of a single human being, and it is certainly true that individuals, working by themselves, can have very good ideas indeed. In many cases, however, even better ideas can be generated by small teams, for that enables ideas to be explored and enriched. In my experience, groups of about six people work well, especially when they each have different backgrounds, and different perspectives. I'll examine this in more detail in chapters 9 and 10; suffice it to say here that the more varied the perspectives around the table, the richer the enrichment. The mistake is often made that the only people around the table are the 'experts', so, for example, the design engineers get together to design a new product. When experts assemble, the result is that either everything converges on what is already the conventional wisdom, or two 'experts' divert the entire discussion by arguing, loudly, and for a long time, about who is the 'bigger expert'. Far better to gather together some design engineers, some manufacturing people, some representatives of the sales force – and, most importantly, some representatives of the customers who are actually going to use it!

It is also my experience that well-constructed workshops, lasting from a few hours to, say, two or (at most) three days, can generate a huge number of ideas, certainly more than enough for the organisation to work with over the next many months. That's the justification of my assertion just now that 'creativity can be done very effectively in a short time'.

That might come as a surprise, for two reasons.

The first reason is that it seems to be in conflict with the story of Kepler: he didn't solve the problem of the orbit of Mars in a few hours at a workshop; he struggled over more than five years. Yes, the story does extend over that time – but not all that time was devoted to creativity, as can be appreciated by considering Kepler's story in the context of figure 2.2.

Those five years relate to Kepler's journey (or 'sleepwalk', to use Arthur Koestler's vivid description) across the entire space depicted by figure 2.2, and all the recursions he took as he started again and again after having wandered down any number of blind alleys. The bulk of the time was in whatever development (green) zone he happened to be in at any particular moment, multiplying long numbers together 'by hand', checking calculated positions of Mars against Brahe's observations, drawing intricate diagrams.

Within all this, it didn't take much time to formulate Kepler's key questions – questions such as 'Is it possible to fit the five Platonic solids within the six celestial spheres?', 'Might the orbit of Mars be egg-shaped?', and 'What is the simplest closed curve that matches Brahe's observations of the orbit of Mars?', which, as we saw in the Prologue, were the critically important acts of creation. But the work required to answer these questions – and then discovering that the hypotheses behind the first two of these questions were wrong – did indeed take years. Kepler's story does not therefore negate my statement that 'creativity doesn't take much time', and, as we shall see in the chapters 9 and 10, by using a well-crafted process, it is possible to generate a host of ideas, including some truly stunning ones, in workshops of duration from a few hours to a few days.

The second reason concerns the experience that we have all had, in which an idea is there, in the back of our minds, but not quite right, over perhaps quite a long elapsed time, as if the idea is somehow being 'incubated'. This 'background thinking' is real, and makes an enormously valuable contribution to creativity – as discussed, for example, in Henri Poincaré's 1910 essay 'Mathematical Creation', in which he lucidly describes how he made his discoveries associated with the mathematical constructs known Fuchsian functions or automorphic forms [20].

But while an idea may be 'incubating' in the back of our minds, we are doing other things, from taking a bath to giving lectures, from going for a walk to solving a different problem altogether. So to equate that total elapsed time to the actual time spent on the 'creativity' associated with that particular idea is, I suggest, misleading.

I therefore return to my assertion that very many ideas can be generated in a relatively small amount of time, especially at a creativity workshop, which might generate ideas in their hundreds. These ideas are likely to be of all scales, and of very varied quality: in any list of, say, 100 ideas, most, on reflection, will be recognised as poor, some 'interesting', and just a very few – maybe 3, perhaps 5 – 'promising'. But at the end of the workshop, with the walls covered in flip-charts and post-it notes, how can you identify the 'promising' ones, and distinguish them from the rest?

And so to…

2.4 Process 2 – Evaluation

… for that's the objective of the second process identified in the Target Diagram – evaluation, or, rather better, 'wise evaluation': to identify the very best ideas wisely.

Of the four processes shown in the Target Diagram, wise evaluation is to me both the most important and the most difficult. It's difficult because a judgement has to be taken on inevitably limited evidence – the question 'is this idea good or bad?' requires firstly, a wise understanding of what 'good' and 'bad' actually look like, and secondly an ability to see into the future to anticipate not only what needs to be done to take the idea successfully through development and implementation, but also the extent to which 'the world' is a better, or indeed worse, place once the idea has been implemented. This requires forecasts and estimates, often involving hopes and wishes. There is no 'truth', no 'hard evidence', for the idea being evaluated is no more than a concept described in a briefing paper or as a set of diagrams. The idea cannot be 'touched' or 'felt' (other than, say, as a physical model), yet a judgement needs to be made as to whether to invest in that idea – or not. Which is why wise evaluation is difficult. But not impossible, especially when following the guidelines to be presented in chapters 14 and 15.

And that word 'invest' is the clue as to why evaluation is so important. As we have seen, the effort, time and resources required for really effective creativity is modest, and the number of people involved can be quite small. But once an idea is taken into development and implementation, the effort, time and resources that might be required can be vastly more.

To illustrate... A small team, say six people, can quite quickly generate some truly great ideas, and another small team can judge that idea wisely, evaluating it as a 'yes'.

What happens next depends on what the idea actually is, but as an example, suppose that the idea is for a new product, such as a washing machine. Here are just a very few of the activities that might need to be carried out: design engineers need to solve all the questions about exactly how the product can be built; new equipment is needed in the factory, so suitable suppliers have to be identified, contracts negotiated, machines installed and commissioned, existing staff trained and new staff hired; the accountants have to do all the costings and the commercial team the pricing; marketing material has to be designed and produced, for both the retailers and the general public; the sales team have to be trained; lawyers have to work out the details of warranties; service engineers need to be trained; call centres have to be given instructions as to how to deal with customers' questions... As can be imagined, to take the idea through development and implementation draws on the resources of the whole organisation. And the risk increases too. Generating ideas is relatively risk-free. But development and implementation can be very risky indeed, perhaps putting the entire enterprise on the line if the idea, once implemented, were to turn out to be less successful than had been hoped.

The transition from creativity to development is therefore also a transition from the small group to the entire organisation; from the expenditure of few resources to the commitment of possibly huge resources; from low risk to high. That's why wise evaluation is important, for it is this process that 'guards' the boundary between low-resource-and-risk creativity and high-resource-and-risk development, so ensuring that resources are not squandered on weak ideas, and protecting the organisation against taking on too much risk.

In commercial organisations, evaluation decisions are usually taken by more senior managers, increasingly so the greater the investment being sought; in academia, some evaluation decisions are taken locally, but the most important as regards research are usually taken by the agencies that award the grants, deciding whether or not to fund the various research proposals that are submitted. Some organisations have very detailed 'rule books' on how evaluation is done, and on the evidence that must be submitted to those taking the decision; others are less rule-bound; still others have a very simple rule – 'if it's the boss's idea, it's great; if it isn't, forget it'. That last rule is quite clear, and everyone understands it. Furthermore, if 'the boss' has a continuing stream of genuinely good ideas, the organisation can thrive, even though the 'imagineers' whose ideas have been ignored have long since left. But if 'the boss' has bad idea, or is just stuck…

Overall, wise evaluation is, as I have said, both important and difficult, of which much more in chapters 14 and 15.

2.5 Processes 3 and 4 – Development and Implementation

An idea approved by the evaluation process has received organisational endorsement for development, in which the idea is transformed from a proposal, a model, a drawing or a concept into something real. And, as we have seen, this can often require a considerable amount of time, and many different resources. And it always requires hard work, lots of it. Also, although development and implementation are depicted in figure 2.1 as separate, sequential, activities, with a 'hard' boundary between them, the reality can often be more fluid, with development and implementation merging into one another.

The skills required to carry out development and implementation depend very much on the specifics of the idea: taking the design concept for a new opera house through to the opening night of the first production is hugely different from a PhD student working all hours of the night trying to debug that programme. In so far as there are relevant generic skills, they are those of project management, for which there are many existing effective methodologies, standards of good practice, books, training programmes and experts.

I have little to contribute in these areas, and so please refer elsewhere.

2.6 The Target Diagram and skills

As we have seen, the Target Diagram – the sequence creativity, evaluation, development and implementation – identifies the activities that are needed to transform an idea into reality. Accordingly, as we have seen, the diagram represents the overall 'meta-process' of innovation, with creativity as the very start.

Given that each of the four processes of creativity, evaluation, development and implementation are different, what are the skills of the people who are most well-suited to each of the coloured zones?

This question invites generalised answers, to which there will always be specific counter examples, so – at the risk of over-generalisation – I offer some thoughts here.

In the central red zone of creativity, the key attributes are a flexibility of mind, free of the dogmatism of 'that's how we've always done things'; a thirst for learning; a willingness to listen to other people, attentively; a spirit of curiosity, and a joy in asking questions. And, very importantly, the ability to imagine, to envisage 'in one's mind's eye' something that is not 'in front of one's nose'.

In my experience, some people find it very hard to imagine something that is not in front of them. That's not a criticism or a judgement; it's just the way things are. Some people can naturally run fast, some can't; some people can readily imagine something not in front of them, even something they've never ever seen, others find it much harder.

Being very comfortable imagining things, however, is a vital attribute of someone who can make a useful contribution to a creativity workshop: after all, what is an idea other than something that doesn't exist yet?

And although evaluation appears to be hard-nosed and down-to-earth (which it is), it also benefits from imagination too – but a rather focused imagination: the ability to imagine, richly and in detail, what the 'world will look like' once the idea has been successfully implemented. This is all about envisaging the 'size of the prize', as well as identifying the adverse issues that are invariably present, for no idea is 'perfect'.

Take, for example, the phrase 'unintended consequences', that is often used – especially by politicians – to explain away something that has gone wrong after a government policy, or perhaps a new system, or perhaps a new organisation structure, has been implemented. The intent is to convince the listener that whatever happened to cause things to go wrong was some arbitrary random event, something out-of-the-blue that no one could ever have foreseen. This is rarely the case. Most alleged 'unintended consequences' are not evidence of unknowable act of nature; rather, they are almost always evidence of very poor thinking when the idea was evaluated, and even poorer development and implementation.

In addition to having the ability to imagine, the 'wise evaluator' must be thorough and disciplined, and very attentive to detail; to cut through the inevitable advocacy of those who favour the idea, and the resistance-to-change of those who don't; to be good at spotting what *isn't* in the presentation, so identifying what people are trying to hide; to have a balanced sense of risk; to understand the numbers; to recognise, and allow for, the uncertainty of forecasts.

That's a sophisticated mix of skills, many of which are acquired over time. So those who can evaluate wisely tend to be older, well-experienced, and – very importantly – trusted to do the job wisely. Wisdom is a good word here.

Development is very different, and associated with words such as pragmatism, focus, problem-solving. That's the archetypical skill set of the engineer – the practical person who can get things to work.

And implementation is all about organising people, meeting budgets, hitting deadlines, delivery – the conventional attributes of the 'manager'. And because those particular skills are often highly valued in many organisations, the people who have them tend to get promoted to more senior positions, and ultimately to the top. So it's no surprise that many Boards of Directors are composed, often exclusively, of those with a proven ability to deliver, on time, within budget. Skills that are all about implementation, and quite different from the skills that are best-suited to the central red zone of creativity. The difference in skill-set, and mind-set too, between the 'deliverers' and the 'imagineers' can create much organisational tension, especially in organisations where those who hold the senior positions, and the power, are 'deliverers'. So the 'deliverers' determine the organisation's culture, and sometimes the 'imagineers' get up and go.

Which can be very bad for the organisation.

Yes, I have drawn generalised, perhaps even crude, pen-portraits of the attributes best suited to each of the four zones. But there is much truth there.

And one such truth is that it is very rare, extremely rare, for any one person to be good and effective across all four zones. The 'imagineer' might not be sufficiently disciplined to 'get things done' in implementation – or worse, might want to 'return everything to the drawing board' because she's just had this great idea for an improvement... Likewise, a really effective 'deliverer' might be too impatient to participate effectively in a creativity workshop – 'Why are we doing all this pfaffing around? Why don't we just get on with it?'

This too can lead to organisational tensions. The 'big boss' (who is of course a 'deliverer'), who demands that 'I must have a blockbuster idea on my desk, no later than Friday mid-day!'. The 'imagineer' who won't let go, and seeks to remain 'in charge' right the way through and ends up disgruntled at being side-lined; the inventor who tries to turn an idea into a business, only to end up with a failed business, and a lot of debt.

These are just some of the reasons why managing innovation, making the entire Target Diagram 'hum', is difficult, and why so few organisations are really good at it. Figure 2.4 summarises the story-so-far, and I'll explore some more in chapters 16, 17 and 18, in which I examine the importance of the organisation's overall culture.

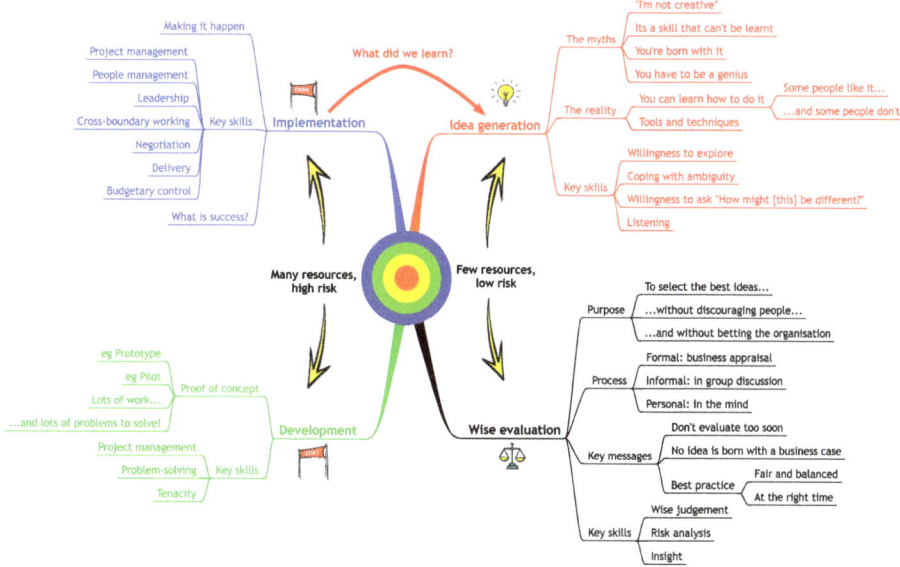

Figure 2.4. The target diagram in context.

References

[1] www.london.edu/-/media/files/publications/bsr/business-heroes-ray-kroc.pdf?la=en
[2] www.nationaltechcenter.org/mcdonalds-statistics-and-interesting-facts/center/—summary-of-mcdonalds-statistics-and-interesting-facts
[3] Gaynes R 2017 The discovery of penicillin – new insights after more than 75 years of clinical use *Emerg. Infect. Dis.* **23** 849–53
[4] www.healio.com/news/endocrinology/20120325/penicillin-an-accidental-discovery-changed-the-course-of-medicine
[5] www.dpag.ox.ac.uk/about-us/bscsc/about-us/john-burdon-sanderson
[6] www.medicinenet.com/penicillin_history/definition.htm
[7] Gould K 2016 Antibiotics: from prehistory to the present day *J. Antimicrob. Chemother.* **71** 572–75
[8] Wainwright M 1989 Moulds in ancient and more recent medicine *Mycologist* **3** 21–3
[9] For example, 'Purge me with hyssop, and I shall be clean: wash me, and I shall be whiter than snow', *The Bible, Book of Psalms*, ch 51, verse 7
[10] Tompsett R R 1996 The discovery of penicillin *Proc Bayl. Univ. Med. Cent.* **9** 37–8
[11] Chen C 2003 On the shoulders of giants *Mapping Scientific Frontiers: The Quest for Knowledge Visualization* (London: Springer)
[12] http://news.bbc.co.uk/local/oxford/hi/people_and_places/history/newsid_8828000/8828836.stm
[13] www.acs.org/content/acs/en/education/whatischemistry/landmarks/flemingpenicillin.html
[14] www.nationalarchives.gov.uk/education/resources/florence-nightingale/
[15] www.sciencemuseum.org.uk/objects-and-stories/florence-nightingale-pioneer-statistician
[16] www.historyofinformation.com/detail.php?entryid=2929
[17] Lilienfield D E 2007 Celebration: William Farr (1807–1883)—an appreciation on the 200th anniversary of his birth *Int. J. Epidemiol.* **36** 985–87 and www.york.ac.uk/depts/maths/histstat/passionate_stat.htm
[18] https://commons.wikimedia.org/wiki/File:Nightingale-mortality.jpg
[19] McDonald L 2001 Florence Nightingale and the early origins of evidence-based nursing *Evid. Based Nurs.* **4** 68–9
[20] Poincaré H 1910 Mathematical creation *Monist* **20** 321–25

IOP Publishing

Creativity for Scientists and Engineers
A practical guide
Dennis Sherwood

Chapter 3

The six domains of creativity

3.1 Creativity is not just about 'the better mousetrap'

Another widespread myth is that creativity is either about the so-called 'creative industries', such as film, television, music, drama, and these days the development of computer games, or about the invention of the 'better mousetrap'.

Not so.

For a more insightful understanding of the 'domains' that can truly benefit from creativity, figure 3.1 shows an enrichment of the Target Diagram of figure 2.1.

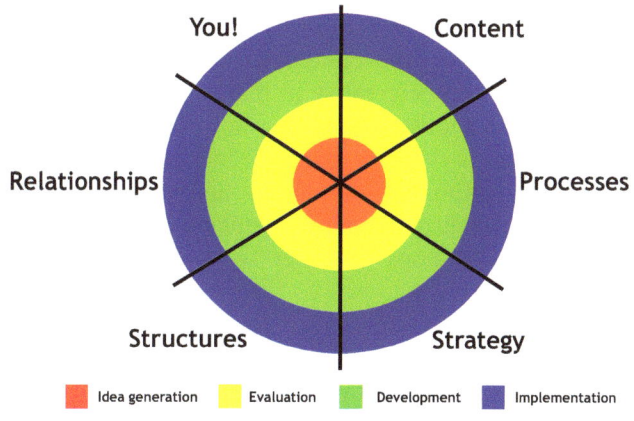

Figure 3.1. The six 'domains' of creativity and innovation.

Each of the six 'domains' represents a different context for benefiting from creativity, as symbolised by the red zone, and the resulting innovation, in which the idea generated within the red zone becomes real, having been evaluated as 'good' and then successfully developed and implemented.

3.2 Content

Starting at the top right, content. This refers to creativity directly related to the 'day job', what you do. This varies according to context, and so some examples are:

- For a product designer, creativity is the generation of the idea for the new product; evaluation, the judgement that the design is good and the product likely to be profitable; development is making the design work in practice, and everything required for manufacture; implementation is taking the product to market and selling it.

- For a research scientist, creativity is the idea for a research project, often expressed as a question; evaluation is the allocation of the required funds, as, for example, determined by an agency that awards grants; development is the actual research; implementation is the publication of academic papers and presentations at conferences.

- For a visual artist, creativity the core idea; evaluation, judging it to be worth the time to paint it; development, the actual painting; and implementation, the final framing, and perhaps exhibiting it in a gallery, selling it, or just hanging it on the studio wall.

- For an author, the content is about writing. So creativity corresponds to having an idea for, say, a novel; evaluation is the author's decision that, yes, the idea is good; development is the actual writing; and implementation is the contract with a publisher…

- …and for the publisher, the content is provided by the author, and so publishers don't need to be creative in that sense – save for having ideas for works they might commission and for how the book, and the cover, are designed; evaluation is the very heart of their business, in deciding whether or not a given manuscript should be accepted; development is about printing, binding and distribution; implementation is about organising sales of the book through shops or other ways of accessing the appropriate markets and customers.

As can be appreciated, the 'content' segment of the Target Diagrams for the author and the publisher are linked, this being a particular instance of a common occurrence, and a variant of the fractal diagram, figure 3.2.

3.3 Process

The second domain, process, concerns not what you do, but how you do it.

One example, appropriate to anyone involved in teaching at any level from the infant reception class to post-graduate programmes, is the development of on-line, remote teaching methods, to supplement, or perhaps replace, conventional face-to-face teaching, as became particularly relevant during the prolonged lockdowns in 2020 and 2021 associated with the control of the Covid-19 pandemic. The outcome required – a well-taught student – is largely the same as conventional face-to-face teaching, but the process of delivery is very different.

To achieve this requires much creativity in generating ideas as to how to put often complex concepts across 'at arm's length'; how to enable effective discussion and exploration between teacher and learner, and amongst a group; how to assess students reliably and fairly; and many other aspects of teaching too. Each of these ideas need to be evaluated wisely, leading to the development of the relevant materials, and the subsequent implementation of the successful delivery of the programme. That's a lot of work!

Another frequently-met example is cost reduction, in which the objective is to achieve a given outcome, but consuming fewer resources, and possibly being achieved more quickly, therefore incurring a lower cost.

3.4 Strategy

An area where many organisations gain value from creative thinking concerns their strategy. We all know the cliché that 'if you keep doing the same thing in the same way you end up in the same place'. But in fact that's the good news: quite often, you can end up in a far worse place as your competitors move past. Although what-you-do-now is familiar, comfortable, and probably works well, for how far into the future will that continue to be true?

For a corporation, strategic creativity is about the product range, markets, the cost structure, and long-term competitive advantage; for a government department, say, education, health or justice, this is about how the corresponding services can be delivered in an effective, efficient and timely manner as the demographics and technology change; for research teams, it is about the 'grand challenges' and having the capability – and the ideas – to address them.

As figure 3.1 implies, strategic creativity follows those now-familiar four steps: strategic ideas are generated and then evaluated, and the best are developed and implemented. What is different, as compared with content and process creativity, is the 'size of the creative canvas' on which the organisation's vision of the future is being painted: the ideas are bigger, bolder, and over longer time scales; they carry much higher risk, and so need to be evaluated more thoroughly and wisely; development and implementation are more complex. But the processes are just the same.

3.5 Structures

Here, the focus of attention is not what we do (content), nor how we do it (process), but rather the organisational structures within which all this happens. A proposal to 'consolidate' two departments is an example, so instead of a college or university having a mathematics department and a statistics department, each with its own Head of Department, secretariat, marketing materials for prospective students and all the rest, the idea (red zone) is to form a single maths and stats department, with consequences such as a single Head, one secretariat, one set of brochures.... At a higher level, the idea might be to merge two nearby campuses, with all the consequences of that. And structural change is endemic in the public sector, as anyone who has worked in the National Health Service in England knows all too well.

In principle, structural creativity and innovation should follow the same overall process as all the other domains: creativity, in which the idea – say, to merge – is proposed; wise evaluation, in which the idea is carefully scrutinised to identify whether it is a good idea or a bad one; development, in which all the problems are ironed out and resolved; and then implementation, in which the merger actually happens, those new brochures are printed, and the new signs are placed on the door.

Unfortunately, the reality is often very different. Structural changes such as mergers are largely driven by politics and power, and those people who play politics and wield power are usually senior, and haven't read this book. So a new Chief Executive is appointed, and decides he has to do something. And the easiest thing to do is to change the organisation structure from one based on products (for example, 'Pumps' and 'Valves'), to one based on geographical regions (say, 'Europe' and 'Pacific'). Because he is the boss, the idea is by definition 'good' (so much for wise evaluation), and because everything needs to be done by Christmas, development and implementation are compressed into a mad scramble to meet an artificial deadline.

Perhaps I exaggerate. But only a little. Structural change is fraught with practical, political and personal difficulties, and is very often botched – a totally tragic example being the partition in 1947 of the former British Raj into the two successor states of India and Pakistan.

3.6 Relationships

No one lives in total isolation: even a Stylite, atop a pillar, has to interact with the people on the ground, if only to receive food. We all have relationships. Some are long-term, such as in families; some every-day interactions, such as with work colleagues; some transient, as with the driver of the bus I rode the other day.

Creativity as regards relationships is all about exploring the question 'If I, or we, were to do something differently, might 'the world be a better place'?' Some contexts in which that might be an 'interesting' question are, for example, with 'the boss', with peers, and with those more junior, all of which are relationships within any organisation; and, beyond an organisation's boundaries, with peers elsewhere, with academic collaborators, with regulators, with customers, with suppliers...

And one of the most important contexts is the team. What, precisely, is a 'team'? Is it the group of people who happen to be in the room when the boss says, 'Hey! You, you and you. Form a team, and fix [whatever]!' ? Or is it something more than that? And what is a 'high-performing team'? Is it the same as 'just any-old team'? Or is it something different? And if it is different, what are the differences? And if, right now, our team is behaving in a way which might be closer to 'any-old', in contrast to 'high-performing', what has to be different to move things up a notch or two?

Those are all questions relevant to relationship creativity. And some of them are mighty tough...

3.7 You!

...but not as tough as the questions associated with this last domain, shown in the diagram as 'You!' – which also stands for 'Me!' and 'Us!'.

An example. The person who is always late for meetings. That's not 'the world's biggest problem' by any means, but it does serve as a case study here.

The convenor of the meeting has a choice. The easiest option, of course, is for the convenor to do nothing at all. That doesn't ruffle any feathers, or precipitate a confrontation best avoided. But it does mean that the person who is always late will continue to be always late, and that might set a precedent for others to be late too.

The alternative is for the convenor to 'have a word' with the latecomer, and initiate a conversation about why being on time is courteous to everyone, for their time is valuable too...

For a convenor whose natural style is dominating and aggressive, that's a 'no brainer' – but that's unlikely in this story, for anyone attending a meeting convened by such a person will probably always be on time, or not be in the team at all. But for the convenor who is naturally consensus-seeking, who shies away from conflict, who worries that the exercise of any authority is likely to bring the accusation of being authoritarian, this situation is a nightmare. And a recurrent one too, at every meeting. Especially when sensing the irritation of the others in the meeting, others whose time is being wasted.

What to do? The easy option is to do nothing. But the wise option...

Taking the wise option, though, is hard. Really hard. And before that action can be taken, the convenor has to have the idea of the wise option of 'having a word' (creativity, red); think about it and decide that, yes, that's the right thing to do (wise evaluation, yellow); pluck up the courage to do it (development, green); and then actually do it (implementation, blue). That's creativity in the 'You!' domain – the creativity to change your mind. Especially when that's tough.

The convenor, of course, is not the only actor in this vignette. The other is the person who is habitually late. That person too can undergo a mind change. But for that to happen, the person must generate the idea of actually being on time (creativity, red); come to terms with the fact that being on time might be the right thing to do (evaluation, yellow); organise things and keep an eye on the clock to get to the meeting on time (development, green); and then, when the alarm actually goes, choose not to answer that email at that instant but to go to the meeting to arrive on time (implementation, blue). That's creativity in the 'You!' domain – the creativity to change your mind. Especially when that's tough.

Yes, that is a trivial example. But everyone reading this will get the message.

People often ask me 'What is the most wonderful example of creativity you know?'.

And those people are expecting the reply, depending on their interests, to be 'Sydney Opera House', 'Picasso's *Guernica*', 'The discovery of the structure of DNA', 'The concept of zero', whatever.

So they're usually very surprised to hear 'The most powerful example of creativity I know is when someone, especially someone in a senior position, changes their mind. For good reason.'

3.8 The importance of the organisational culture

My depictions of the Target Diagram so far have represented the four processes of innovation as a stand-alone concept, albeit, in practice, rather more complex than a simple 'target' but both fractal and recursive. The reality, of course, is that creativity and innovation do not 'stand alone'; rather, they are embedded within an enterprise or organisation, and every enterprise and organisation has a culture, a culture that exerts subtle, but powerful, influence on how people behave, what people say, and how they react to organisational rules.

The culture is never written down, but everyone within the organisation can sense it. Yes, these days, many organisations have 'values statements' seen on posters along the corridors and in offices. But all those abstract nouns – 'Integrity', 'Excellence', 'Teamwork' and the rest – though usually chosen with good intent, suffer from two drawbacks. The first is that they are quite obvious – especially since it would be an unusual organisation indeed that has posters everywhere trumpeting opposites such as 'Fraudulent', 'Dullness', or 'Every man for himself' (and for any organisation that chose this one, it would certainly be 'man'!). And secondly, these cosy-cuddly abstract nouns are, by themselves, meaningless. What counts are the actual day-to-day behaviours, attitudes, decisions, not the words. And there are many organisations whose values statements include 'teamwork', and where the poster of the boat, powered by those eight strong rowers acting in perfect unison, is there for all to see on the cafeteria wall. But where the reality is 'every man for himself'.

An organisation's culture forms the backdrop to everything that happens, including creativity and innovation, as captured in figure 3.2 in which the Target Diagram is positioned within the 'envelope' of the organisation's culture. And since an organisation's culture is often hard to define specifically, and is rather 'fuzzy', if not nebulous, I represent that envelope as a cloud.

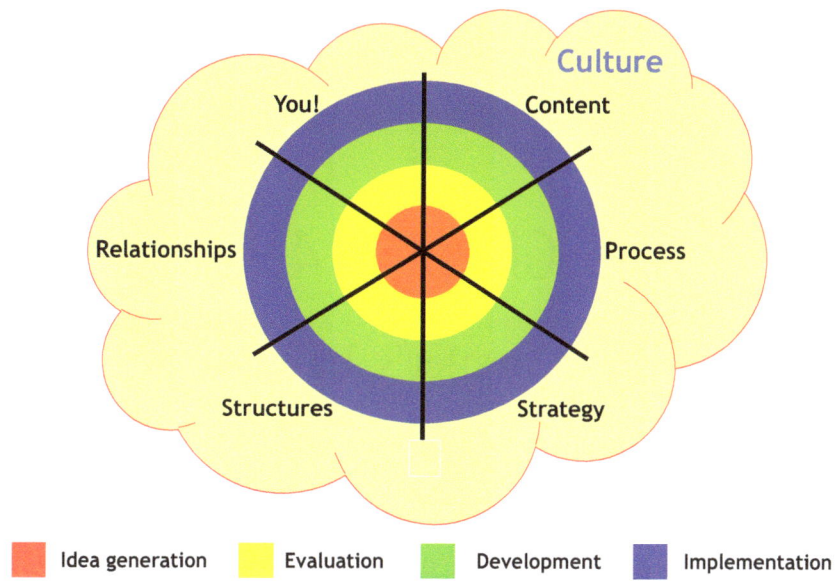

Figure 3.2. Creativity and innovation happen within the 'envelope' of the organisation's culture.

Some examples of the influence of an organisation's culture on creativity have been alluded to several times so far in this book, for example, the discussion of 'permission' towards the end of the Prologue, and the cameo about 'My first brainstorming workshop' in chapter 1. Furthermore, the way in which any organisation carries out evaluation, the yellow zone, is often deeply cultural – is not an evaluation process described as 'If it's the boss's idea, it's great; if not, forget it' a cultural statement as much as a procedural one?

The way in which creativity and innovation actually happen in any organisation is very largely determined by its culture. Some cultures stifle them; in other cultures they flourish. And the cultures in which creativity and innovation flourish are not 'accidents' that just 'happened'. Rather, they are carefully crafted and maintained, as I shall describe in detail in chapters 16, 17 and 18.

Before we get there, though, we need to dig more deeply into that word 'creativity'. The sound-bite 'creativity is having an idea' is helpful but very limited, and the Target Diagram has broadened our understanding of creativity in an organisational context.

But an important question remains. What, fundamentally, is 'creativity'?

IOP Publishing

Creativity for Scientists and Engineers
A practical guide
Dennis Sherwood

Chapter 4

Koestler's Law

4.1 Arthur Koestler's definition of creativity

We've already met Arthur Koestler, in connection with his book *The Sleepwalkers* [1], which is my primary source for the story about Kepler. Throughout his life, Koestler was very interested in creativity, and wrote four books about it. The first was *Insight and Outlook* [2], published in 1949, in which he set the scene for his later works; the second, *The Sleepwalkers*, appeared ten years later, and has its origins in Koestler's recognition that the period from Copernicus to Newton was, from a scientific point-of-view, hugely creative, and of great significance in transforming our world from that of the Middle Ages to that in which we live today. Koestler was therefore intrigued by questions such as 'Why did this happen in the period from around 1550 to 1650?', 'Why did it happen in Northern Europe?', 'How did it happen?'. *The Sleepwalkers* presents his answers, and along the way, tells wonderful stories about Copernicus, Brahe, Kepler, Galileo and Newton, as well as many other less well-known figures too. It is an interesting, engaging, enthralling and hugely informative read. And very well-written too: Koestler was an accomplished journalist, and was nominated for the Nobel Prize in Literature three times – but did not receive the award, much to his (understandable) disappointment.

Koestler's fourth book, *The Ghost in the Machine* [3] of 1967, is now rather dated, but his third, *The Act of Creation* [4], published in 1964, is, in my view, a gem, and the key inspiration for this book.

The Act of Creation take a broad perspective, and explores the nature of creativity, as manifest in literature, drama, poetry, visual art, humour, and science, asking 'What, fundamentally, is creativity?', 'Is creativity different in different contexts, such as when writing a great novel, as compared with, say, the discovery of a fundamental law of nature? Or are there some common underlying principles?'

And here are the words used by Koestler to describe what he thinks creativity is, and how it works:

> *The creative act is not an act of creation in the sense of the Old Testament.*
>
> *It does not create something out of nothing; it uncovers, selects, re-shuffles, combines, synthesises already existing facts, ideas, faculties, skills.*
>
> *The more familiar the parts, the more striking the new whole.*

To me, these words are a deep, insightful, fundamental definition of creativity. They underpin everything in this book, and although they don't have the same status or significance as Kepler's Laws, Newton's Law of Gravitation, or the Laws of Thermodynamics, I refer to them as 'Koestler's Law of Creativity' nonetheless.

4.2 The 'Eureka moment' myth

Koestler's first sentence dismisses two widely-believed myths: the myth of the 'Eureka moment', and the myth of the 'creative person'.

Although the 'Eureka moment' makes a great story, and an even better feature film – the genius inventor suddenly magics up a great new device, the genius scientist runs to her desk to write down the great equation, the genius musician dashes to the piano and plays the great symphony – the much more prosaic truth is that creativity is the outcome of very hard work, over often very long periods of time. It is true, as we saw in the story of Kepler, that when at last 'everything comes together', that moment can be euphoric, ecstatic, triumphal. But that's the end-point, not the process overall. So forget the 'Eureka moment', and get down to the hard, relentless, painstaking work.

Implied in that first sentence is also the busting of the myth of the 'creative person' – the person, as portrayed in those feature films, who always has long hair, is often self-obsessed, and who pursues a careless, if not carefree, lifestyle.

Certainly, some historic 'creatives' comply with this image: Salvador Dali actively played up to it; Beethoven had long(ish) hair and was cantankerous; Einstein, especially in his older years, looked exactly as the caricature of a 'mad scientist' is supposed to look. But by far the majority of people who have ideas, and good ideas, look like you and me.

That's because it's people like you and I that have ideas; it's people like you and I that are creative. As we have seen, creativity is a natural activity of a human brain. All humans have brains. So all humans can be creative. Certainly, not all humans are equally good at it, just as some people can naturally run faster than others. But – physical disabilities apart – we can all run, and if I want to run faster, I can practice running, do some training, and get a little bit faster. I may never enter the Olympic Games, but I will improve. Creativity is very similar. Yes, some people will be naturally better at it than others. But we can all participate, and by training and practice, we can all become better. All of us. And this book can help, for it will both convince you that, yes, *you* can do it, and show you *how* it can be done.

4.3 'But I'm not a creative person'

Unfortunately, the belief that creativity is something you are either born with, or not, is very widely held. In fact, this belief is half-true: *everybody* is 'born with it', for every human being is born with a brain and so every human is intrinsically creative. The untrue half is the 'or not' part, for there are no 'nots'. But many people believe there are people who truly are 'uncreative', and one of them might be me. So many people lack confidence in their capability to be creative, and lacking that confidence, do not seek out opportunities – or possibly actively avoid opportunities – to be creative. With the result that they don't do anything creative. Which reinforces my self-belief that I am not creative, and so I don't seek opportunities... so setting up a vicious circle of non-creativity from which I can never escape... a vicious circle reinforced at work, for my boss thinks that I'm not a creative person, so she never asks me for my ideas, so I never give them, so validating her belief – trapping her in a vicious circle too.

If that anecdote about the vicious circles rings true, that's because it is. And they are indeed vicious, and they are indeed traps. The best way to escape from them is to believe that 'yes, I am creative!', for, yes, you are. And I hope that this book will help.

4.4 Creativity is all about patterns

And so to Koestler's second sentence:

> *It (creativity) does not create something out of nothing; it uncovers, selects, re-shuffles, combines, synthesises already existing facts, ideas, faculties, skills.*

There's something missing! The word 'new'. And not only is the word 'new' missing, it is explicitly *denied*, not once but twice: firstly, in the statement 'does not create something out of nothing' and secondly, the phrase '...already existing...'. Surely, creativity is about the new, the out-of-the-blue, which is *exactly* all about 'something out of nothing'? And if – as Koestler seems to be suggesting – creativity uses 'already existing facts...', then there is nothing new on the planet!!! I think my brain has just exploded...

In chapter 1, I mentioned, almost in passing, that 'there is a significant philosophical problem as to what the word 'new' actually means'. And it is this insight by Koestler that identifies what that problem is. Since the 'already existing facts...' are 'already existing', they can't be 'new'. Yet there is something about creativity in which something 'new' emerges. How does that happen? That's quite a puzzle. And it is in resolving that puzzle that our understanding of creativity moves from the superficial to the fundamental.

To explore this, let's firstly take an example of true creative genius. The music of Beethoven. No one would dispute that his entire output is a role model of creativity; the only argument would be about which particular work is the greatest, the most profound, the most creative. So choose whichever work you wish, and play it in your mind. And if Beethoven is not your thing, no matter; Duke Ellington, Bob Dylan, George Gershwin... choose whomever you consider to be your exemplar of great musical creativity.

Now think about what the-composer-of-your-choice actually *did* to create such great music.

To help, here are four things none of them did:

- Invent any of the notes or tones.
- Invent a 'musical grammar', different from the conventional major and minor scales.
- Invent a notation for music, different from the conventional five lines, with conventional note symbols for crotchets, quavers and the like.
- Invent any hitherto unknown musical instruments.

That's quite a list of musical things that the most creative musical composers on the planet didn't do.

So what *did* they do?

They used the pre-existing notes and tones, within pre-existing frameworks of major and minor musical scales, to craft *new patterns and sequences of those notes*, which, if written down, used pre-existing conventions of symbols to represent the music, which can then be performed on pre-existing instruments. Yes, all musical instruments have been invented by someone, but usually not by the same person who wrote the music – with some very few exceptions such as John Philip Sousa, the composer of those stirring marches, who commissioned a manufacturer of brass instruments to make the sousaphone so he could have some rich bass notes in his marching bands. Overall, though, in my view, those four bullet points are valid.

Now read Koestler's second sentence again, especially 'combines existing facts' – where in this case, the 'facts' are the notes, the 'grammar', the notation and the instruments, and what is being 'combined' are the pre-existing notes and tones, within the pre-existing formalisms of scales, notation and the rest.

The notes in Beethoven's music pre-existed him, and were 'there', available to him to use. But the *patterns that he crafted* had not been heard before; no one hitherto had put those pre-existing notes together in the sequences, and in the combinations, that we now label as the 'Moonlight Sonata', the '7th Symphony', 'Fidelio'.

Likewise, Bob Dylan used exactly the *same* notes as those used by Beethoven, to craft *different* patterns, that we now label as 'Blowin' in the Wind', 'Quinn the Eskimo', 'Subterranean Homesick Blues'. And – something Beethoven didn't do – Bob Dylan wrote the words too, and good words they are: so good that he was awarded the Nobel Prize in Literature in 2016 [5].

The same applies to every composer, creating music in the Western tradition. The music of other traditions – Indian classical music, for example – is based on a different 'grammar', but all composers within that tradition executed the equivalent process: the formation of different patterns, using the same components, within the 'rules' of that tradition.

This resolves the problem of the word 'new'. Although the *notes* are all the same, the resulting *patterns* are different. And the first performance of the 'Moonlight Sonata' was the first time that total pattern had ever been heard, despite the fact that each individual note, by itself, had been heard countless times. Similarly, when Bob Dylan first sang 'Blowin' in the Wind', that too was the first time that pattern had ever been heard, despite the fact that each of the notes can also be found in the 'Moonlight Sonata'.

In so far as creativity is about the discovery of the new, what is 'new' is the final 'pattern'. Prior to the formation of that pattern, each of the individual components necessarily pre-existed, independently, separately, and not within that pattern, and so in that sense, nothing is new – everything is a re-combination of pre-existing components.

'Novelty' is therefore a matter of level: in any particular context, at the level of the overall 'pattern', the combination might be new, in that any particular complete pattern might never have been compiled before; at any lower level, each component within that pattern must have pre-existed.

This insight has great significance, for it demystifies creativity, and gives some important clues as to how to design a process by which ideas can be generated deliberately, on demand, whenever and wherever a new idea would be useful.

Koestler's Law tells us that since the result of creativity is a combination of pre-existing components, then the result of decomposing that pattern will be a set of separate components, each of which existed in its own right before the combination had been formed. Looking at that the other way around, if all those components are 'on the table' in front of me, side-by-side, and in the right configuration, then I have the opportunity to assemble them. And so discover the resulting idea.

To make that real. Suppose that in 1605, someone other than Kepler, who happened to be interested in astronomy, had, on their table, four sheets of paper. One was a diagram like that shown in figure P.2; a second had written on it $x/y = 0.00429$; a third, $\theta = 5°\ 18'$; and the fourth, secant $5°\ 18' = 1.00429$. Let us further imagine that this person also had some knowledge of the geometrical properties of the ellipse. At that moment, that person is looking at all the components required to generate the idea that the orbit of Mars is an ellipse with the Sun as focus. Those components are present, they exist; but they are still 'in separate compartments', they have not yet been assembled into a meaningful pattern, a pattern that would be truly revolutionary. But there they are, and the person looking at them has the opportunity to assemble them and make that discovery.

It is of course quite possible that this person, staring at those four sheets of paper, just looks at them, doesn't put the pattern together, and moves on; perhaps the person sees the papers, is struck by the numerical coincidences, but – although possessing knowledge of the geometry of ellipses – just happens not to be thinking about ellipses at that time, and so only says 'that's odd', and then does something else. The opportunity was there, but was missed – and this happens frequently: in 1953, the structure of DNA was proposed in a very short paper, authored by James Watson and Francis Crick, published in *Nature* [6]. All of the components required to do this had been published in a variety of journals over the previous several years; all were available to all scientists around the world; and in 1953, the structure of DNA was a 'hot topic', with many individuals and teams working on it, some very eminent indeed. But, as we shall see in the next chapter, it was the Watson–Crick team that put the right components together in just the right way to get the right answer. Many others could have done exactly the same thing. But they didn't.

The message to all researchers is very important. Subject to what I call the 'da Vinci problem' which I will discuss in chapter 6, Koestler's Law says this: if you are searching for an idea, the components needed to discover that idea are out there, now. Probably in different places, and probably established at different times. But they're 'out there', somewhere, now. All you have to do is to find them, and then put these pre-existing components together in the right pattern.

So rather than being a process of 'out-of-the-blue-lightning-striking', creativity is a process of 'collecting' and 'assembling'. That makes the process far more practical and amenable. Anyone can gather things, anyone can put things together in different patterns. It's rather like playing with Lego: if I choose the right pieces from all those strewn on the floor, I can build that rocket ship. And the more pieces I have available, the more exciting the resulting rocket ship will be. In research, of course, the components aren't plastic bricks. They are the aggregate of all my knowledge. And so the more knowledge I have, the greater the likelihood that I will be able to select specific components of that knowledge, form 'interesting' patterns, and generate great ideas.

That's why knowledge is fundamental to creativity, and why the more knowledge you have, the more creative you can be. And Kepler was a beneficiary of that. If he did not have knowledge of the geometry of the ellipse, he would have missed the clues associated with the three other components, that $x/y = 0.00429$, that $\theta = 5° \, 18'$, and that secant $5° \, 18' = 1.00429$. The more you know, the more creative you will be. And the more access you have to knowledge, the greater the likelihood of discovering ideas. Which explains why high-performing teams, within which each individual's knowledge can be pooled with the knowledge of everyone else, are likely to be more creative than the same individuals, working by themselves and therefore able to draw on only their own knowledge.

4.5 'Bisociation' and 'thinking aside'

As noted towards the end of the Prologue, Arthur Koestler refers to the fundamental process of creativity, in which hitherto unrelated concepts come together in a meaningful way, as 'bisociation' – a term he introduced in his first book on creativity, *Insight and Outlook*, and developed and enriched in *The Act of Creation*. For bisociation to happen, two (or more) previously unrelated concepts need to be in one's mind at the same time – just as we saw in the example of Kepler.

One way in which this can happen is by (often accidental) observation: you are thinking about something, and then you happen to notice something completely unrelated. And then…

That, for example, is how George de Mestral got the idea for Velcro: he was thinking about how to fasten a garment whilst walking through some woods, and noticed, as he brushed past some vegetation, that some seeds were sticking to his coat (and his dog's coat too!). He looked more closely, and saw the little hooks on the seed interlocking with the fibres on his coat, and his dog's fur… [7]

The invention of the bar code is a similar story. In 1948, Bernard Silver was a graduate student at Drexel University, and, walking along a corridor, accidentally overheard a conversation between the Dean of Engineering and the boss of a local supermarket chain about how to identify, quickly and easily, goods at the checkout. Bernard then related this conversation to his friend, Joe Woodland. A few months later, in January 1949, Woodland was sitting on a beach in Florida, thinking about how products might be identified, idly running his fingers through the sand. Then, when he looked down, he saw a stripey pattern… [8]

Yes, as we saw in the Prologue, observation is important. And being lucky too. But there's a huge difference between merely *seeing* something, and *observing* it, recognising its significance.

Another way this can happen is by what Koestler, in *The Act of Creation*, calls 'thinking aside', a process in which we think about whatever-it-is rather differently from the way we usually think about it, so increasing the likelihood that something else, perhaps unexpected, comes into our minds that might make that special bisociative connection. Koestler's 'thinking aside' is therefore related to familiar terms such as 'looking at things from another angle', 'taking a different point of view', and, as coined by Edward de Bono in 1967, 'thinking laterally' [9].

In discussing 'thinking aside', Koestler acknowledges his debt to a PhD thesis, entitled *Théorie de l'invention* (The theory of invention), by Paul Souriau, submitted to the Sorbonne in 1881, and still currently available as a book [10]. In his introductory chapter, in which he explores the fundamental nature of creativity, Souriau describes how creativity rarely follows a direct path from 'here' to 'there', from 'problem' to 'solution'; rather, to use his phrase, '*Il faut penser à côté*' – '*you must think aside*'. And later in his thesis, Souriau writes, '*L'invention n'est pas une création absolue, mais un combinaison nouvelle d'idées or d'images antérieures*' – '*Invention is not an 'absolute' creation, but a new combination of pre-*

existing ideas or images', which is clearly a pre-cursor of Koestler's Law. Souriau therefore provided two important pre-existing ideas to Koestler's own, personal, act of creation, as manifest in his book of that name.

4.6 The more familiar the parts, the more striking the new whole

> ### It's a *GOAL!!!*
>
> The scorer couldn't believe what was happening, and buried his head in his hands in despair. The goalkeeper was gesticulating wildly, pointing at the post. The other players were wild-eyed, shouting, pushing, looking more like a rugby maul. The referee was talking to the linesman, whilst vainly gesturing the players to untangle their brawl so that the game could re-start. The home fans were in uproar; the visitor's fans jubilant.
>
> The goal had been disallowed.

So what's new about a disallowed goal in football? That happens all the time, and that's what the Video Assistant Referee sorts out.

But there was no Video Assistant Referee that afternoon on 26th October 1889. And yes, there is now, and was then, nothing new about a disputed goal. But what was new was something that didn't happen on the pitch, but in the crowd. As this was all happening, a spectator had an idea. An idea that no one had ever had before – or no one we know about, for it is of course possible that someone else had had the idea, but had done nothing with it. But this time, something was done. Something that reduced the number of 'colourful incidents' associated with all those disputed goals – and made the game of football much fairer as a result. The idea? The goal net.

That day, Everton were at home to Accrington Stanley. Both teams continue to play in the English league to this day; Everton ranking rather more highly than Accrington. The game was played at Anfield, now the home ground of Everton's city neighbours, Liverpool, and the disallowed goal would have been the winner in an Everton 3–2 victory. But the official result was a 2–2 draw.

In the crowd was a 31-year old engineer, John Brodie, and as he watched the mini-riot on the pitch, he was thinking too. 'Suppose that there were a net attached to the posts and the cross-bar, but behind the goal line. If the ball goes within the goal mouth, it will be caught by the net, plain for everyone to see. And if the ball misses, it will just go wherever. That way, there can be no doubt as to whether a goal is scored or not, and all those disputes about which side of the post the ball actually went will no longer take place.' Indeed so. For up to that time, football had been played with just the posts and the cross-bar, leading to any number of arguments over which side of the post the ball had actually travelled.

Brodie subsequently tried a variety of designs, and was granted a patent in 1890. His invention was first used in January 1891, in which, highly appropriately, Everton's Fred Geary became the first player to 'hit the back of the net' to score the opening goal in a 3–0 win over Nottingham Forest. The idea caught on: goal nets were used for the first time in an English Football Association Cup Final in March 1891, and from September 1891, the rules of the English Football Association made nets compulsory [11].

I tell this story for two reasons.

The first is that it is a splendid example of bisociation. One 'fragment' is the 'disputed goal'. The second might have been 'net', had, for example, Brodie been thinking about going fishing the next day. Another possibility, and the one that is usually associated with this story, is 'pocket', as, for example, the pockets in his trousers (he might have been looking for something in his pocket just as the dispute occurred). Either way, the concepts of 'disputed goal', 'net' and 'pocket' all pre-existed the invention of the football net, exactly as Koestler's Law tells us. And if they were in Brodie's mind simultaneously, then…

The second reason is in relation to the third sentence of Koestler's Law: 'The more familiar the parts, the more striking the new whole'. For surely the football net must be one of the most obvious ideas there ever could be. The 'component parts' were very familiar, and anyone reading this story must be thinking 'Why had nobody thought of the football net before?' – especially anyone who had ever seen a billiard table, for nets had been used to catch the balls for many years, as exquisitely shown in a print by Thomas Rowlandson dating from about 1820, reproduced as figure 4.1.

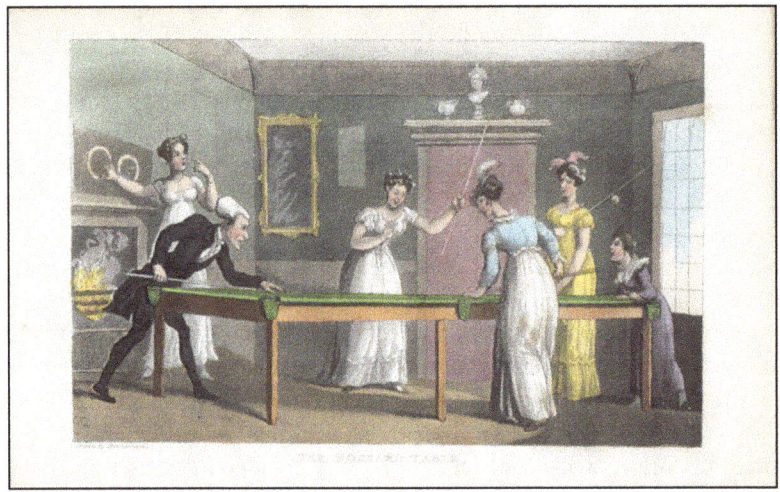

Figure 4.1. A billiard table, around 1820, an aquatint by Thomas Rowlandson. Reproduced courtesy of Collectorsprints.com www.collectorsprints.com/7032/antiqueprint/thebilliardtable.

But they didn't – or if they did, they did nothing with the idea, and so it is Brodie's name that goes down in history.

There are many examples of 'obvious' ideas that took an amazingly long time to be discovered. Whilst I am talking about sport, one further, and in many ways quite similar, story.

Basketball was deliberately invented, in 1891, by James Naismith as a solution to the problem of how to enable young men to exercise indoors during the winter. As the name implies, the objective of the game is to throw a ball into a basket positioned some 10 feet off the ground. The original baskets were those used for carrying peaches, and made of thin strips of wood with a closed base. The closed base performed the useful function of keeping the peaches inside the basket. But is also held the basketball too, so that every time a ball was successfully thrown into the basket, a player had to climb up, retrieve the ball, and then come down so that play could restart. That was tiresome, so someone thought of having a small hole in the bottom of the basket, so that the ball could be dislodged by poking a stick. Then, 15 years later, in 1906, someone had a great idea. To cut a larger hole so that the ball would fall through [12]. Now why didn't I think of that before???

4.7 What Koestler's Law does, and doesn't, do

In my opinion, Koestler's Law gives a powerful definition of creativity, going far beyond, and far deeper, than my sound-bite 'creativity is having an idea', and the dictionary definitions given at the start of chapter 1. And the insight that all creativity is the discovery of a new pattern from pre-existing components is liberating and empowering, for it tells me that if I want to discover a new idea, the components are out there, somewhere, if only I can find them. The components I need might – at least with hindsight – be quite obvious, as with the football net, or they might be much more hidden, as with the components $x/y = 0.00429$ and $\theta = 5° \, 18'$ which were essential for Kepler's discovery of the elliptical orbit, and which could only be identified after much laborious work.

But somehow, the assembly of pre-existing components seems to be a more accessible task than 'having an idea out-of-the-blue'. And I don't have to be a 'special person', born with the gift of 'creativity'. In principle, anyone can put Lego bricks together. I can do it. And you can do it too – as I will show in chapter 9.

For that is one thing that Koestler's Law doesn't do: it doesn't tell you *how* to be creative, what you can actually *do* to discover a great new idea tomorrow. Yes, Koestler's Law certainly does tell us that any idea is a new pattern of pre-existing components, and so, given any idea, we can then deconstruct it and identify what those components are. Which is fine after-the-event, but less helpful beforehand: yes, the 'right' components are out there, somewhere, mingled amongst a huge number of other potential components, any combination of which could, in principle, give the 'right answer'. But which is the right combination? And how do I find it?

Perhaps we need to try every possible combination, and, once any trial 'pattern' has been formed, to test it to see if it works. In human terms, such a task is impossible, for there are far too many combinations to test. But for a search that can be carried out by a computer, this is possible, as exemplified by the way in which Turing and his colleagues used the 'bombe' at Bletchley Park to decipher Enigma-encoded messages [13]. The algorithms available to the Bletchley Park team were of course relatively simple, and the computing machines slow. Today, and looking ahead, machines are vastly faster, and perhaps machine learning algorithms open the possibility of making the search process much more efficient. At the time of writing (2021), computers are already composing music [14], once the preserve of 'creative talent'. In the future it is more than likely that the use of artificial intelligence will enable computers to be as 'creative' as any human.

That, however, does not stop humans from being creative, even if the machines are being creative too. And the time at which machines will be sufficiently creative is still some way into the distance, so we poor humans are not out of a job quite yet. That still leaves the question of 'how?', which Koestler's Law doesn't answer. But chapter 9 will.

There's something else, too, that Koestler's Law doesn't do, and which is really important. Koestler's Law identifies creativity as the formation of a new pattern from existing elements. Which is true. But in his definition, Koestler makes no statement about the *quality* of the resulting pattern. And some patterns are much, much, better than others.

To illustrate the significance of this, consider music. Yes, I can access exactly the same musical notes that Beethoven used, as, for example, made available by the keys on a piano. A standard piano has 88 keys, corresponding to 88 notes, some very high, some very low. So let me restrict myself to, say, 36 around the middle of the keyboard. Suppose I wish to compose a piece of music of, say, 16 notes. That's a very short piece of music, but it could be a key theme or riff. I can select each of the 16 notes in 36 different ways, for I can repeat notes if I wish, and so – ignoring any further variations such as different durations for each note, or loudness or softness – that results in 36^{16} combinations of 16 notes, or about 8×10^{24} different sequences.

The current population of the Earth is about 8×10^9 people, and there are about 5×10^5 min in a year. Assuming that about one quarter of any individual's time is spent sleeping, the maximum number of 'working minutes' in principle available to the global population each year is therefore some 3×10^{15}. If each pattern took about ½ minute to play, the total number of patterns that could be played in any one year by the entire global population, assuming no human being did anything else at all, is therefore about 6×10^{15}. But as I have just estimated, the number of possible patterns of 16 notes is about 8×10^{24}, so that implies it will take about $(8 \times 10^{24}) / (6 \times 10^{15}) \approx 10^9$ years to play them all.

That time of $\sim 10^9$ years should be considered in the context of the current estimates that the planet was formed about 4.5×10^9 years ago, and *homo sapiens* evolved about 2×10^5 years ago.

So if I happen to go to a piano now, and randomly select 16 notes, there is a reasonable probability that the resulting pattern would be one that no human being has ever created before, a musical pattern that is genuinely new. There is also a very high probability, from a musical point-of-view, that my particular pattern will be awful. Truly bad. But that pattern would comply with Koestler's Law. It would be an 'act of creation' in his terms. Just a very poor one.

From a practical standpoint, generating patterns from existing components for the sake of generating patterns is a waste of time and resources. And although, as discussed in chapter 1, words such as 'valuable', 'useful' and 'good' were deliberately and explicitly excluded from my sound-bite definition of creativity as 'creativity is having an idea', if we are thinking of how to design a process of 'deliberate creativity', in which we can harness Koestler's Law and use its insights to discover something new, I suggest it makes good sense to design that process with a view to maximising the likelihood that the resulting idea will indeed be valuable, useful and good rather than to expend time and energy on discovering ideas that are the opposite.

Koestler's Law makes no reference to the goodness or badness of any pattern: in Koestler's terms, any pattern that is new is creative, whether good and useful, or bad and useless.

In the chapters that follow, I address these limitations, and will explore the issue of the 'goodness' of a pattern, building on Koestler's Law to design a process that enables new, and potentially good, ideas to be discovered, deliberately.

But just before I do that, the next chapter tells some more stories, from a variety of fields, that show how Koestler's Law provides a compelling explanation of literally everything.

4.8 The Koestler Challenge

Which brings me to the 'Koestler Challenge'. If you can find anything, from any field, that 'breaks' Koestler's Law, please let me know. For a Law is valid only as long as there are no counter-examples – so finding one is important, for that would identify a fatal flaw. The absence of finding one is important too, for whilst there are no known counter-examples, the 'Law' continues to be, at the very least, a good working hypothesis.

I have given the 'Koestler Challenge' to a large number of people – everyone, in fact, that I have ever met, and with whom I have discussed creativity. So far, no valid counter-example has been found. Yes, I've had many suggestions, but on closer investigation, Koestler's Law has been found to hold. But there might be a counter-example out there, somewhere. And so if you can find one, do let me know – please email me at dennis@silverbulletmachine.com.

Arthur Koestler

Source: © The Estate of Athol Shmith, Courtesy Kalli Rolfe Contemporary Art.

Darkness at Noon is a chilling read. Incarcerated in prison in solitary confinement, and expecting to be executed, the central character, Nikolaj Rubashov, discovers he can communicate with a neighbouring prisoner by tapping on the intervening wall. As the novel progresses we learn of Rubashov's past, his beliefs and his relationships, all in the context of his times and what it feels like to be fearing imminent death [15].

Written whilst Koestler was in Paris in 1940, the setting is Russia in 1938, at the height of the Stalinist purges. But the work is not just the fruits of Koestler's imagination. While he was a journalist covering the Spanish Civil War, Koestler was arrested, imprisoned, and sentenced to death. In February 1937, as his cell door slammed shut, he was not to know that some five months later he would be released in a prisoner exchange. Koestler's life was nothing if not dramatic, colourful, vivid.

Koestler was born in Budapest in 1905, within what was then the Austro-Hungarian Empire, ruled by the King-Emperor Franz-Joseph I of the same House of Habsburg as Rudolf II, whom we met in the story of Kepler. Having studied engineering in Vienna (to an extent: he was expelled for non-payment of fees), Koestler travelled and took up journalism and writing: between the 1930s and his death in 1983, his books ranged over themes from politics to the paranormal, from propaganda (he worked for the UK Ministry of Information during the Second World War) to the history of science. For his 'outstanding contribution to European culture', he was awarded the 1968 Sonning Prize (other recipients being, among others, Bertrand Russell, Albert Schweitzer, Karl Popper, Ingmar Bergman, Renzo Piano and Orhan Pomuk [16]); in 1972 he was made a Commander of the British Empire [17].

His contributions to creativity are primarily to be found in four books: *Insight and Outlook* (1949), *The Sleepwalkers* (1959), *The Act of Creation* (1964) and *The Ghost in the Machine* (1967). *Insight and Outlook* was in essence superseded by *The Act of Creation*, the key source not just for this chapter but for much of this book too.

References

[1] Koestler A 1990 *The Sleepwalkers* (London: Penguin Books)
[2] Koestler A 1964 *Insight and Outlook* (London: Macmillan)
[3] Koestler A 1967 *The Ghost in the Machine* (London: Hutchinson)
[4] Koestler A 1964 *The Act of Creation* (London: Hutchinson)
[5] www.nobelprize.org/prizes/literature/2016/dylan/lecture/
[6] Watson J D and Crick F H 1953 Molecular structure of nucleic acids: a structure for deoxyribose nucleic acid *Nature* **4356** 737–38
[7] www.velcro.com/blog/2016/11/an-idea-that-stuck-how-george-de-mestral-invented-the-velcro-fastener/
[8] www.channel4.com/news/barcode-inventor-who-transformed-commerce-has-died
[9] Bono E de 1974 *The Use of Lateral Thinking* (London: Penguin Books)
[10] Souriau P 2013 *Théorie de l'Invention* (Paris: Hachette Livre BNF)
[11] https://sportinglandmarks.co.uk/liverpools-19th-century-contribution-to-goal-line-technology/
[12] https://stadiumfreak.com/facts-about-the-history-of-basketball/
[13] https://en-academic.com/dic.nsf/enwiki/436621-British_bombe
[14] https://scienceline.org/2021/01/ai-music-tools/
[15] https://lithub.com/the-eerily-prescient-lessons-of-darkness-at-noon/
[16] https://event.ku.dk/sonning_prize/recipients/
[17] Scammell M 2010 *Koestler: The Indispensable Intellectual* (London: Faber & Faber)

IOP Publishing

Creativity for Scientists and Engineers
A practical guide
Dennis Sherwood

Chapter 5

Some more examples of Koestler's Law

5.1 Literature

One of the greatest novels in the English language is Jane Austen's *Pride and Prejudice*, which opens with these words:

> *It is a truth universally acknowledged, that a single man in possession of a good fortune, must be in want of a wife.* [1]

Jane Austen did not invent any of those words, any more than Beethoven invented any musical notes; but Jane Austen certainly crafted the most wonderful linguistic patterns.

Jane Austen is an example of a novelist (an interesting word, implying 'new'), using combinations of words to tell powerful stories; poets take linguistic pattern formation one step further, combining the meanings of the words with additional features such as rhythm and rhyme. And one step beyond that is to combine linguistic rhythm and rhyme with musical tones and structures: the creative genius of the great songwriters such as John Lennon, Paul McCartney and Bob Dylan – even if some of the rhymes might have been used before…

> **Subterranean Homesick Blues, Bob Dylan, 1965**
>
> > Better jump down a manhole
> > Light yourself a candle
> > Don't wear sandals
> > Try to avoid the scandals
> > Don't wanna be a bum
> > You better chew gum
> > The pump don't work
> > 'Cause the vandals took the handles.
>
> **Up at a Villa – Down in the City, Robert Browning, 1855**
>
> Look, two and two go the priests, then the monks with cowls and sandals,
> And the penitents dressed in white shirts a-holding the yellow candles,
> One, he carries a flag up straight, and another a cross with handles,
> And the Duke's guard brings up the rear for the better prevention of scandals:

For sources, see [2] and [3].

Browning might have thought of 'vandals', but chose not to use it; had he thought of 'manholes', he certainly would have rejected it on the grounds that it doesn't (strictly) rhyme, and he probably would not have wished wish to 'break the rules' of conventional mid-Victorian poetry. Sometimes, however, 'breaking the rules' is what creativity is all about…

Some authors, however, have invented words: the Reverent Charles Dodgson, for example, who under the pseudonym Lewis Carroll was the author of the highly imaginative and creative *Alice in Wonderland*, invented words such as 'boojum' (in *The Hunting of the Snark*) and 'chortle' (in *Through the Looking Glass*) [4]; Shakespeare invented very many words, one of which is 'gloomy', first used in Act IV of *Titus Andronicus* – '*Forced in the ruthless, vast, and gloomy woods*' [5].

These words were indeed new, but if we go to a lower level, they are each patterns of letters in the English alphabet, letters that already existed. Koestler's Law once more. And inventing the word 'gloomy' wasn't too difficult. Shakespeare wanted a word to describe dark, perhaps rather sinister, woods. Although the adjective 'gloomy' was invented by Shakespeare, the noun 'gloom' was already in use, as were many noun-adjective pairs describing atmospheric conditions, such as rain-rainy, sun-sunny, wind-windy, fog-foggy. So gloom-gloomy.

5.2 Art

Piet Mondrian did not invent red, yellow or blue paint. But he did combine them into patterns – such as his *Composition with Red, Yellow, Blue, and Black* of 1921, reproduced as figure 5.1 – which many find quite beautiful.

Figure 5.1. Mondrian's Composition with Red, Yellow, Blue, and Black, 1921.
Source: Wikimedia: 'Composition with Red, Yellow, Blue, and Black' https://commons.wikimedia.org/wiki/File:Piet_Mondriaan,_1921_-_Composition_en_rouge,_jaune,_bleu_et_noir.jpg.

Some people do not find Mondrian's art beautiful at all, choosing words such as 'boring, 'simplistic', 'child-like', 'formulaic'. The words any person decides to use to describe Mondrian's art are not so much describing the art as describing their personal feelings about what they see, and whether what they see conforms to their view of what 'good' art should look like. This is a matter of judgement and evaluation, of which more in future chapters. But in the context of this chapter, Mondrian's art – and the art of every other painter – complies with Koestler's Law as being patterns of pre-existing components, the various coloured pigments.

Painting is also full of examples of artists who 'broke the rules', and who now are regarded as creative geniuses. Claude Monet's *Impression, soleil levant*, exhibited in Paris in 1874, was greeted by much criticism for breaking the rule that all paintings must be 'finished', and the term 'Impressionism' was first used in derision [6]; Picasso's Cubist work was described as 'schizophrenic' and 'satanic' [7] for breaking the 'rule' that any depiction of the human face requires that each eye must be either side of the nose. But even though Monet and Picasso were 'revolutionaries' at their respective times, they too 'stood on the shoulders of giants': Monet's predecessor JMW Turner was nothing if not a proto-impressionist [8], and Picasso was strongly influenced by Cézanne and the indigenous art of Africa (to name just two) [9].

5.3 Chemistry

All non-nuclear chemistry is a vivid example of Koestler's Law. No mining engineer invents the iron. No laboratory chemist, no industrial chemist, no pharmaceutical chemist invents those carbon atoms, oxygen atoms or nitrogen atoms. The atoms are there and – nuclear reactions apart – always have been there, always will be there. What chemistry does is to rearrange them: to decompose larger patterns into fragments; to rearrange those fragments into different patterns, some of them being patterns that have never been formed before (that's the new drug); and then rearranging them again. But the fundamental components – the atoms – never change.

In an essay entitled *Carbon* in his book *The Periodic Table* [10], Primo Levi vividly describes the 'story' of a specific, individual, identifiable, single carbon atom within a molecule of calcium carbonate within some limestone rock. The rock remained deep within the Earth's crust for a million years or more, until, one day, some 150 years ago, that rock is mined, and burnt in a limestone kiln, transforming that carbon atom into carbon dioxide, drifting in the atmosphere. Later, in the miracle of photosynthesis, that carbon atom becomes part of a sugar molecule in a green plant. The green plant – a vine – incorporates that molecule within its fruit, a grape; a grape that is harvested, and made into a delicious wine, with just the right degree of sweetness, a sweetness attributable to its sugar content. And one of those sugar molecules contains that original, specific, carbon atom. The carbon atom in the sugar molecule in that glass of delicious, sweet, wine you drank with your evening meal last night. So, right now, that same carbon atom is part of you.

That's Koestler's Law.

And it also applies to nuclear chemistry. All those sub-atomic particles are 'there' too.

5.4 How chemistry made impressionist art happen…

I've just talked about chemistry, and previously about art; combining these together, here is a Koestler's Law story about why, without chemistry, and some technical ingenuity, we would never have had the benefit of gazing on Monet's paintings of water-lilies.

My starting point is a question: what is the difference between the art of a renaissance master such as van Eyck, Michelangelo or Raphael, and that of a late 19th century master such as Monet, Renoir or van Gogh? Take a look, for example, at the paintings reproduced in figure 5.2: on the left, a van Eyck from 1434, on the right, a Monet from around 1898.

Figure 5.2. Left: Jan van Eyck's *The Arnolfini Marriage*, 1434; right: *Water Lilies, Pink*, Claude Monet, 1897–99.
Sources: Wikimedia: 'Arnolfini Portrait' https://upload.wikimedia.org/wikipedia/commons/3/33/Van_Eyck_-_Arnolfini_Portrait.jpg and 'Water Lillies Pink' https://www.wikiart.org/en/claude-monet/water-lilies-pink-1899.

There are of course many differences: style, context, subject, patronage, to name just a few.

But there are two, perhaps less obvious, differences that are of relevance here. The first is the colour palette: the Impressionists had a much wider range of luminous colours to work with than their renaissance predecessors [11]. And the second is that the Impressionists were able to work in the open air, and so could capture the essence-of-the-moment and the vibrancy of natural light; by contrast, van Eyck, Michelangelo, Raphael and their contemporaries worked primarily indoors, in studios.

It's all about the paint. In the 17th Century, the pigments used by artists were sourced directly from natural materials, for example, black, from burnt animal bones; red from crushed cochineal insects, or from mercury sulphide contained within the natural mineral cinnabar; the richest blue, ultramarine, came from the mineral lapis lazuli, sourced from mines in what is now Afghanistan. Sometimes, some rudimentary 'cooking' was required, such as the production of 'lead white' from metallic lead and vinegar. The manufacture of each pigment, in a form that the artist could use, required skill, and, once made, usually needed to be applied quite quickly, before it dried. The final pigments were therefore often made in small quantities, in a space adjacent to the studio, so they could be used when still fresh [12].

But by the first decades of the 19th Century, knowledge of chemistry and chemical methods was much improved, so enabling molecules that did not exist in nature to be synthesised. Some of these were highly coloured and made good pigments for painting, such as cobalt blue (discovered in 1802) [13], emerald green (1808) [14], chrome yellow (around 1816) [15], French ultramarine (1826) [16], zinc white (marketed in 1834) [17], and cadmium yellow (marketed after 1840) [18]. Then, in 1856, 18-year old William Perkin noticed a dark sludge at the bottom a vessel in which he had hoped he would make quinine, but had failed. As we saw in the Prologue, he didn't – as many others would – throw the sludge away; rather, he worked out what it was, and recognised that not only was mauveine a beautiful new colour, but it also could dye textiles [19]. He took out a patent, and mauve-coloured fabrics became hugely fashionable. This increase in demand stimulated the search for new, brightly coloured molecules – for example, fuchsine, aniline yellow, Bismarck Brown – and also the growth of the companies, such as Bayer and BASF, that manufactured them. Most of these new colours were used for dying cloth, but artists benefited too with the introduction of new pigments such as alizarin red, first synthesised in 1868 [20].

Whilst the chemists were developing new pigments, in 1841 an invention was patented that is often overlooked, but which transformed the way artists work. The paint tube [21]. By enabling pre-mixed paints to be stored in small quantities, in a secure container with a resealable cap, artists were freed from their studios [22].

The inventor, John Rand, was an American portrait painter who wanted to paint outdoors. At the time, paints could be carried in a pig's bladder tied with string and which could be punctured to release the paint, or in a recently-invented syringe-like device. Both of these were clumsy to use. Rand then had an ingenious idea: a small tube, sealed at one end and with a removable and replaceable screw top at the other, made from a material that did not react with the contents, and that could be easily crumpled to allow the contents to be squeezed out – for, example, thin tin or lead. As a by-the-by, the toothpaste tube, exactly the same concept but with different contents, was not introduced until 1879 by Dr Washington Sheffield, a Connecticut dentist – after his son, Lucius, on a visit to Paris, had seen some artists at work! [23]

Some other technical inventions helped too, such as the collapsible easel that could easily be carried; also, since tube paint was rather thick and heavy, new brushes had been introduced with stiffer bristles held by a metal 'ferrule' in a range of different widths and shapes, so enabling a greater variety of brush strokes with which to form those wonderful three-dimensional textures [24].

Overall, synthetic chemistry, the paint tube, the portable easel and the stiff brush all contributed to making Impressionist art what it is, and without any one of those... well, who knows? Or, to use the words of one the greatest of the Impressionists, Renoir:

Without paints in tubes, there would have been no Cézanne, no Monet, no Sisley or Pissarro, nothing of what the journalists were later to call Impressionism. [25]

5.5 …and how physics has facilitated contemporary art

John Wynne is a contemporary artist whose preferred medium is not visual but aural, for example, his *Installation for 300 speakers, Pianola and vacuum cleaner* of 2009, exhibited at London's Saatchi Gallery [26] and winner of the 2010 British Composer Award for Sonic Art [27]: an *Installation* which he would have been unable to create had physicists not studied the science of sound, and engineers not invented any number of acoustic devices.

John Wynne is just one example of a contemporary artist whose work relies on science; another is the visual artist David Hockney. Since 2019, Hockney has been living in Normandy, where he has painted many pictures capturing the local landscape, and how it evolves as the seasons change, including 220 completed during the Covid-19 pandemic in 2020. But 'painted' is the wrong word, for Hockney did not use paint, pigments, and palettes, nor brushes and easels. He used an iPad, 'painting' with a stylus on a flat screen [28].

To produce his wonderful images – of which just two are shown, joined together, in figure 5.3 – Hockney didn't have to concern himself with the physics of semiconductors and LEDs. But Hockney, and all of us, are beneficiaries of the physicists who did.

Figure 5.3. David Hockney, at the Orangerie museum in Paris on 7th October 2021, in front of a small section of his 80-metre-long artwork, 'A year in Normandy', formed from a series of pictures, each of which was created during 2020 using an iPad, and then printed [29].
Sources: From photograph by Thomas Coex/AFP, reproduced by permission of Getty Images.

A key component of the iPad is the liquid crystal flat touchscreen display. Liquid crystals go back a long way, to around 1888, when Friedrich Reinitzer and Otto Lehmann identified that below 145.5 °C the cholesterol derivative, cholesteryl benzoate, is solid, and above 178.9 °C, a clear liquid. Between those temperatures, however, the material flows, and so is liquid, but it is cloudy, and the molecules are arranged not randomly as in a 'normal' liquid, but more systematically, rather like a crystal [30]. The ordering of the molecules within this 'liquid crystal' state results in an interaction with polarised light, and the overall fluidity of the material allows that ordering to change, as the result of, for example, a local electric field.

The flat screen of the iPad exploits the ability to change the arrangement of the liquid crystals in response to an electrical signal, thereby causing a change in the polarisation of transmitted light. But getting that to work effectively took a long time and much research, as well as depending on a host of previously-discovered technological inventions, not least semiconductors, transistors, printed circuits and fast switches. Some milestones on the journey from 1888 are Daniel Vorländer's synthesis during the 1920s and 1930s of many molecules that form liquid crystals [31], the invention in 1927 of the electrically switched light valve by Vsevolod Fredericks [32], and the work of George Heilmeier at the RCA laboratories in the 1960s on operational flat panel displays [33]. Small LCDs were incorporated into wrist watches [34] and electronic calculators [35] in the 1970s, and the first commercial colour LCD television was manufactured in Japan in 1984, but with a very small screen, just two inches across the diagonal [36]. Since the late 1990s, flat screens have become much larger and cheaper, and therefore more widely used, and the iPad itself was launched in 2010 [37]. Along the way, Pierre-Gilles de Gennes was awarded the 1991 Nobel Prize in Physics for his work showing that 'phase transitions in such apparently widely-differing physical systems as magnets, superconductors, liquid crystals and polymer solutions can be described in mathematical terms of surprisingly broad generality' [38].

As well as being based on the physics of liquid crystals, the iPad is also touch-sensitive, such that the position of the stylus on the screen acts as an input signal. The position of a specific location on the screen's surface, as identified by a finger or a stylus, can, in principle, be determined in a number of ways, for example, optically (the stylus blocks the transmission of infrared light across the surface of the screen), acoustically (the touch triggers an acoustic signal that is detected by appropriately located sensors), by measuring the resistance of the screen material (which is changed by the compression attributable to the touch), or, most commonly, by changes in the capacitance of the screen material resulting from the (very small) current flow through the finger or the stylus [39].

The first capacitative touchscreens were developed in the mid-1960s [40], and in 1972, a patent was filed for an optical touchscreen [41]. For the most part, performance remained unsatisfactory and generally unreliable until the 1990s, when IBM, Apple and others launched various hand-held devices with stylus touchscreens. And, as always, further developments made touchscreens cheaper, bigger and more reliable, until touchscreens became sensitive enough, and the software powerful enough, to enable one of the world's greatest living artists to create such captivating non-painted paintings.

Yes, David Hockney owes much to physics! And for an update on some of the most recent advances in the technology of screen displays, see section 13.13, 'Newton's Rings and flat screens'.

5.6 History, politics, philosophy, and economics

Although this is primarily a book about science and engineering, the fundamental principles of creativity in general, and of Koestler's Law in particular, apply to every discipline. This section therefore cameos four examples from the humanities, drawing on the explosion of creative intellectual endeavour that took place in Europe and North America in the 18th Century during 'The Enlightenment' [42]. And three of the examples feature documents from one single year, 1776.

The first, published on 17th February 1776, is Volume I of Edward Gibbon's magisterial six-volume history, *The Decline and Fall of the Roman Empire* [43]. It was an immediate hit, and had six printings in that year – a feat of which any author would be proud, let alone an author of a history book. The Western Roman Empire, of course, had collapsed more than a millennium before Gibbon's book, and the Eastern Empire three hundred years earlier in 1453 [44], so you might expect that there was not much that might be fresh to write about in the 1770s. Gibbon's work, however, was truly innovative in that he was a stickler for drawing – as far as he was able – on original sources, rather than relying on second- and third-hand re-tellings [45]. This is archetypal Koestler's Law, combining 'already existing facts', as recorded in original sources, in a manner that had not been done hitherto.

Gibbon, 'the greatest of Enlightenment historians' [46], was not the first, nor the last, to adopt a creative approach to telling a historical narrative. Every society preserves its past in myths, legends and folk-tales; the ancient Greeks Herodotus and Thucydides, were the first (known) to attempt to record 'the facts' [47]; the Victorian Thomas Carlyle advocated the 'Great Man' theory of history, in which single individuals (almost invariably men!) shape events [48]; these days, there is much more emphasis on 'the people' – for example, David Olusoga, Professor of Public History at the University of Manchester, has presented several television programmes focusing on a single house in a city such as Liverpool, Leeds and Newcastle, relating the stories of the 'ordinary' – and sometimes quite extra-ordinary – men, women and children who lived there over the years [49]; another television series traces the ancestors of a particular person, one example being the British actress Dame Judi Dench, who discovered that one of her nine-times great-grandfathers was a Danish noble, Steen Bille. In 1544, Steen's sister, Beate married another Danish noble, Otte Brahe. Their second child, Tyge, has featured earlier in this book. But under his anglicised name, Tycho [50]!

Back to 1776. When, in Philadelphia on 4th July, the delegates to the Second Continental Congress, are approving the text of one of history's landmark documents, the *United States Declaration of Independence* [51]. No multi-volume treatise this; merely 1,333 words, of which just one of its memorable statements is its assertion that

> *"We hold these truths to be self-evident, that all men are created equal, that they are endowed by their Creator with certain unalienable Rights, that among these are Life, Liberty and the pursuit of Happiness"* [52].

Those who crafted the Declaration [53] – notably the future presidents Thomas Jefferson and John Adams, and also Benjamin Franklin (who will appear again soon) – were masterful wordsmiths, yet even these titans drew on the work, and words, of others. Tom Paine was hugely influential, especially his 47-page pamphlet *Common Sense*, published in Philadelphia in January 1776, which was certainly known to the Declaration's authors [54]; John Locke too, whose *Two Treatises on Government*, published in 1689, argues strongly for people's natural rights to life and liberty [55]; and the Scottish philosopher Henry Home, Lord Kames, in his 1751 *Essays on the Principles of Morality and Natural Religion*, wrote 'While man pursues happiness as his chief aim...' [56]. So the *Declaration* is indeed a 'synthesis of already existing ideas', as Koestler would say. Furthermore, as well as having its own antecedents, the *Declaration* is itself, directly or indirectly, the precedent for the independence declarations of some 100 other nations, such as Venezuela (in 1811), Vietnam (1945) and Kosovo (2008) [57].

And talking of Lord Kames, it turns out that he was instrumental in securing the appointment of Adam Smith as a lecturer at the University of Edinburgh in 1748 [58], prior to his 1751 appointment as Professor of Logic at the University of Glasgow [59]. That's the same Adam Smith who wrote two profoundly significant books, *Theory of Modern Sentiments* (published in 1759) [60], and *An Inquiry into the Nature and Causes of the Wealth of Nations* (published on 9th March 1776 [61], so that's my third example), which together are the foundation of free-market economics [62].

Adam Smith was truly innovative, advocating a view of economics very different from the orthodoxy of the time, mercantilism [63]. One of the ideas Smith explores – and strongly advocates – is the 'division of labour', in which complex tasks are broken down into individual steps, with each step being performed by a specialist – perhaps someone with particular skills, perhaps a locality with access to particular resources – so enabling increased productivity. And as an example he describes the 18 steps required for the manufacture of pins [64].

Smith's work drew on many sources, not least Sir William Petty's *Another Essay in Political Arithmetick Concerning the Growth of the City of London* of 1682, which identifies the key principles of the division of labour using the example of watch-making: 'In the making of a Watch, If one Man shall make the Wheels, another the Spring, another shall Engrave the Dial-plate, and another shall make the Cases, then the Watch will be better and cheaper, than if the whole Work be put upon any one Man.' [65] Furthermore, descriptions of pin manufacture are to be found in a number of French writings that pre-date *Wealth of Nations*, notably Diderot's 1755 *Encyclopédie* [66]. And two others who influenced Smith were Francis Hutcheson, Smith's former teacher [67] (and who wrote of 'unalienable rights' - so linking him to the *Declaration of Independence* [68]), and also the philosopher David Hume [69]. A Koestler 'synthesis' once more, with the added coincidence that Hume died in our year of interest, on 25th August 1776.

The intellectual environment of the Enlightenment was truly vibrant, and thinkers and scholars across Europe and North America were in contact as much as communications in that era allowed. In 1765, Smith travelled in France, where he met François-Marie Arouet, better known by his pen-name, Voltaire (there is no first name, rather like some of today's star footballers!), who was then 70 years old, and at that time living in Fernet (now Fernet-Voltaire), just inside France and some 10 miles from Geneva [70]. Those 70 years had been eventful. In 1717, aged 22, Voltaire was incarcerated for 11 months in the notorious Bastille prison as punishment for writing a satirical poem that had upset Philippe d'Orléans, the Regent of France on behalf of the then seven-year old King Louis XV [71]. That was not Voltaire's only visit to the Bastille: following an altercation in 1726 with a French nobleman, the Chevalier de Rohan, he was imprisoned again, but released after three weeks having agreed to being exiled to England, where he remained for three years [72].

And so it was that on 28th March 1727 Voltaire happened to be in London, and present at the funeral of Sir Isaac Newton [73]. Voltaire was amazed. How could a 'commoner' receive so lavish a funeral? With the Lord High Chancellor, two dukes, and three earls as pall bearers? [74] In his native France, only the highest-born would be honoured in this way. What was it about the society in England that enabled this to happen?

Such thoughts stimulated Voltaire to think hard about politics, society and fundamental human rights – thoughts that would determine the content of his writings over the next several decades, thoughts that would influence both contemporaries as well as philosophers, historians, economists and sociologists of the future; thoughts that were expressed in prose and poetry, in plays and polemics [75].

Furthermore, Voltaire himself was much influenced by Newton – especially the way in which Newton combined experience and verifiable experimental evidence with more abstract – and in Newton's case, often mathematical – theories, all based on reason. And it was 'Voltaire's crowning achievement to amalgamate [Newton's] ideas with the theories of Descartes and John Locke to create his own blend of understanding' [76]. 'Amalgamate'. 'Blend'. Pure Koestler.

Which takes us back to science in general, and Newton in particular...

5.7 Newton's Laws of Motion and Gravitation

Newton's Laws of Motion, and of Gravitation, were first published in 1687 in his book *Mathematical Principles of Natural Philosophy*, often known as the *Principia*, but it is certain that he had determined them over the preceding two decades [77]. These laws draw together the threads running through the development of science from Copernicus to his own time, and Newton is famous for saying 'If I have seen further, it is by standing on the shoulders of giants' [78] – a metaphorical, and alternative, statement of Koestler's Law. And, as a by-the-by, an example of Koestler's Law itself, for Newton was not the first to make a statement of this form: a 12th Century monk, Bernard of Chartres is attributed to have said, 'We are like dwarfs on the shoulders of giants, so that we can see more than they, and things at a greater distance, not by virtue of any sharpness of sight on our part, or any physical distinction, but because we are carried high and raised up by their giant size.' [79]

Three of the Newton's giants were Brahe, Kepler and the less well-known astronomer Jeremiah Horrocks, who was the first to determine that the orbit of the Moon was elliptical and to observe the transit of Venus across the face of the Sun, and whom Newton acknowledges in the *Principia* [80]. As discussed in the Prologue, Kepler had discovered that the intensity of light diminished as the square of the distance from its source, but had mistakenly assumed only a simple inverse relationship for the force between the Sun and Mars. He was also unsure as to the nature of this force, describing it as a 'broom', sweeping the planets in their orbits. Newton built on this foundation, and recognised that Kepler's First (elliptical orbits) and Second (equal areas in equal times) Laws could be explained by an inverse square law describing a force directly between the Sun and each planet, a force of exactly the same nature as the force that caused an apple to fall from a tree (to draw on that myth), this being the force of gravity.

A fourth giant was Galileo Galilei, who, as far back as the 1590s, had been studying motion, and who established that, in the absence of a force, a body would maintain its velocity unchanged, but that in the presence of a force, a body will accelerate or decelerate. He further showed that acceleration is independent of mass, and that objects tend to resist changes in motion – so identifying the concept of inertia [81].

And yet another giant was the French polymath René Descartes, who made monumental contributions to many fields, not least philosophy, analytic geometry and mathematical physics. In his book *Principles of Philosophy*, published in 1644, we find, amongst many other things, 'each thing, as far as in its power, always remains in the same state' and 'all movement is, of itself, along straight lines', statements which are very close to Newton's First Law of Motion, 'A body will continue in its state of rest, or uniform motion in a straight line, unless acted on by a force' [82].

Newton's Law of Gravitation, and his three Laws of Motion, all have their precedents; that said, it was Newton's genius to bring those precedents together to create his great synthesis (to use one of Koestler's words once more).

5.8 A brief digression – coincidence, co-invention and the zeitgeist

Given the existence of the work of Brahe and Kepler, of Galileo and Descartes, and of other lesser-known figures too, how was it that all these threads came together with Newton's inverse square law of gravitation? Why did no one else do this?

Well, perhaps someone else did. A contemporary of Newton was Robert Hooke, the scientist after whom Hooke's Law, about the linear extension of a spring with force, is named. Hooke and Newton knew each other well, and one of Hooke's main interests was gravity – a subject on which he and Newton were known to have discussed together. In 1686, Hooke read an early copy of Newton's *Principia*, and became enraged, claiming that Newton had stolen the idea for the inverse square law from him. Hooke demanded he be recognised as the originator of the idea; Newton refused. And the rivalry, and acrimony, continued until Hooke's death in 1703 [83].

Newton was also involved in another dispute over what would now be called intellectual property rights – his long-running argument with Gottfried Liebniz concerning who had invented (or is it discovered?) calculus [84]. And a later example of two towering geniuses coming up with the same idea at the same time was in the 1850s when Charles Darwin and Alfred Wallace proposed their very similar theories of evolution by natural selection – but without the personal bitterness [85].

So who did discover the Law of Gravitation – Hooke or Newton? Should Liebniz or Newton be hailed as the discoverer (or is it inventor?) of calculus? And which of Darwin or Wallace was the true originator of evolution by natural selection?

The answers lie within Koestler's Law: creativity is the formation of new patterns from pre-existing components. And since any idea, from 'small ideas' to *BIG IDEAs* like gravitation, calculus and evolution, must be formed from pre-existing components, then anyone who can identify those components is in a position to assemble them in just the right way. If an important component is missing, then no one can do it, so Kepler was not in a position to discover what were to become Newton's Laws of Motion since he could not benefit from the work, for example, of Descartes.

And if some key components are accessible to only a single person, then only that person can be the discoverer, as was the case for Kepler and the elliptical orbit. The key components were within the observations of Tycho Brahe, and nowhere else, and only Kepler and Brahe's immediate heirs had access. But although Brahe's observations contained all the required components to discover the ellipse, those components were very well 'hidden', and could only be 'set loose' by prodigious, painstaking computational labour – which Kepler alone was willing to carry out. Kepler was also secretive, a loner; the only person with whom he might collaborate and discuss ideas, Brahe, was dead. Only Kepler was in a position to discover the elliptical orbit of Mars, and he did not share his thinking. So his claim to be the discover of the elliptical orbit of Mars is uncontested.

The context of the Law of Gravitation is very different. Kepler himself had talked of the 'sweeping force' between the Sun and Mars, and had suggested a simple inverse relationship between the strength of that force and distance. As over the course of the next few decades, many others investigated – both experimentally and intellectually – how bodies move, why they move, and the important distinction between motion at constant speed and accelerated motion. Notions of cause and effect, and of force, were becoming more sophisticated, as were mathematical techniques and methods. And people corresponded, published, and – very importantly – not only talked to one another, but recognised the value and significance of collective discussion. So, for example, in England in the 1640s, Robert Boyle (after whom Boyle's Law is named) was an active participant in a community known as 'The Invisible College' – which included Sir William Petty whom we met in section 5.6 – who gathered together to discuss scientific and philosophical matters [86]; a similar group known as 'Académie Montmor' met in Paris in the 1650s [87]; London's Royal Society, still the UK's premier association of scientists, was founded in 1660 [88]. Hooke was appointed 'Curator' of the Royal Society in 1662 and Secretary in 1677 [89], placing him at the hub of the scientific community; Newton became a Fellow in 1672, and was the president from 1703 until his death in 1727 [90].

Newton and Hooke therefore met, knew each other, and talked to one another. The must have talked about gravity, about their hunches, their ideas. And at first, those hunches and ideas would be incomplete, not quite right. But as a result of these discussions – by exploring these ideas in the 'space' between them – these hunches and ideas become enriched, as we saw in section 1.2. At some point, everything 'works' – and it is quite possible that everything 'worked' for both Newton and Hooke around the same time. Yes, one may have published his thoughts before the other, and the 'other' might have felt aggrieved for not being given the explicit recognition he felt he deserved, providing the backdrop for an argument along the lines of 'Don't you remember that conversation in which I said [whatever], which was really important? Why didn't you acknowledge that?!?'.

Koestler's Law, however, provides a framework for understanding this, and for appreciating that to attribute a single person as 'the' originator of an idea is not only fraught with problems, but potentially dangerous and pernicious too. If 'I' am fearful that 'you' might 'steal' my ideas, I become reluctant to speak to you, to explore ideas, to share ideas. I retreat into my shell; you don't have the benefit of any interaction; the potential of 'us' is diminished.

Similarly, a PhD student, aware of the university's statutes that might say something like 'The degree of PhD is awarded to students who have proven their ability to carry out original research', might think, 'My work needs to be 'original'. That means 'mine'. So I have to do everything myself, and I need to be really careful about what I talk about, and to whom I talk. Especially my supervisor. I can't risk someone leaning forward when I present my thesis saying, 'Mmm... Is that idea about [whatever] really 'original'? I seem to remember a conversation we had a year or so ago when I think I mentioned exactly that...'. No. Can't risk that. Better to do everything myself...'.

For one of the most fundamental insights from Koestler's Law is that, unless someone is operating in truly exceptional circumstances, as Kepler did, all creativity is, *necessarily*, a communal activity. No single person can *ever* be 'the originator' of an idea. And that applies to Kepler too: he certainly could not have discovered the ellipse without Brahe's observations, and even without the results of whoever did the work to compile those trigonometric tables showing that secant $5° \ 18' = 1.00429$. There's also the possibility that, before his death, Brahe had a conversation with Kepler, perhaps over a drink, and said, 'I've just had an idea! Kepler, have you ever considered the possibility that the orbit of Mars might be an ellipse?', to which Kepler might have replied, 'An ellipse? I didn't think you could be so stupid! How could a tetrahedron fit neatly within an ellipsoid of revolution? Impossible! What a dumb idea!' Of course I made that up. But it might, just might, have happened.

So when concepts such as gravitation, calculus, evolution by natural selection, are 'in the air', in the 'zeitgeist'; when the Koestler fragments and components are gradually being identified; when people are beginning to fit those fragments and components together in partial patterns, with some working better than others; when many, very able, minds are thinking about the same things, then it cannot be a 'surprise' that different people find the 'right' pattern at about the same time. On the contrary, far from being a surprise, we should expect it.

So no wonder that Hooke and Newton both had a 'claim' to the Law of Gravitation; Newton and Liebniz to calculus; Wallace and Darwin to evolution by natural selection. And what a great pity that in some of those cases, disputes arose, with both parties saying 'It's mine!'. No idea can ever be 'mine'. Or 'yours'. All ideas are 'ours'. Which should give patent lawyers a big headache, but that's another story...

And talking of stories, here are some more, all of which are archetypes of Koestler's Law, that all creativity is about the formation of patterns from pre-existing components, and that all creativity is a team activity.

The first is about that very icon of creativity, the electric light bulb; that's followed by stories about Casa Battló in Barcelona, the DC motor, Sydney Opera House, and finally two of the very greatest scientific achievements of the 20th century, special relativity, and the discovery of the structure of DNA.

5.9 The light bulb

Who invented the light bulb?

Most people will answer 'Thomas Edison'.

And yes, on 1st November 1879, Edison applied for a patent for an electric light bulb [91] of a form that would be recognised today, and using as the source of luminescence a carbonised filament connected to platinum wires. US patent No 223,898 was duly granted on 27th January 1880 [92]. Those were the days! The interval between filing a patent and having one granted is now much, much longer...

But almost a year earlier, on 18th December 1878, Sir Joseph Swan had demonstrated an electric lightbulb, of a rather different shape from those used today [93], but using as the source of luminescence a carbonised cotton thread filament – that's cotton thread, treated with sulphuric acid and then strongly heated, but without catching fire, rather like the formation of charcoal from heated wood – connected to platinum wires. Further demonstrations took place on 17th January and 3rd February 1879, but Swan did not take out a patent – on the grounds that all the components he had used were all in the public domain, a true Koestler inference! Swan did, however, take out UK patent 4933 on 27th November 1880 [94], ten months after Edison's US patent, and the drawing associated with Swan's patent – as reproduced in figure 5.4 – looks just like the light bulbs that were in use for the next 100 or so years until the introduction of LEDs [95].

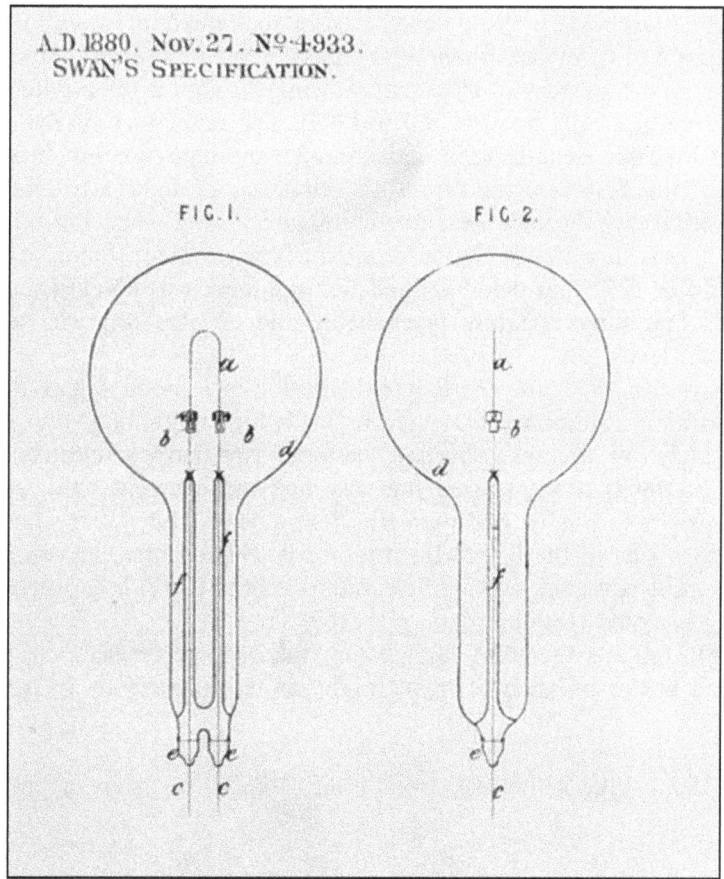

Figure 5.4. Swan's UK Patent 4933, 1880.
Sources: Reproduced courtesy of Herman H J Lynge & Søn A/S, Copenhagen www.abebooks.com/Electric-Lamps-Letters-Patent-Invention-IMPROVEMENTS/15853802035/bd.

Swan's invention was also featured in an article in the July 1879 issue of *Scientific American*, which Edison certainly knew about, for he references it in a notebook in 1880 [96]. It is therefore very likely that Swan's invention influenced the design of Edison's US patent filed in November 1879 and granted in 1880.

Swan began to manufacture light bulbs in Britain in 1880, prompting representatives of Edison to come across the Atlantic with the intention of suing

Swan for infringement of Edison's US patent. They soon realised, however, that in suing Swan, Edison would run the risk that his own patent would be invalidated on the basis of Swan's 1878 invention. So instead of there being a court case between Swan and Edison, in 1883 they decided to join forces to create the Edison and Swan Electric Light Company Limited [97], which, despite a number of mergers and take-overs, continued to manufacture light bulbs under the Ediswan brand into the 1950s [98].

In fact, this story goes back way before both Edison and Swan. One starting point is 12th March 1761, the date of a letter to Benjamin Franklin (whom we met in section 5.6) from one Ebenezer Kinnersley, who describes, amongst many other things, that a current of electricity flowing through a brass wire caused the wire to become not only hot but red too [99]. The redness is attributable to the emission of long-wavelength light, and although the light was not bright enough to read by, this a demonstration of the emission of light attributable to the passage of electricity through a conducting filament. And while I'm talking about Franklin, it was his (really quite dangerous) experiment flying a kite in a thunderstorm in 1752 that demonstrated that lightning was an electrical phenomenon [100]. Lightning certainly gives light, and so this too was a clue that electricity can give light.

Fast-forward to 1809, and the Royal Institution in London, where the Director of the Laboratory, Humphry Davy (from 1812, Sir Humphry Davy) reproduced lightning safely by arcing a current between two carbon electrodes, having previously (in 1802) demonstrated the very first incandescent light, achieved by flowing electricity through a platinum wire that glowed white [101]. Davy was able to do this by virtue of the Royal Institution's possession of what was at the time the world's most powerful source of electric current: a (very) large version [102] of the recently invented (1800) Voltaic pile [103].

From 1802, it was therefore known that the passage of electricity through a wire resulted in the emission of light, bright enough to read by in the otherwise dark.

So why did it take a lifetime, from 1802 to 1878, to invent a practical light bulb?

Because to do so, (very) many practical problems had to be solved. For example, the filament. Davy used platinum wire, which was expensive. Not only that: as the current flowed, the wire became hot, causing it both to vapourise and to oxidise. So that highly expensive material just disappeared. And there's also the problem of the power supply: relying on Voltaic piles is very limiting.

Through most of the 19th century, different people solved these various problems at different times. So, for example, in 1841, Frederick de Moleyns lessened the problem of oxidation by placing the filament, formed from a combination of platinum and carbon, within a glass envelope from the interior of which some of the oxygen had been removed by the creation of a partial vacuum [104]. And in 1845, John Starr was granted a patent for a light bulb using a filament of carbon alone, also within a partial vacuum [105]. Another vital contribution was made by Hermann Sprengel in 1865, with his invention of a more powerful vacuum pump [106], so that much more oxygen could be extracted, thereby considerably extending the time until the filament burnt out.

But it wasn't until 1878 that all the right components came together in the right way, enabling the Koestler Law puzzle to be put together by Swan in England, and by Edison in the United States in 1879. Another example of the zeitgeist.

So who invented the light bulb? Frederick de Moleyns has quite a good claim, for his patent of 1841 was the very first for a light bulb that actually worked. But it didn't work for long enough, and in no way was it a commercial product. Starr's carbon-only filament was another important advance, and Sprengel's vacuum pump was used by both Swan and Edison. Both Swan and Edison, independently, set up companies to manufacture bulbs, which they chose to merge in 1883, after Edison had decided not to sue Swan for infringing his 1880 US patent – is this a tacit acknowledgement by Edison of Swan's contribution?

Given a story that lasted a lifetime, from Davy's demonstration in 1802 to the formation of the Edison-Swan company in 1883, then the conclusion must be that no single person invented the light bulb. Rather, many people over many years made important contributions, giants on whose shoulders Swan and Edison simultaneously stood.

The story of the light bulb is a story of Koestler's Law *par excellence* – and also a great story to illustrate the fractal Target Diagram, figure 2.2.

5.10 Casa Batlló

In 1904, Josep Batlló y Casanovas commissioned the Catalan architect Antoni Gaudí to refurbish his home in central Barcelona. The work was finished in 1906, and the result is one of the most spectacular townhouses in the world [107]. Figure 5.5 shows just the façade visible from the street; the interior is a creative marvel too [108]. How did Gaudí imagine it all?

Look firstly at the roof. It is scaly, reptilian even. The tiles are not the familiar terracotta colour, but rather a mixture of blues, purples, reds. And the scaly effect has been achieved by laying the tiles differently – rather than side by side in straight lines, they are side by side with the corners downwards. The shape of the roof is arched, and looks rather like the 'bicorne' hat that Napoleon is associated with [109] – and hats are worn on the head, at the top of the body…

But what is that hole towards the right of the roof? Perhaps the shape of the roof is not a hat, but a dragon, with the tail to the left and the head to the right – so the hole is underneath the neck. But why should a dragon be on a building in Barcelona? Could it be because Saint Jordi is the patron saint of the province of Catalonia, of which Barcelona is the capital? Saint Jordi might not be familiar to those outside Catalonia. Jordi, however, is the Catalan version of the English name George. And it was St George who slew that dragon. With a sword [110]. Might that tower to the left of centre be the sword piercing the dragon?

Figure 5.5. Casa Batlló, Barcelona.
Source: Wikimedia 'Casa Battlo image' https://upload.wikimedia.org/wikipedia/commons/b/bf/Casa_Batllo_Overview_Barcelona_Spain_cut.jpg.

In front of the windows are balustrades like no others. Conventionally, balustrades are made of stone, wood or metal, perhaps forming a continuous low wall, or a barrier of slender columns (the 'balusters') topped by a horizontal handrail. These balustrades aren't like that at all; they are huge versions of the kind of mask worn at the Carnival in Venice [111], masks that hide the identity of the wearer by shading the eyes. Do the windows of a house represent its eyes?

Like the roof, the façade of the house is brightly coloured – in fact, the surface contains thousands of ceramic and glass pieces arranged like an impressionist painting. And it was in the late 1890s and early 1900s that Monet was painting his garden, and his water lilies… [112].

Yes, Casa Batlló is truly unique. But all the individual components that Gaudí combined together pre-existed. Just as Koestler's Law says. So perhaps identifying all those components and putting them together wasn't so difficult. More surprising, perhaps, is that Gaudí's sponsor let him do it!

5.11 The DC electric motor

In 1820, the Danish physicist Hans Christian Ørsted observed that the needle of a compass deflected when an electric current passed through a nearby conductor. This was the first demonstration of electromagnetism, but it was not an accidental discovery: Ørsted had been searching for a link between electricity and magnetism for about two years. He realised that an electric current generated a magnetic field, and at first believed that this field was radial, directly away from the conductor – but after a few months of experimentation, he established that the magnetic field was in fact circular around the wire [113].

Then, in France, André-Marie Ampère discovered that two parallel electrical conductors experience a mutual force of attraction or repulsion, depending on whether the currents are flowing in the same, or in opposite, directions. This force is attributable to the interaction of the (unseen) magnetic fields around both wires, and the presence of that force implies that if one conductor is fixed whilst the other is free to move, it will. That established a connection between electricity and motion, a connection resulting from magnetism [114].

Very soon thereafter, in 1821, Michael Faraday designed the ingenious apparatus shown in figure 5.6.

The two vessels each contain mercury. On the left is a fixed vertical wire and a bar magnet loosely hinged at the bottom; on the right, a fixed bar magnet, and a wire loosely hinged at the top. When the electrical circuit is closed, the bar magnet on the left rotates about the fixed wire; simultaneously, the wire on the right rotates about the fixed magnet [116]. The motions on each side are caused by the interaction of two magnetic fields: one associated with the bar magnet and the other the field caused by the current flow through the wire. The resulting rotary motions are the first example of machine which rotates as a consequence of a current flowing through a conductor within a magnetic field – or, more simply, a DC electric motor.

The 1821 experiment showed that in the presence of a magnetic field, a current flowing through a wire causes the wire to move; ten years later, in 1831, Faraday showed the opposite effect – that when a loop of wire is made to move within a magnetic field, a current flows in the conductor. Faraday had invented the dynamo [117]. This was an important breakthrough, for until that time, the only sources of electricity were batteries, or clumsy machines that held static charges. To create electricity by rotating a coil within a fixed magnet was much more reliable, and so in the mid-1800s, the technology of generation advanced quickly, and dynamos became progressively more powerful.

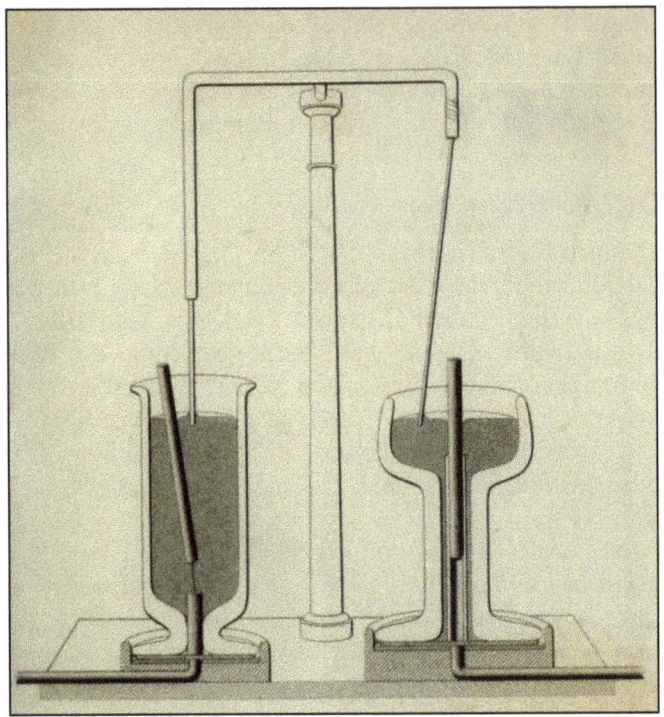

Figure 5.6. Faraday's 1821 'Homopolar Motor'.
Source: Based on Faraday's paper published in 1822 [115].

In the early 1870s, the most powerful DC dynamos were manufactured by the company run by the Belgian engineer Zénobbe Gramme [118], as shown in figure 5.7.

And in 1873, Gramme was in Vienna to participate in that year's International Industrial Exposition. Whilst setting up the company's exhibit, Gramme's colleague, Hippolyte Fontaine, inadvertently joined the terminals of two dynamos together. And when one dynamo was switched on, and – as intended – generated a current, Gramme noticed that the other dynamo rotated 'all by itself' [119]. As a result of Fontaine's inadvertent error, Gramme re-discovered what Faraday had first observed more than 50 years earlier – that when a current flows through a conductor within a magnetic field, the conductor moves.

So was born the commercial DC motor.

As Koestler said, 'creativity it is a new pattern of pre-existing components, and the more familiar the parts, the more striking the new whole'. And in this case, the key components are identical: physically, a DC dynamo and a DC motor are the same, for the difference lies not in the structure, but in the way that structure is used. For a dynamo, the coil is rotated, within a fixed magnet, and a current flows through the coil; for a motor, the current flows through the coil, which then

Figure 5.7. A Gramme dynamo, c. 1873. The hand-crank rotates a coil within the 'horseshoe' permanent magnet, so driving an electrical current through the trailing wires.
Source: Wikimedia 'Vue d'ensemble de la machine Gramme' https://commons.wikimedia.org/wiki/File:LaNature1873-345-MachineGramme.png.

rotates within the fixed magnet. Despite the simplicity of that concept, those who were building dynamos in the mid-1800s didn't notice it, until Gramme stumbled over it in Vienna as a result of Fontaine's mistake.

5.12 The impossible building

Many cities are 'defined' by the visual profile of their architecture, perhaps a group of buildings – for example, the skyline of Manhattan – or an individual structure, such as the Eiffel Tower in Paris. So if you think of Australia in general, or Sydney in particular, what comes to mind? Perhaps the bridge over Sydney Harbour. More likely, Sydney Opera House, and the distinctive shape of the 'sail roof' as captured in figure 5.8.

Situated on a peninsular jutting out into Sydney Harbour close by the bridge, the Opera House opened in 1967, and its dramatic, elegant, imaginative profile immediately invokes the wings of a sea-bird or the sails of a yacht, both images so appropriate to the building's location. Maritime locations were indeed familiar to the architect, Jørn Utzon, the son of a naval architect and yacht designer, and who grew up in the Danish port city of Aalborg [120]. The profile visible today, however, isn't quite the same as Utzon's original idea, as depicted in his entry to the international competition run by the government of New South Wales in 1956, one drawing from which is reproduced as figure 5.9.

Figure 5.8. Sydney Opera House.
Source: Wikimedia 'Sydney Opera House Side View' https://commons.wikimedia.org/wiki/File:Sydney_opera_house_side_view.jpg.

Figure 5.9. The original 1956 design – notice the absence of ribs within the 'sails'.
Reproduced from Jørn Utzon's Competition Entry, page 7, © State of New South Wales through the State Archives and Records Authority of NSW 2016 [121]. NSW State Archives: Department of Public Works; NRS 12825, Competition drawings submitted by Jørn Utzon to the Opera House Committee, 1956. NRS-12825-1-[SZ112]-[7] | West elevation National Opera House, Sydney, Australia, 1956.

The original design is somehow sleeker, lower; the actual structure more upright. Why is there a difference?

The Opera House competition attracted 233 entries [122], and in January 1957, Jørn Utzon's design was declared the winner. Soon thereafter, Arup & Partners were appointed consulting engineers [123], commissioned to work closely with Utzon to bring the design to reality. To do that, Arup's engineers had to keep three requirements in mind:

- The final structure had to maintain the aesthetic of the original design.
- The roof had to maintain its own weight safely, and withstand high winds.

- The cost had to be 'sensible', for the project was being funded by the government of New South Wales, with an original budget of (Australian) $7 million [124].

Central to the design is that wonderful sail-roof, originally envisaged as a sequence of free-standing reinforced concrete shells [125]. The engineers, however, quickly established that free-standing shells with those shapes were structurally unsound, and suggested using a framework of pre-cast concrete ribs to support the roof surface. Furthermore, the engineers discovered that the shape of a yacht's sail is mathematically highly complex, with a different radius of curvature at each point on the surface. As a consequence, they were unable to determine an equation that would correctly represent that surface, and that could be used to define the shape of the structure's ribs. In the absence of that equation, it would not be possible to fabricate the mould required to form the concrete segments from which each rib could be constructed. So the direct representation of a billowing sail had to be abandoned too, and replaced by a more geometrically regular, and mathematically simpler, shape that would still be analogous to a sail, but much more easily manufactured.

Determining what that shape should be proved to be a problem – a problem that would not be solved for some four years, and, when it was solved, the solution was discovered in a most unexpected way. And over those four years, twelve geometries were investigated, based on a variety of mathematically well-defined curves such as the parabola and the ellipse. Each possibility required many calculations of shear forces, bending moments, tensions, compressions and loads, and Arup were pioneers in the use of computers for engineering structural analysis [126]. Despite all this work, however, no satisfactory solution was found. It looked as if this structure was impossible to build.

But in October 1961, Utzon discovered the answer. Not by advanced mathematics. Not by using a powerful computer. But by having breakfast, and noticing something 'interesting'. For there is a story that he was eating an orange, and happened to cut a wedge rather like that shown on the left-hand side of figure 5.10. And he recognised the shape [127].

An orange is a sphere, the simplest of geometrical shapes. If the 'sails' could be made from segments cut from the same sphere, then the ribs can be defined precisely, and all would have the same radius of curvature. That enables them to be manufactured (relatively) easily, and at a (relatively) controllable cost. The 'Key to the Shells' plaque, located on the Opera House concourse, shows what those segments are, and those are those are the shapes of the 'sails' we see today.

The story of the orange is a vivid example of Koestler's 'bisociation' – the accidental juxtaposition of the orange segment and the problem of the 'sails'. But did this actually happen? No one really knows, and different sources give different explanations of how the spherical segment solution was discovered. Indeed, there is a similar story about how Eero Saarinen – who was one of the judges for the Sydney Opera House competition – had a similar idea whilst eating not an orange, but a grapefruit, for breakfast. In fact, there are two stories, one

Figure 5.10. An orange segment, and the 'Key to the Shells' plaque.
Source: Wikimedia 'Shells, Sydney Opera House, Australia' https://commons.wikimedia.org/wiki/File:Shells,_Sydney_Opera_House,_Australia.jpg.

associated with the shell of MIT's Kresge auditorium (opened 1955) [128], and another with the TWA flight centre (now the TWA hotel) [129] at New York's Idlewild (now Kennedy) airport (opened 1962), as shown in figure 5.11.

Figure 5.11. Eero Saarinen's Kresge Auditorium (left) and TWA flight centre (right).
Sources: Wikimedia 'MIT Kresge Auditorium' by Madcoverboy under a CC BY-SA 3.0 licence https://commons.wikimedia.org/wiki/File:MIT_Kresge_Auditorium.jpg and 'TWA Flight Center building at John F Kennedy International Airport, NYC in 2004' by Dmitry Avdeev under a CC BY-SA 3.0 licence https://commons.wikimedia.org/wiki/File:New_York_-_John_F._Kennedy_International_(Idlewild)_(JFK_-_KJFK)_AN0619542.jpg.

In the end, all Sydney Opera House's design and structural problems were solved, resulting in a true masterpiece. The statistics are impressive: 940,840 ceramic tiles were carefully laid side-by-side (rather than overlapped, as on a conventional roof) on 3,382 'tile lids' [130], themselves laid on the concrete ribs, formed from 2,194 pre-cast segments. And there are 6,225 square metres of glass

[131] too – for a comparison, a standard football pitch is about 7,140 square metres [132].

Some other statistics are singularly dramatic too. The final cost was some 15 times the original estimate, and the building opened in 1973, ten years later than the original intention [133]. Nor is the drama confined to the numbers. Jørn Utzon resigned in 1966 [134], and was never to set foot in Australia again. Nor was his vision fulfilled. Although the exterior, and in particular those 'sails' are (reasonably) true to his original concept, Utzon's design for the interior was totally scrapped: after Utzon returned to Denmark in 1966, a new architect, the Australian Peter Hall, was appointed to finish the project [135].

5.13 Special relativity

On 30th April 1905, Albert Einstein's submitted his PhD thesis 'A New Determination of Molecular Dimensions' to the University of Zürich [136]. It was approved on 27th July, and a slightly revised version was subsequently published by the leading German-language journal, *Annalen der Physik*, on 8th February 1906. And in the meantime, during 1905, *Annalen der Physik* published four other papers, all solely authored by the 26-year-old Einstein. The first, published on 9th June, was his analysis of the photoelectric effect, the phenomenon in which an electric current can be caused to flow by the absorption of light by certain materials; the second (18th July), a mathematical analysis of Brownian Motion, and a discussion of how this provides evidence for the existence of atoms and molecules; the third (23rd September), was the description of the key principles of special relativity, including length contraction and time dilation; and the fourth (21st November), the exposition of the equivalence of mass and energy, and the derivation of probably the most famous equation in science, $E = mc^2$ (although, in the paper, this relationship is expressed rather differently) [137].

Each of these four papers is a masterpiece; together they form a collection unparalleled in the history of science; and it was the first, on the photoelectric effect, for which Einstein was awarded the 1921 Nobel Prize in Physics [138] – he was never awarded a Nobel prize for special relativity, or for his subsequent development of general relativity [139].

Einstein's main 1905 paper on special relativity (the third in the sequence), has no references, so it might be thought that special relativity was the brain-child of the archetypal 'lone genius'. But was it?

Despite the absence of references, the paper does mention James Clerk Maxwell and Hendrik Lorentz by name, as well as ending with the sentence:

"In conclusion I wish to say that in working at the problem here dealt with I have had the loyal assistance of my friend and colleague M. Besso, and that I am indebted to him for several valuable suggestions." [140]

The names of Maxwell, who unified electricity and magnetism, and of Lorentz, who shared the second Nobel Prize in Physics in 1902 [141], might be familiar, but who is 'my friend and colleague M Besso'? That question does have an answer: since 1904, Michele Besso (an engineer) and Einstein had been work colleagues at the Swiss Patent Office; they had met each other by chance some six years earlier at a music party which took place in Zürich whilst Einstein was a student at the Federal Polytechnic Institute (now ETH), and they had remained friends ever since [142].

Maxwell and Lorentz were indeed important providers of pre-existing ideas which Besso helped Einstein combine together, but these are just three of the many links in the chain of events that led to $E = mc^2$.

Newton himself is a starting point, for he, along with Galileo and Descartes, recognised the fundamental principle of relativity, that certain measurements are not 'absolute', but contingent on the circumstances of the observer. So, for example, someone standing at the side of a straight road might measure the speed of a horse, moving away, as u m s^{-1}. At the same time, an observer in a carriage, travelling along the road at a constant speed of v m s^{-1} as measured by the roadside observer, and moving in the same direction as the horse, will measure the speed of that horse as u' m s^{-1} such that $u' = u - v$. This is the first time derivative of the 'Galilean (alternatively, Newtonian) transformation', $x' = x - vt$, in which the relative speed v m s^{-1} of the two observers is, as already stated, constant. Importantly, if the horse were to accelerate, then both observers would measure the same acceleration, and so infer the presence of the same force. So despite the fact that the measured speeds are different, 'the laws of physics' – which are very much associated with forces and accelerations – are the same to both observers. In more technical language, Newton's Laws of Motion, and many other laws too, are 'invariant with respect to a Galilean transformation'. A feature of the Galilean transformation, however, is that time, and time intervals, are the same for all observers, even those moving with constant speed relative to one another: $t' = t$ always and for everyone. To Newton and his contemporaries, whereas measures of distance and speed were 'relative', time was 'absolute'.

Fast forward to the early 1860s, when, in a series of papers published in 1861, 1862 and 1865, James Clerk Maxwell, as already noted, drew on the work of Ørsted, Faraday, Ampère and Gauss (and others too), unifying electricity and magnetism in the four 'Maxwell Equations', which themselves can be combined to show that associated electric and magnetic fields can propagate as waves. The resulting 'wave equation' enabled the speed of those waves to be determined from other known physical constants, giving a value of approximately 3×10^8 m s^{-1}. This was very similar to the then current estimate of the speed of light, leading Maxwell to conclude that light itself must be an electromagnetic wave [143].

That speed, $c = 3 \times 10^8$ m s^{-1}, is the speed as measured by an observer who is stationary with respect to the source of light. If, however, the observer and the source of light are moving away from one another at a constant relative speed v m s^{-1}, then the Galilean transformation implies that the observer will measure the speed of light as $c' = c - v$. The speed of light will therefore differ according to the relative motion of the source and the observer.

The most important light source, of course, is the Sun. Relative to which the Earth moves. And in 1887, Albert Michelson and Edward Morley carried out a series of experiments using an instrument (briefly described in section 13.2) sufficiently sensitive to measure the change in the speed of light attributable to the relative motion of the Earth and the Sun, as predicted by the Galilean transformation. But their result was puzzling. No speed difference could be detected [144].

This presented a paradox, with the implication that the physics of Galileo, Newton and mechanics was incompatible with the physics of Maxwell and light. Over the following years, this led to something of an intellectual scramble, as different people proposed a range of different ideas, all attempting to explain the result of the Michelson-Morley experiment, and all contributing concepts that Einstein would bring together in his theory of special relativity, by which that paradox was finally resolved.

One such idea was already on the table: in 1887, a few months before the Michelson-Morley experiments just noted but in response to an earlier study Michelson and Morley had carried out in 1886 [145], Woldemar Voigt (who coined the modern meaning of the word 'tensor', the entity that plays such an important role in general relativity) published a paper [146] concerning the mathematics of the wave equation when there is relative movement at a constant speed v m s^{-1} between the wave source and an observer – the phenomenon known as the Doppler effect, as commonly experienced with sound waves. Voigt discovered something that would turn out to be highly important – although he probably didn't appreciate its significance at the time: a transformation under which the wave equation is invariant, implying the speed of light would be independent of any relative movement of the source and the observer, as Michelson and Morley were about to confirm. Voigt's transformation includes the quantity $\gamma = 1/\sqrt{(1 - v^2/c^2)}$, and also requires that time is no longer absolute, but varies with the (constant) relative speed v such that $t' = t - vx/c^2$ [147]. The identification that $t' = t - vx/c^2$ introduces the concept of relativity as regards time, and Voigt's transformation anticipates what was later to become known as the Lorentz transformation – key elements in special relativity, and identified when Einstein was just 8 years old [148].

Then, in 1889, George Fitzgerald suggested that the length of a moving body might contract in the direction of motion [149], and three years later, in 1892, Lorentz – apparently independently of both Voigt and FitzGerald – proposed the transformation now associated with his name; he also demonstrated that the Maxwell equations are invariant with respect to it, implying that the speed of light is a constant regardless of the relative motion of a source and an observer, confirming Voigt's findings and as verified by Michelson and Morley. Lorentz developed his approach in further publications in 1895, 1899 and 1904 [150], and also discusses a concept he called 'local time', whereby time measurements in a moving frame depend on the speed of that frame relative to another frame. This implies that time measurements are relative, not absolute, as indeed had been strongly argued by Ernst Mach in his book, *The Science of Mechanics*, published in 1883 [151] and, as we have seen, established by Voigt. Lorentz, however, didn't explore the consequences of that profound thought – rather, he regarded the time transformation as a mathematical construct, as required to ensure the invariance of Maxwell's Equations.

Joseph Larmor subsequently enriched Lorentz's work and established time dilation [152], and around the turn of the twentieth century, Henri Poincaré applied his most formidable intellect to offer a deeper insight into 'local time', taking it beyond a mathematical artefact into an appreciation of what, fundamentally, 'simultaneity' means, and how time might be perceived and measured by different observers [153]. And it was Poincaré who first expressed the Lorentz transformation in the form in which we know it today, and indeed coined the phrase 'Lorentz transformation' [154].

Then, in a lecture delivered on 24th September 1904 at the St Louis Congress of Arts and Sciences, Poincaré articulated the principle of relativity in this English translation of the original French:

"The principle of relativity, according to which the laws of physical phenomena must be the same for a stationary observer as for one carried along in a uniform motion of translation, so that we have no means, and can have none, of determining whether or not we are being carried along in such a motion." [155]

This lecture is intriguing, and full of gems, not least:

"From all these results, if they were to be confirmed, would issue a wholly new mechanics which would be characterised above all by this fact, that there could be no velocity greater than that of light (because bodies would oppose an increasing inertia to the causes that would tend to accelerate their motion, and when approaching the velocity of light, this inertia would become infinite), any more than a temperature below that of absolute zero." [156]

Furthermore, in two papers, published in June and July 1905 respectively, prior to Einstein's two papers on special relativity, Poincaré asserted that relativity was a general law of nature, and predicted gravitational waves too [157], a prediction that would not be verified until over a century later, in 2016, as will be described in section 13.2.

So by 1905, when Einstein published his special relativity papers, the key 'components' (to use Koestler's word) that Einstein 'combined' were all present: the mathematics of the Lorentz transformation had been formulated by Voigt and Lorentz, and rather more philosophical concepts such as length contraction, time dilation and the relativity of time had been tabled and explored to one extent or another by, for example, FitzGerald, Larmor, Lorentz and Poincaré. But whereas much of that work was focused on addressing the consequences of the Michelson-Morley experiment, Einstein took a more theoretical approach, investigating the dynamics of moving bodies starting from a single fundamental principle: that the speed of light must be measured as the same value regardless of any (constant) relative motion of the light source and an observer. This is pure Koestler, '… combining and synthesising already existing facts…' into a truly wonderful, and indeed striking, 'new whole'.

But what contribution did Michele Besso make? And what were his 'several valuable suggestions'? In 1922, at a lecture in Kyoto, Einstein recalled that:

> "… a friend of mine living in living in Bern helped me by chance. One beautiful day, I visited him and said to him: 'I presently have a problem that I have been totally unable to solve. Today I have brought this 'struggle' with me.' We then had extensive discussions, and suddenly I realised the solution. The very next day, I visited him again and immediately said to him: 'Thanks to you, I have completely solved my problem.'
>
> My solution actually concerned the concept of time. Namely, time cannot be absolutely defined by itself, and there is an unbreakable connection between time and signal velocity. Using this idea, I could now resolve the great difficulty that I previously felt.
>
> After I had this inspiration, it took only five weeks to complete what is now known as the special theory of relativity". [158]

For many months, Einstein had been thinking about time, about simultaneity, and about how different observers might measure time intervals, and he was familiar with the work of others such as Poincaré, and also Ernst Mach, whose writings had been introduced to Einstein by Besso [159].

But it was a muddle, things just didn't fit. Until, in that conversation with Besso, the pieces of the Koestler's Law puzzle all suddenly came together… [160].

5.14 The structure of DNA

5.14.1 DNA and X-ray crystallography

On 10th December 1962, King Gustav VI Adolf of Sweden presented the Nobel Prize in Physiology or Medicine to Francis Crick, James Watson and Maurice Wilkins [161] 'for their discoveries concerning the molecular structure of nucleic acids and its significance for information transfer in living material' [162]. The name of Maurice Wilkins is less well-known; those of Crick and Watson more so. But almost everyone recognises the double helix of DNA.

The discovery of the structure of DNA is regarded by many as one of the greatest scientific achievements of the 20th century. So what did Crick, Watson and Wilkins actually do?

That DNA plays the key role in heredity is itself a Koestler's Law story, and so, for the current purposes, I just mention the contributions made by Frederick Griffith and Oswald Avery in the 1920s, 1930s and 1940s who proved that DNA, rather than some cellular protein, was the transmitter of hereditary information from one generation to the next [163]. Chemically, DNA is a long polymer, containing four 'nucleotide bases' (the two 'pyrimidines' adenine and guanine, and the two 'purines' thymine and cytosine), a sugar (deoxyribose), and some phosphate components too [164].

Also central to this story is the analytical method known as X-ray crystallography. Very briefly, when a fine beam of X-rays is targeted on a crystal, the beam does not just pass through or bounce back. Rather, the incident beam is split into an array of many exit beams, all of different intensities, and at different angles. This process is known as 'diffraction', and the exit beams may be detected and recorded, for example, on a photographic plate, resulting in a 'diffraction pattern' [165]. This phenomenon was first observed by Max von Laue in 1912, and in 1913, William Bragg (then a professor at the University of Leeds) and his son Lawrence Bragg (then a research student at University of Cambridge) showed how the diffraction pattern could be analysed and interpreted to give information about the structure of the crystal, and also the molecules from which the crystal is formed. Von Laue was duly awarded the Nobel Prize in Physics in 1914 [166], the Braggs in 1915 [167] – Lawrence, to this day, being the youngest-ever winner of a science prize at 25 [168], and second only (at the time of writing, 2021) to Malala Yousafzai, who was 17 when awarded the 2014 Peace Prize [169].

Finally by way of background, the interpretation of an X-ray diffraction pattern, and the determination of the corresponding structures, was, until very recently, both an experimental and computational marathon. Mathematically, a diffraction pattern is the square of the Fourier transform of the product of a function defining the shape of the crystal and the convolution of two other functions, one defining the three-dimensional structure of the crystal lattice, and the other defining the structure of the molecule from which the crystal is formed [170]. So analysing the intensities of a set of spots on a photographic plate to get back to the molecule's structure is a lot of work. And many Nobel Prizes have been awarded to those who have succeeded [171].

5.14.2 Photo 51

X-ray photographs play a significant role in this story, and 'Photo 51', shown in figure 5.12, is central.

This photograph of the X-ray diffraction pattern of DNA was taken in May 1952 [172] by a PhD student named Raymond Gosling [173], whose supervisor, Rosalind Franklin [174], was a research associate at the Medical Research Council Biophysics Unit at King's College, London, the Deputy Director of which was Maurice Wilkins [175]. And it was this photo that Wilkins showed Watson, without Franklin's knowledge, during a visit by Watson to Wilkins on 30th January 1953 [176]. As we have seen, Kepler described his realisation that the orbit of Mars is an ellipse as if he had been awoken from a sleep; Watson is more dramatic, for, as soon as he saw the photo, 'my mouth fell open and my pulse began to race' [177]. For he recognised that what he was looking at was the tell-tale X-shape of the diffraction pattern of a helix.

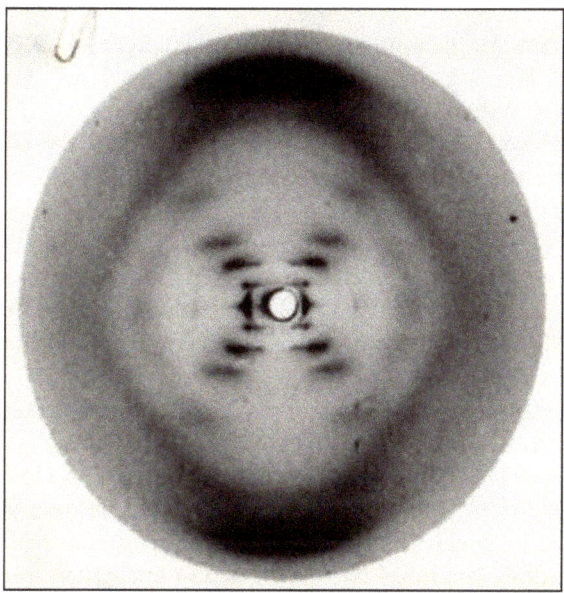

Figure 5.12. Raymond Gosling's 'Photo 51' of the X-ray diffraction pattern of DNA.
Reproduced courtesy Ava Helen and Linus Pauling Papers, Oregon State University Libraries http://scarc.library.oregonstate.edu/coll/pauling/dna/pictures/sci9.001.5.html.

5.14.3 Florence Bell

It turns out, though, that Photo 51, taken in 1952, is not the first X-ray diffraction pattern of DNA, nor the first to show the characteristic X-shape. That photograph, shown on the left of figure 5.13, was taken in 1938 by Florence Bell [178], a research student of William Astbury [179] at the University of Leeds (where William Bragg had been in 1913 [180]).

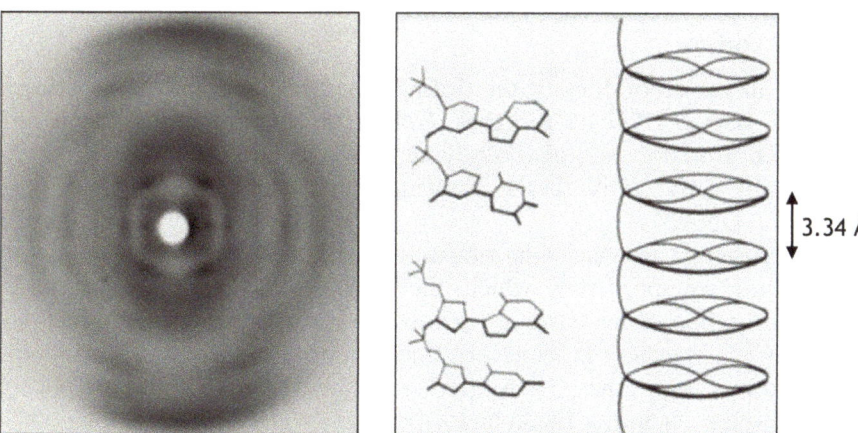

Figure 5.13. One of Florence Bell's X-ray diffraction photographs of DNA, and the suggested 'pile of pennies' structure, 1938.
Reproduced courtesy Leeds University Library, Special Collections, MS 2229; https://explore.library.leeds.ac.uk/special-collections-explore/650413, pages 146 and 150.

Why was William Astbury interested in the structure of DNA and the use of X-ray diffraction? That's a 'mini' Koestler's Law story too. Astbury had been a research student of William Bragg, and so was well-versed in the technique. In 1928, he was appointed as a Lecturer in Textile Physics at the University of Leeds [181], a leading centre of research into the physical and chemical properties of naturally occurring fibres such as wool, as might be expected in a university located in the city at the heart of Britain's wool industry. During the 1930s, although the role of DNA as the carrier of genetic information was not yet clear, it was becoming established that the molecule is long and fibrous, rather like the protein keratin, the key structural component of wool, and much studied by Astbury during the 1930s.

Putting those 'Koestler's Law fragments' together, the result was Bell's PhD thesis [182], and two papers, co-authored by Bell and Astbury – the first published in *Nature* in April 1938 [183] and the second presented at a Cold Spring Harbour Symposium in July 1938 [184]. In these papers, DNA is described in terms of a columnar structure in which the nucleotide bases are stacked above one another, 'like a pile of pennies', separated by 3.34 Å (0.334 nm) [185], but – crucially – not in the form of a helix: in 1938, the mathematics of the diffraction pattern of a helix had not yet been done, and so Bell and Astbury did not recognise the X-shape as being the 'signature' of a helix. That 1938 paper included something else important too – a measurement of the density of DNA as about 1.62 g cm^{-3}. Density is a measure of how much mass there is in a given volume, and this measurement was to be critical in proving that DNA could not be a single chain, but either two or three chains, somehow intertwined [186].

The Second World War then intervened, and afterwards, Astbury continued studying DNA, with further, clearer, X-shaped photographs, such as the one illustrated in figure 5.14, taken by his new student, Elwyn Beighton, in May and June 1951 [187], about a year before Raymond Gosling's 'Photo 51'of May 1952.

5.14.4 Sven Furberg

Meanwhile, in the laboratory of the distinguished scientist and X-ray crystallographer J D Bernal [188] at Birkbeck College in London, a Norwegian PhD student, Sven Furberg was also studying the structure of DNA using X-ray crystallography [189]. His thesis, presented in 1949, proposes two possible structures for DNA.

One is a structure in which 'the ribose rings and the phosphate groups form a flattish central column, from which the purines and pyrimidines stand out perpendicularly. Successive bases are pointing outwards in opposite directions…'.

The other is a structure in which the nucleotides 'are piled in a central column directly on top of each other, 3.4 Å apart, with the ribose rings and the phosphate groups in a spiral enclosing the column' [190].

This 'spiral' structure spaces the nucleotides by 3.4 Å (0.34 nm), sensibly in agreement with Bell and Astbury's measurement of 3.34 Å, and enriches their 'pile of pennies' idea by adding that all-important word 'spiral'. Strictly speaking, a spiral is a line drawn on the surface of a cone, and so the radius changes

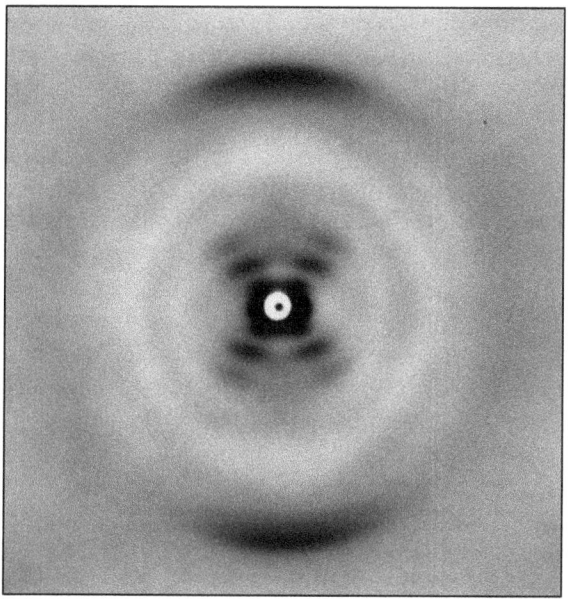

Figure 5.14. Elwyn Beighton's X-ray diffraction photograph of 28th May 1951, which is very similar to Raymond Gosling's 'Photo 51' of May 1952, figure 5.12.
Reproduced courtesy Leeds University Library, Special Collections, MS 419 Box C.7; www.kerstenhall.com/2020/07/21/photo-51-and-the-road-not-taken/.

depending on the position on the cone; by contrast, a helix is a line drawn on the surface of a cylinder, and so the radius remains constant. Although Furberg refers to a 'spiral', there is no doubt that he meant 'helix', making Furberg to be the first to suggest a helical structure for DNA – but only a single helix, not a double one. And, importantly, he explicitly identifies that the nucleotides are on the *inside*, and the phosphate and sugars on the *outside*.

Furberg's idea that the structure of DNA might be helical was not a direct inference from X-ray diffraction, for he did not say, 'Because the X-ray diffraction pattern shows a prominent X-shape, the structure must therefore be helical'. And the reason he did not do that was because he did not know that: as we shall see, it was not until 1952 that the mathematics associated with the diffraction of X-rays by helical structures was first published. Rather, Furberg identified the possibility of a helical structure by 'chemical dead reckoning'. He knew much about the geometry of the key components – the sugar molecule, the phosphate group, and the four nucleotides – and he also knew much about how different chemical groups form bonds, and how they are directed in three-dimensional space. Drawing on this knowledge, he was then able to deduce how a complex molecule might be structured, so that each of the components preserved their own geometries, and that the various chemical bonds were in the right orientations.

The overall result gave two possibilities, the 'flattish column with the nucleotides standing out perpendicularly' and the structure in which the nucleotides 'are piled in a central column directly on top of one each other … with the ribose rings and the phosphate groups in a spiral enclosing the column'. Though theoretically possible, the first was a blind alley; the second was an important step along the right track, but at the time Furberg did his research, he did not know that. The academic paper resulting from his work, published in 1952, therefore presents both suggestions without expressing a preference [190].

Furberg's suggestion of a helical structure for DNA is another example of the zeitgeist, for whilst Furberg was working on his thesis in London, perhaps the greatest chemist of the age was recovering from a cold. In April 1948, Linus Pauling [191] was visiting Oxford, and fell unwell. He had been working on the structure of the fibrous protein keratin (which had also been a research interest of Astbury), and, whilst in bed, scribbled the chemical structure on a sheet of paper. As he was idly fiddling with the paper, he happened to fold it so that the two-dimensional paper formed a three-dimensional shape. The paper had become a cylinder, and what had hitherto been a flat structure now appeared as a helix. When wrapping the paper more and less tightly, Pauling noticed that a helix of a particular diameter resulted in a configuration that enabled what chemists know as 'hydrogen bonds' to be formed at regular intervals along the resulting structure, so stabilising the helical shape [192], as shown in figure 5.15.

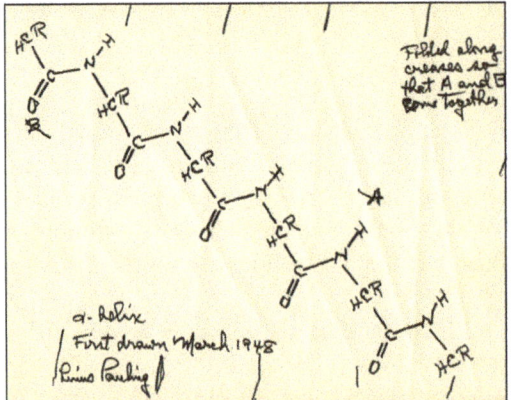 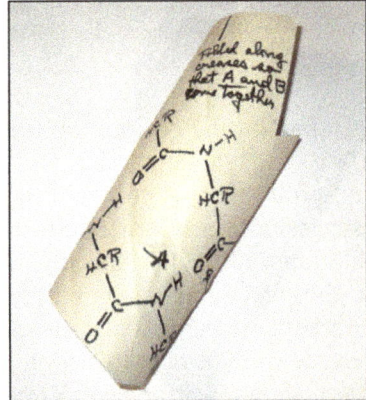

Figure 5.15. Linus Pauling's folded paper, 1948, showing the formation of hydrogen bonds between each C=O and the fourth-next N-H (as indicated by the asterisks), so stabilising the helix.
Left image reproduced courtesy Ava Helen and Linus Pauling Papers, Oregon State University Libraries https://paulingblog.wordpress.com/2011/03/09/the-alpha-helix/; right image, author's photograph.

This accidental bisociation of 'structure of keratin' and 'folded paper' led to the discovery of a fundamental biological structure known as the 'α-helix', as described in a series of papers published in 1951 – a bisociation which contributed to Pauling's award of the Nobel Prize in Chemistry in 1954 [193]. That original sheet of folded paper can now be seen in the Nobel Prize Museum in Stockholm, within their collection 'Cultures of Creativity' [194], which uses the context of the

Nobel Prizes (in all fields) to address the questions 'What is creativity and how can creative activity best be encouraged? Which is more important to the creative process: the individual or the environment?' – questions that, I trust, this book helps answer!

5.14.5 The diffraction pattern of a helix

So, back to DNA... for it's around this time that Francis Crick [195] enters the story, albeit obliquely. In 1951, Vladimir Vand [196] was a research fellow at the University of Glasgow. He was working on the mathematics of the X-ray diffraction patterns of helical structures, fundamental to which is the computation of the Fourier transform of a helix. Seeking some advice, Vand contacted the most eminent crystallographer in the country, Lawrence Bragg, now Sir Lawrence Bragg, and head of the Cavendish Laboratory of Physics at Cambridge. Bragg referred Vand's request to one of the local experts, William Cochran [197], who in turn discussed the matter with Francis Crick, at that time a PhD student. The result was the definitive paper on the mathematics of diffraction by helical molecular structures, known as the 'CCV' (Cochran, Crick and Vand) paper, published on 10th September 1952 [198]. The mathematics shows that the diffraction pattern is defined by a sequence of the squares of Bessel functions of progressively higher orders, symmetrically above and below a horizontal mid-line. When captured on a photographic plate, this pattern is X-shaped, such that the angle between the 'arms' of the X is determined by the radius of the helix (the greater the radius, the narrower the angle) and the spacing between the successive horizontal rows by the distance between successive 'layers' of molecules within the helical structure (the more widely spaced the rows, the narrower the distance between the layers) [199].

5.14.6 James Watson

James Watson [200] enters the story now too. In 1950, Watson, aged 22, gained a PhD in zoology, and became a post-doctoral researcher at the University of Copenhagen, where he became increasingly interested in DNA and its structure. That led him to join Cambridge's Cavendish Laboratory, which included a distinguished team using X-ray crystallography to study biological molecules, notably John Kendrew [201], and Max Perutz [202], both of whom were to join Crick, Watson and Wilkins in Stockholm in 1962 to receive their Nobel Prizes, in Chemistry [203] – Kendrew for the structure of the protein myoglobin; Perutz, for haemoglobin. Watson arrived in Cambridge in October 1951, and soon met Crick – a most fortunate bisociation of skills: Crick's first degree was in physics, and he was confident in mathematics; Watson's knowledge related primarily to biology, biochemistry and genetics.

During 1952, Watson and Crick were in touch with, among many others, Maurice Wilkins and Rosalind Franklin, and, as noted earlier, in January 1953, Watson visited Wilkins and saw Photo 51. At which point, to use Watson's own words once more, 'my mouth fell open and my pulse began to race' [177].

Watson knew the CCV paper on the mathematics of the diffraction pattern of a helix, so he recognised that Photo 51's X-shape was cast-iron proof that DNA was a helix – evidence also plainly visible in the photographs taken by Elwyn Beighton at Leeds in 1951, but as yet unseen by Watson, and also in the 1938 paper by Astbury and Bell. And over the following weeks, Watson worked closely with Crick exploring a variety of possible models. One key question was 'two helices or three?'. Watson and Crick knew from Florence Bell's density measurements that DNA was 'too heavy' to be just a single helix, but whether there were two or three helices was at that time still unresolved. Or not quite – as well as showing a very prominent X, Photo 51 has a further feature which is rather less obvious: the dark spots that 'should' be on the fourth lines above and below the central horizontal are 'missing' (see figure 5.12), and, as was subsequently realised, this absence is evidence for a double helix, not a triple one [204].

5.14.7 The triple helix

Just a few days before Watson was shown Photo 51, Watson and Crick had read a pre-print of a paper that was published on 1st February 1953 [205]. The authors of the paper were the two eminent chemists Linus Pauling (who discovered the α-helix) and Robert Corey [206], and the content of the paper was a proposed structure for DNA. To quote the paper: 'We have now formulated a promising structure for the nucleic acids… [a] structure that is not a vague one, but is precisely predicted…'. And, given the prestige of Pauling and Corey, that 'precisely predicted structure' carried not inconsiderable weight. The structure then described was a *triple* helix, with the nucleotides on the outside, as shown in figure 5.16. Was this right?

Watson and Crick immediately thought 'no'. But to say 'that's wrong' is not enough. They needed to propose an alternative; an alternative that was demonstrably better…

5.14.8 The missing Koestler's Law fragment

Pauling and Corey's triple helix structure was wrong because they were missing a key Koestler's Law fragment – a fragment that was available to anyone in 1952, so long as they were looking in the right direction and were alert enough to spot it. For this fragment was not in the fields of crystallography, physics, chemistry, chemical structures or mathematics, but in the fields of biology and biochemistry – Watson's home territory.

By 1952, the role of DNA as the carrier of genetic information from one generation to the next was well established. How this transfer of information was achieved was still unknown, and so many biologists and geneticists were hard at work trying to discover the mechanism. A plausible explanation was 'something to do with the molecule', and while the crystallographers were looking at the structure of the molecule as a whole, biologists and biochemists were looking at the biological function, and the details of the biochemistry.

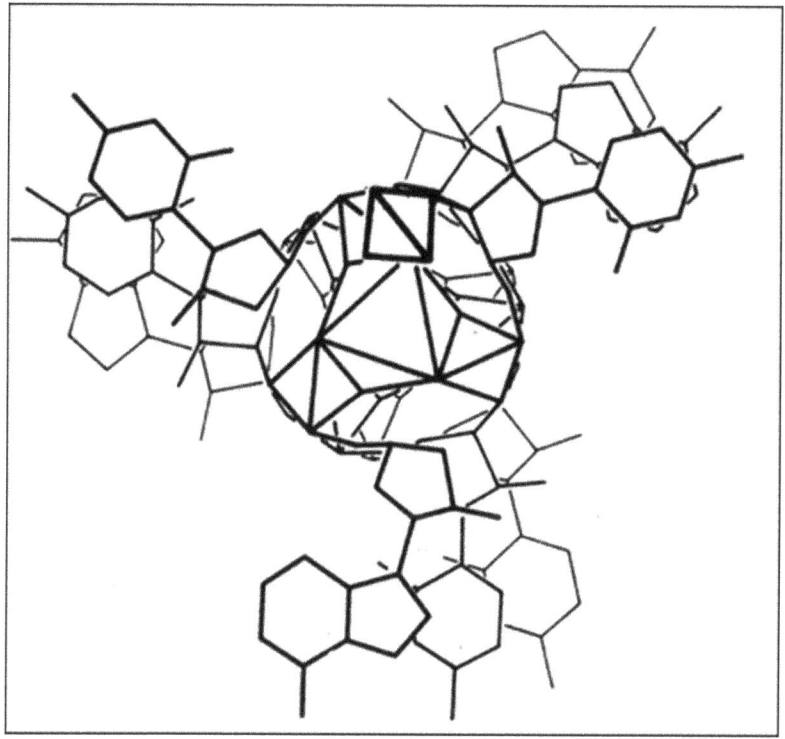

Figure 5.16. The triple helix structure proposed by Pauling and Corey, 1953. Reproduced courtesy Ava Helen and Linus Pauling Papers, Oregon State University Libraries http://scarc.library.oregonstate.edu/coll/pauling/dna/papers/1953p.9–092.html.

One of these researchers was Erwin Chargaff [207], working at Columbia University. For some years, biochemical studies of DNA had shown that each DNA molecule is a polymer – an assembly of a series of identical, or near identical, units called 'monomers' – formed from phosphates, the sugar deoxyribose, and four nucleotides, adenine (A), cytosine (C), guanine (G) and thymine (T). Furthermore, quantitative analyses had shown that the amounts of each of these nucleotides in any strand of DNA were about equal: A = C = G = T [208]. That suggests that DNA is either a largely random sequence of the same numbers of each nucleotide, or a polymer of A-C-G-T monomers.

Over the period from the late 1940s to the early 1950s, Chargaff carried out some very careful and detailed studies of the chemical structure of DNA from a variety of sources, publishing his findings in a series of papers, the first of which was published in 1948 [209], as well as presenting them at a number of conferences and on some lecture tours too. No, said Chargaff. The equation A = C = G = T is not true. What is true is that, for any strand of DNA, A = T and C = G, always, but not that A = C = G = T.

The twin relationships A = T and C = G imply that DNA might have *two different 'monomers'*: one a combination of A and T; the other a combination of C and G. But the numbers of As and Ts, and of Cs and Gs, are independent: a molecule of DNA could, theoretically, be all A-T, or all C-G. But fundamentally, A and T are paired together, as are C and G. Always.

The idea of 'base pairs' was therefore in the public domain from about 1948, with the evidence becoming increasingly convincing in 1949 and 1950. Evidence available for anyone to read and think about. But only if they looked in that direction, and were alert enough to spot its significance.

Pauling and Corey had not done that, for their triple helix structure did not incorporate anything in particular about the nucleotides, nor that very specific A = T, C = G finding, subsequently to be known as 'Chargaff's First Rule' [210].

Nor had Wilkins, Franklin or any of their team noticed it either.

But Watson did. For Watson was a biologist, and he had been present, with Crick, at a lecture given by Chargaff in Cambridge in May 1952 [211]. So he knew he had to incorporate that finding into any proposed structure of DNA. To do so, he (literally) 'played' with models. Using cut-outs that each represented the molecular shape of each of the four nucleotides, he tried to fit them together, this way and that, to see if any of the resulting patterns made any sense. Which they didn't [212].

5.14.9 Jerry Donohue

Not, that is, until Jerry Donohue [213], an American from Caltech, who just happened to be on a sixth-month stay in Cambridge, and just happened to share an office with Watson and Crick, made a significant contribution. On 27th February 1953, four weeks after Watson had seen Photo 51, Donohue noticed Watson trying to fit the nucleotides together. Donohue was a chemist, and knew that T and G can each have two different forms, known as the 'enol' and the 'keto', in which one hydrogen atom is located in a different position. And these different forms have different molecular shapes. When Donohue saw the shapes being used by Watson, he recognised these as enols, and so he suggested to Watson that he might try the ketos.

So Watson made some more cut-outs, in the alternative keto shapes, and...

...there it was. A way of fitting A and (keto) T together, side-by side, and C and (keto) G too, with each pair being stabilised by the formation of hydrogen bonds, the same chemical force that Pauling had invoked to discover the α-helix. There was no way in which A or T would associate with C or G: the only possibilities were an A-T pair and a C-G pair [214]. And when Watson fitted the cut-outs together in the right way, he saw that the side-by-side combination of A and T has *exactly* the same overall width as the side-by-side combination of C and G, implying that the base pairs could be stacked, like 'a pile of pennies' (as Astbury and Bell might say) within an overall helix (as Furberg might say), as verified by Photo 51 (taken by Gosling, Franklin's student) – and by the photographs taken by Bell in 1938 and Beighton in 1951 too.

On the morning of Saturday 28th February 1953, they cracked it.

Watson and Crick then went to a nearby pub, The Eagle, to celebrate – as commemorated on a brass plaque, reproduced as figure 5.17, that can be seen inside the Eagle to this day.

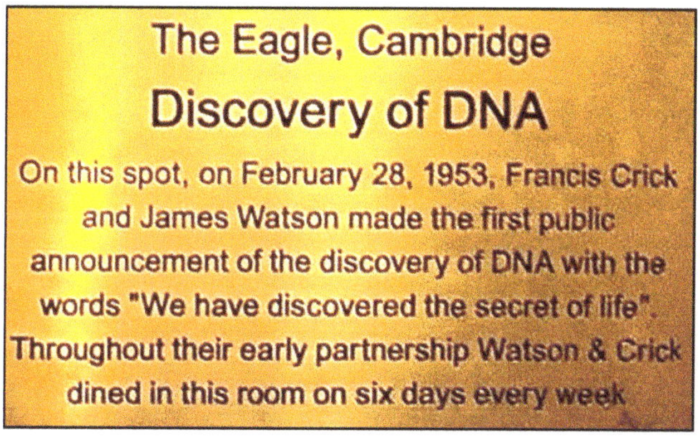

Figure 5.17. The plaque on the wall of The Eagle.
Reproduced courtesy of Greene King www.greeneking-pubs.co.uk/discover/historic-pubs/the-eagle/.

5.14.10 The Nobel Prize

Put all that together, and there emerges a structure for DNA as a *double* helix, where each strand is a sequence of A, C, T and G, such that if, at any point, the nucleotide on one strand happens to be, say, a T, *then the nucleotide on the opposite strand must be an A*, or if a C then a G. And since the overall width of each A-T, C-G pair is always the same, the helix is a perfect cylinder, all along its length.

And there's more too. If the double helix 'unzips' so that each strand separates, the sequence of nucleotides is preserved, but each, now separate, strand becomes a template for a new strand: where ever there is a C on the 'old' strand, there must be a G on the 'new' one; wherever a C, a G; wherever an A, a T; wherever a T, an A. One 'parent' double helix structure can therefore replicate to form two identical 'daughter' double helix structures, as shown in figure 5.18, so preserving the 'message' contained in the sequence of nucleotides and transmitting that message from generation to generation…

At a stroke, everything came together. The X-shape of the diffraction photographs implies that DNA is a helix. Chargaff's Rule implies base pairs and a double helix, as is consistent with the known density, and that missing spot on the fourth row of the diffraction photograph. And biologically, the structure makes

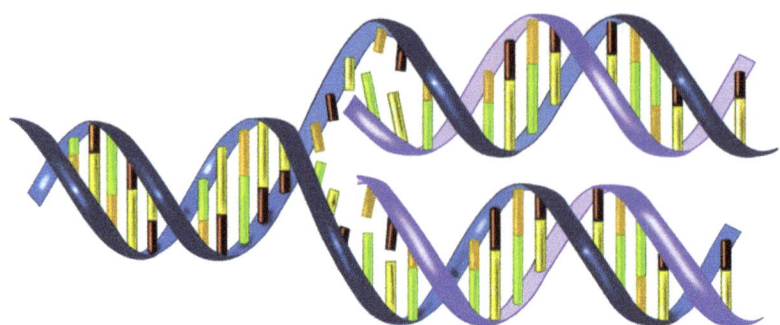

Figure 5.18. Schematic representation of the replication of DNA showing how a 'red' (A) always pairs with a 'yellow' (T), and a 'green' (C) with an 'orange'(G).
Source: https://quizlet.com/122422264/flashcards.

great sense, for it provides a totally plausible explanation of *how* heredity actually works.

Watson and Crick published all this in a paper of fewer than 900 words, taking up about a single page in the 25th April 1953 issue of *Nature*, and with just six references [215]. And in those 900 words, they firstly dismiss Pauling's triple helix as being 'unsatisfactory for two reasons' (which they give), and then 'put forward a radically different structure', the double helix. To support their idea, they cite Furberg's structure, Chargaff's Rule A = T, C = G, and the X-ray photographs of Wilkins and Franklin, who are thanked, and Astbury, who is referenced. Jerry Donohue is thanked too. And the paper also contains the almost throw-away sentence, 'It has not escaped our notice that the specific pairing we have postulated immediately suggests a possible copying mechanism for the genetic material'. Indeed.

And that paper was how James Watson and Francis Crick, alongside Maurice Wilkins, came to shake hands with King Gustav VI Adolf of Sweden that December day in 1962.

Not on the platform were Rosalind Franklin (she had died in 1958, and one of the Nobel rules does now allow posthumous awards), nor Raymond Gosling (who had actually taken Photo 51), nor William Astbury (he too had died, in 1961), nor Florence Bell (who took the very first X-ray photograph in 1938), nor Jerry Donohue (who gave Watson the key idea to make base pairing work), nor Erwin Chargaff (who discovered the critical rule that A = T and C = G).

5.14.11 A moment to reflect

That's a long story, and there are many further features that I have omitted. But it does capture the key events associated with one of the landmark discoveries of the twentieth century, one of the most creative acts in the history of science.

So what did Watson and Crick actually *do*?

The short answer, and the long one too, is 'not a lot'. Unlike Kepler, who laboured, really laboured, for years, Watson and Crick did almost no work at all. Yes, Crick contributed to the CCV paper, but that was initiated by Vand's enquiry to Bragg. And – certainly according to Watson's own account as recorded in his book *The Double Helix* – Watson didn't do that much work either, but spent most of his time, when not playing with models, going to conferences and enjoying himself.

What they did not, and actually did, do was exactly what Koestler's Law says: they did 'not create something out of nothing'; rather, they 'uncovered, selected, re-shuffled, combined, synthesised already existing facts, ideas, faculties, skills'. In spades. And as they themselves recognised: one of Watson's subsequent roles was to become firstly Director, then President, then Chancellor, of the Cold Spring Harbor Laboratory on Long Island, New York, and these are some words to be found on their 'exploratorium' website [216]:

"Norwegian scientist Sven Furberg's DNA model – which correctly put the bases on the inside of a helix – was one of many ideas about DNA that helped Watson and Crick to infer the molecule's structure. To some extent, they were synthesizers of these ideas. Doing little laboratory work, they gathered clues and advice from other experts to find the answer."

As explicit a statement of Koestler's Law as might be imagined.

And the overall story contains many embedded Koestler's Law mini-stories, such as how Pauling discovered the α-helix by bisociating 'structure of keratin' with 'folded paper', and the bisociation of Watson's skills with Crick's. Neither could have solved the structure of DNA by themselves, but together...

Watson and Crick were the beneficiaries of a lot of luck too. What would have happened had Vand not contacted Bragg about the Fourier transform of a helix? Or if Cochrane had not discussed it with Crick? Or if they had not shared an office with Donohoe? Yes, Watson had gone to Cambridge deliberately, so he helped luck happen in that sense. But even so... Watson and Crick surely were in the right place at the right time.

Also, given that all the Koestler's Law fragments were 'there', waiting to be assembled into the right 'pattern', why was it Watson and Crick who put all the right fragments together in the right way, rather than anyone else?

Florence Bell and William Astbury did not have a chance. Their work was too early, for not all the fragments were available. In particular, there was no mathematical theory of the diffraction pattern of a helix, so the X-shape so clearly visible on Bell's photograph could not have been fully interpreted – yet they were able to identify the stacking of the nucleotides, as well as getting the right result for the intervening spacing. Furthermore, in the late 1930s, knowledge of the chemical composition of DNA was rudimentary, and no more than $A = T = C = G$ at best.

It might be argued that Bell and Astbury might have done the mathematics to discover the X, or worked with a mathematician, for the mathematical knowledge required to determine the Fourier transform of a helical structure should certainly be within the capabilities of a mathematically-confident graduate student. But they didn't, nor did they have any particular reason to, for there were no particular clues that the structure might be helical.

Sven Furberg had far better reason, for he was the first to suggest a helical structure. So he must have wondered what the diffraction pattern of a helix might be. But perhaps not. X-ray crystallography was at that time, and still largely is, about how to infer a molecular structure by analysing the intensities of the spots on a photograph, and working back to the structure. One of the breakthroughs in the DNA story was to *think the other way around*, for what Cochrane and Crick did, triggered by Vand, was to ask, and answer, the question 'If a structure is helical, what would the diffraction pattern look like?'. Perhaps Furberg did ask that question, and then ignored it, or asked it and got stuck with the mathematics; or asked it and didn't know who to approach for help; or asked it, approached someone, and then got rejected. We'll never know. And even if he had done the mathematics, his PhD thesis was presented in 1949, just around the same time that Chargaff presented his First Law at his lectures in Europe, which did not happen to include a visit to London, so Furberg was denied the key information that A = T and C = G.

For Pauling, Wilkins and Franklin, however, the situation was very different. By 1952, they had access to all the information that was available to Watson and Crick – and arguably, especially as regards Wilkins and Franklin, more so for they were living-and-breathing DNA every minute of every day. But either they didn't spot the biochemical clue provided by Chargaff, or they ignored it – as evidenced, for example, by Pauling's triple helix which makes no reference either to DNA's chemistry (which is something of a surprise since Pauling was a chemist) or its biology (perhaps less of a surprise, save for the glaringly-obvious fact that DNA's significance is all about, and nothing but, the biology). And as for Maurice Wilkins and Rosalind Franklin, who knows? The only thing we do know is that they didn't.

The other key player is Erwin Chargaff. Why didn't he spot the answer? Chargaff, of course, was not an X-ray crystallographer, so he did not spend his days looking at spotty photographs and thinking about Fourier transforms. True. But is that relevant? For there are two key features about the structure of DNA: the double helix, and the base pairs.

To me, that DNA is a double helix is 'interesting', in that it is an elegant structure. But what's *important* about DNA is base pairing, for it is the pairing of bases that enables replication, and it is the pairing of bases that explains heredity. *Even if the overall structure is not a double helix.* A 'straight ladder' or 'scrunched up ball' would do the same job provided that the integrity of the base pairs is maintained. And – as demonstrated by the structure suggested by Pauling and Corey – a helical structure, of its own accord, doesn't go very far. And yes, Pauling and Corey were wrong in suggesting a triple helix rather than a double one, but it is very easy to imagine the Watson-Crick structure the other way around, with the bases sticking out, but with no base pairing. That's a double helix. But a wrong one. It seems to me that the truly fundamental feature of the structure of DNA is base pairing; the double helix feature just makes the structure more pleasing.

It was Chargaff, working with just a few colleagues, who discovered the one, most fundamental, fact about DNA – that A = T and C = G. From that information alone, he had the opportunity to discover *how DNA works*, which is important, even if he didn't discover the double helix.

In my view, it is Chargaff who, for whatever reason, missed the great opportunity, an opportunity for which he had sufficient, and the right, Koestler's Law components, but which he was unable to put together into the right pattern. Even so, his contribution to the determination of the double helix was profound, yet he was not on that platform in Stockholm in December 1962.

But then no one has ever claimed that Nobel Prizes are fair.

5.15 DNA – a final word

Every human being on the planet is a different, unique individual. And very different from an oak tree, a jellyfish, a mushroom. But every living entity, from a bacterium to a giant sequoia, has, in every cell, molecules of DNA. And all those molecules of DNA are composed of phosphate, the sugar deoxyribose, and the four nucleotides adenine, thymine, cytosine and guanine, nucleotides that form the base pairs A-T and C-G. For completeness, I mention a fifth nucleotide, uracil, that is similar to thymine and can form an A-U base pair with adenine, but the occurrence of uracil in DNA is very rare.

One of the triumphs of molecular biology, following the discovery of the structure of DNA, was the elucidation of how DNA can store genetic information, so determining the development of a living form from embryo to full sized animal, tree or whatever, and then pass that information on to the next generation.

DNA molecules are, in chemical terms, extremely long chains: human DNA, for example, comprises some 3×10^9 base pairs, and even a 'simple' bacterium has about 5×10^6 [217]. Each one of those pairs can be any one of the four combinations A-T, T-A, C-G and G-C, and so the total number of different sequences that can, in principle, be formed for a human is 4 to the power 3×10^9, which is about 10 to the power 2×10^9, a colossally large number – 1 followed by 2 billion zeros; for a bacterium, the theoretical number of different possible sequences is rather smaller, 'only' around 10 to the power 3×10^6, but still colossal – 1 followed by 3 million zeros.

Those 'colossally large numbers' represent the theoretical number of different messages that can be encoded by a sequence of 3×10^9 elements (for a human) or 5×10^6 elements (for a bacterium), in which each element may take one of four different forms. Very many of those sequences will be genetically meaningless, but a sufficiently huge number will be very genetically meaningful indeed. And that's the explanation of how DNA works: each molecule encodes information which the cell's biochemical mechanisms can 'read' resulting in you, or me, or a tree, or a mushroom. And the reason a tree becomes a tree, a mushroom a mushroom, and you you is because the 'message' encoded in the tree's DNA sequence is different from that encoded in a mushroom's DNA sequence, which in turn is different from the message encoded in your sequence.

The messages are different because the sequences are different. But the DNA in the tree, in the mushroom, and in you are all sequences of *the same four* nucleotides, adenine, thymine, cytosine and guanine.

The huge variety of all living forms, the awe-inspiring creativity of life itself, is the result of different patterns of the same four components.

What better way to end this chapter on Koestler's Law, the fundamental principle that all creativity is the formation of new patterns from pre-existing components.

Thomas Edison

Source: Wikimedia 'Thomas Edison and his early phonograph' https://commons.wikimedia.org/wiki/File:Edison_and_phonograph_edit2.jpg.

That's an 1878 photograph of the 31 year-old Thomas Edison with the second version of his invention, the phonograph – this being another fascinating Koestler's Law story in which a key antecedent is the 'phonautograph', a device, patented in 1857 by Édouard-Léon Scott de Martinville [218], which recorded sound as a trace on smoke-pattered glass, but could not play back. Edison was indeed a prodigious inventor, and is credited with 1093 US patents [219], either in his own name, or jointly with others. Yet he attended school for only a few months, and at age 12 was earning a living selling sweets, fruit and newspapers on a local train.

Edison was not only an inventor of 'things', he also invented the process of invention, notably at Menlo Park [220], New Jersey, between 1876 and 1887, and subsequently in larger premises in West Orange [221], where he created and managed true powerhouses of research and development.

Amongst his many ideas, some did not catch on. Edison's very first patent, US 90,646, issued on 1st June 1869, was for an 'Electrographic Vote Recorder' [222]—a machine enabling voters to register their votes by flipping a switch, so that the votes could be counted electronically, and hence very quickly. But all came to nought when the Chairman of the US Congress Committee appointed to evaluate the device dismissed it, with the withering remark, 'If there is any invention on earth that we don't want down here, that is it' [223].

Despite this rejection, Edison worked on, not only as inventor but as a (very good) businessman too: today's commercial giant General Electric is a direct successor of the Edison Electric Light Company, formed in 1878 to manufacture and market light bulbs [224].

Erwin Chargaff

Source of Chargaff photograph: Reproduced from http://www.nasonline.org/publications/biographical-memoirs/memoir-pdfs/chargaff-erwin.pdf courtesy of American Philosophical Society (2010).

As discussed in the main text, the structure of DNA as a double helix is indeed 'interesting'; but the presence of base pairs is fundamental, for it is base pairing that is the explanation of DNA as the carrier and transmitter of genetic information. And the discoverer of base pairing was the biochemist Erwin Chargaff.

Chargaff is the yet another person to have an association with the Habsburg Empire, having been born in 1905 in Czernowitz [225], at that time the capital of the Austro-Hungarian Duchy of Bukovina. During the First World War, his family moved to Vienna, where he studied chemistry, receiving his PhD in 1928.

After a few years at Yale, Chargaff joined the University of Berlin, but in 1933 was forced by the Nazi government to leave, moving via the Pasteur Institute in Paris to Columbia University, where he arrived in 1935 and remained for the next 40 years. And it was in 1944 that he embarked on his study of the chemistry of DNA, for he had realised that the only way DNA could act as the carrier of genetic information – as had by then been proven, particularly as a result of the work of Oswald Avery – had to be as a result of differences in the chemistry of the DNA molecules of different species.

His research required very careful measurements of the quantities of each of the nucleotides A, T, C and G in his DNA samples. These nucleotides are very similar, so distinguishing between them, and separating them for measurement, proved a major challenge – a challenge resolved by the use of the then-novel technique of paper chromatography (for separation), discovered in 1952 by Nobel Laureates Archer Martin and Richard Synge [226], and the commercial availability (since the early 1940s) of equipment for ultraviolet spectrophotometry (for measurement).

This is itself a Koestler's Law story, for these are two critical components which enabled Chargaff to do his work – work that would have been even more difficult, if not impossible, in the absence of either one.

References

[1] Austen J 2003 *Pride and Prejudice* (London: Penguin Classics) p 5
[2] www.bobdylan.com/songs/subterranean-homesick-blues/
[3] www.bartleby.com/337/1242.html
[4] https://theweek.com/articles/468355/10-whimsical-words-coined-by-lewis-carroll
[5] www.rsc.org.uk/shakespeare/language
[6] https://arthive.com/publications/1812~Pictorial_Louis_Leroys_scathing_review_of_the_First_Exhibition_of_the_Impressionists
[7] www.flavorwire.com/302259/early-exhibition-reviews-of-famous-artists
[8] www.impressionniste.net/turner_whistler_monet_tate_britain.htm
[9] https://montmartrefootsteps.com/picasso-demoiselles-d-avignon-influences/
[10] Levi P 2000 *The Periodic Table,* (London: Penguin Modern Classics)
[11] https://artic.edu/articles/862/color-chemistry-and-creativity-in-monets-water-lilies
[12] www.pcimag.com/articles/86476-a-history-of-pigment-use-in-western-art-part-1
[13] www.webexhibits.org/pigments/indiv/overview/coblue.html
[14] www.webexhibits.org/pigments/indiv/overview/emerald.html
[15] http://webexhibits.org/pigments/indiv/overview/cryellow.html
[16] https://jacksonsart.com/blog/2019/11/05/the-story-of-ultramarine-blue-and-french-ultramarine/
[17] www.webexhibits.org/pigments/indiv/overview/zincwhite.html
[18] www.webexhibits.org/pigments/indiv/history/cdyellow.htmll
[19] https://blog.sciencemuseum.org.uk/mauve-mania/
[20] www.visual-arts-cork.com/artist-paints/colour-pigments.htm
[21] www.aaa.si.edu/collections/items/detail/john-goffe-rand-patent-improvement-construction-vessels-or-apparatus-preserving-paint--c-460
[22] Hurt P 2013 Never underestimate the power of a paint tube *Smithsonian Mag.* www.smithsonianmag.com/arts-culture/never-underestimate-the-power-of-a-paint-tube-36637764/
[23] www.sheffieldpharma.com/our-history/
[24] https://curtisward.com/inventions-that-led-to-impressionism
[25] Quoted in Jean Renoir 2001 *Renoir: My Father* (Engl. transl.) Randolph and Dorothy Weaver (New York: New York Review Books) p 69
[26] www.saatchigallery.com/artist/john_wynne
[27] www.crisap.org/people/john-wynne/
[28] www.latimes.com/entertainment-arts/story/2020-12-16/david-hockney-lockdown-covid-normandy-france
[29] www.musee-orangerie.fr/en/agenda/events/david-hockney-year-normandie
[30] www.liquidcr.ovgu.de/en/history/history.html
[31] www.alternator.science/en/longer/liquid-crystals-the-beautiful-state-of-matter/
[32] www.newvisiondisplay.com/lcd-history/
[33] www.optica.org/en-us/history/biographies/bios/george_h_heilmeier/
[34] https://americanhistory.si.edu/ontime/expanding/images/enlargements/98-4542.html
[35] https://americanhistory.si.edu/collections/search/object/nmah_214337
[36] https://global.epson.com/company/corporate_history/milestone_products/16_et10.html
[37] www.apple.com/uk/newsroom/2010/01/27Apple-Launches-iPad/
[38] www.nobelprize.org/prizes/physics/1991/press-release/

[39] www.explainthatstuff.com/touchscreens.html
[40] https://illumin.usc.edu/touchscreen-an-engineered-harmony-between-humans-and-machines/
[41] https://worldwide.espacenet.com/patent/search/family/022862990/publication/US3775560A?q=pn=US3775560
[42] www.spectator.co.uk/article/the-enlightenment-was-a-many-splendoured-thing
[43] www.theguardian.com/books/2017/sep/04/100-best-nonfiction-books-decline-and-fall-of-the-roman-empire-edward-gibbon
[44] www.washingtonindependentreviewofbooks.com/index.php/features/idecline-and-fall-of-the-roman-empire-i-by-edward-gibbon-an-appreciation
[45] Gibbon E *In Our Time*, BBC radio 4 podcast, 17 June 2021 www.bbc.co.uk/programmes/m000x0v2 and www.richardblake.me.uk/what-i-like-about-edward-gibbon/
[46] Carr E H 1964 *What is History?* (London: Penguin) p 111
[47] www.academuseducation.co.uk/post/greek-historiography-from-herodotus-to-thucydides
[48] www.historytoday.com/archive/head-head/there-still-value-'great-man'-history
[49] www.bbc.co.uk/programmes/b09l64y9 and Olusoga D and Backe-Hansen M (ed) 2021 *A House Through Time* (London: Picador)
[50] www.bbc.co.uk/iplayer/episode/m0010st7/who-do-you-think-you-are-series-18-2-dame-judi-dench
[51] www.constitutionfacts.com/us-declaration-of-independence/dates-to-remember/
[52] www.archives.gov/founding-docs/declaration-transcript
[53] www.constitutionfacts.com/us-declaration-of-independence/drafting-the-declaration/
[54] www.history.com/news/thomas-paine-common-sense-revolution
[55] https://plato.stanford.edu/entries/locke-political/
[56] https://lawliberty.org/american-liberty-and-the-pursuit-of-happiness/
[57] http://ap.gilderlehrman.org/history-by-era/road-revolution/essays/declaration-independence-global-perspective
[58] www.hetwebsite.net/het/profiles/kames.htm
[59] www.universitystory.gla.ac.uk/biography/?id=WH0016&type=P
[60] www.adamsmith.org/the-theory-of-moral-sentiments
[61] www.investopedia.com/updates/adam-smith-wealth-of-nations/
[62] www.adamsmith.org/about-adam-smith
[63] www.econlib.org/library/Enc/Mercantilism.html
[64] https://faculty.babson.edu/krollag/org_site/org_theory/granovet_articles/smith_wealth.htm
[65] https://oll.libertyfund.org/title/hull-the-economic-writings-of-sir-william-petty-vol-2-lf0605-02_head_034
[66] Peaucelle J-L and Guthrie C 2011 How Adam Smith found inspiration in french texts on pin making in the eighteenth century *History of Economic Ideas* **19** 41–67 https://hal.univ-reunion.fr/hal-01403681/document
[67] www.hetwebsite.net/het/profiles/hutcheson.htm
[68] www.libertarianism.org/publications/essays/excursions/philosophy-declaration-independence-part-1
[69] www.prospectmagazine.co.uk/magazine/david-hume-adam-smith-odd-couple
[70] www.adamsmithworks.org/speakings/what-adam-smith-ate-voltaire-and-the-vegetarian-salad

[71] www.smithsonianmag.com/history/how-voltaire-went-bastille-prisoner-famous-playwright-180970854/
[72] www.visitvoltaire.com/voltaire_bio.htm
[73] https://royalsocietypublishing.org/doi/10.1098/rsnr.1994.0024
[74] www.westminster-abbey.org/abbey-commemorations/commemorations/sir-isaac-newton
[75] www.voltaire.ox.ac.uk/about-voltaire/
[76] Watson P 2005 *Ideas* (London: Weidenfeld & Nicholson) pp 527–8
[77] https://weblearn.ox.ac.uk/access/content/group/1803d649-431f-46cd-9428-f1dc63ef43a3/Rare Books/newton.html
[78] In a letter to Robert Hooke dated 5 February 1675 www.newtonproject.ox.ac.uk/view/texts/normalized/OTHE00101
[79] http://oxfordreference.com/view/10.1093/acref/9780191826719.001.0001/q-oro-ed4-00000952
[80] https://mathshistory.st-andrews.ac.uk/Biographies/Horrocks/
[81] https://ccrma.stanford.edu/~jos/pasp/Newton_s_Three_Laws_Motion.html
[82] https://plato.stanford.edu/entries/descartes-physics/#LawsMotiCartConsPrin
[83] www.bbvaopenmind.com/en/science/leading-figures/hooke-the-genius-whose-big-mistake-was-confronting-newton/
[84] https://amsi.org.au/ESA_Senior_Years/SeniorTopic3/3b/3b_4history_2.html
[85] www.bbvaopenmind.com/en/science/leading-figures/wallace-and-darwin-a-pact-for-evolution/
[86] https://technicaleducationmatters.org/2010/12/12/the-invisible-college-1645-1658/
[87] http://users.clas.ufl.edu/ufhatch/pages/02-teachingresources/ClioElectric/1-Electronic%20Texts/Hatch-Academie%20Montmor%20-%20Sci%20Rev.pdf
[88] https://royalsociety.org/about-us/history/
[89] https://royalsociety.org/science-events-and-lectures/2003/summer-science/mr-hooke/
[90] https://mathshistory.st-andrews.ac.uk/Biographies/Newton/
[91] www.archives.gov/historical-docs/todays-doc/index.html?dod-date=1101
[92] www.ipwatchdog.com/patents/US223898-electric-lamp.pdf
[93] https://madeupinbritain.uk/Light_Bulb
[94] www.abebooks.com/Electric-Lamps-Letters-Patent-Invention-IMPROVEMENTS/15853802035/bd
[95] www.shineretrofits.com/knowledge-base/lighting-learning-center/a-brief-history-of-led-lighting.html
[96] Friedel R and Israel P 2010 *Edison's Electric Light: The Art of Invention* (Baltimore, MA: JHU Press) p 68 https://books.google.co.uk/books?id=8U-Naf4DuzMC&printsec=frontcover-v=onepage&q&f=false
[97] https://collection.sciencemuseumgroup.org.uk/people/ap8506/the-edison-and-swan-electric-light-company-limited
[98] www.lamptech.co.uk/Documents/Factory-UK-Ponders End.htm
[99] https://founders.archives.gov/documents/Franklin/01-09-02-0107 experiment 11
[100] www.history.com/this-day-in-history/franklin-flies-kite-during-thunderstorm
[101] www.delmarfans.com/educate/basics/who-invented-light-bulbs/
[102] www.osti.gov/servlets/purl/823202
[103] https://libraries.mit.edu/collections/vail-collection/topics/electricity/the-voltaic-pile/
[104] www.delmarfans.com/educate/basics/who-invented-light-bulbs/

[105] www.lamptech.co.uk/Documents/People - Starr JW.htm
[106] www.nature.com/articles/134280c0.pdf
[107] www.casabatllo.es/en/antoni-gaudi/casa-batllo/history/
[108] https://casabatllo.es/en/antoni-gaudi/casa-batllo/inside/
[109] www.napoleon.org/en/history-of-the-two-empires/paintings/bonaparte-crossing-the-greatst-bernard-pass/
[110] www.casabatllo.es/en/saint-georges-day/
[111] www.costume-ideas.com/2011/05/venetian-masks/
[112] http://art-monet.com/1900_3.html
[113] http://scihi.org/hans-christian-orsted-electricity-magnetism/
[114] www.nist.gov/si-redefinition/ampere-history
[115] Faraday M 1822 *Q. J. R. Inst.* **12** 283–85 plate VII https://biodiversitylibrary.org/item/50623-page/487/mode/1up
[116] https://ri-science.tumblr.com/post/96622046007/this-simple-looking-object-was-made-by-michael
[117] https://spectrum.ieee.org/200-years-ago-faraday-invented-the-electric-motor
[118] www.invent.org/inductees/zenobe-theophile-gramme
[119] www.bie-paris.org/site/en/blog/entry/zenobe-gramme-s-electrifying-discovery-at-expo-1873-vienna
[120] www.sydneyoperahouse.com/our-story/the-architect-jorn-utzon.html
[121] https://records-primo.hosted.exlibrisgroup.com/primo-explore/fulldisplay?docid=ADLIB_RNSW112410845&context=L&vid=61SRA&lang=en_US&search_scope=Everything&adaptor=Local Search Engine&tab=default_tab&query=lsr01,exact,NRS-12825
[122] www.d2architects.co.uk/the-sydney-opera-house-an-architectural-masterpiece-1408.html
[123] www.arup.com/projects/sydney-opera-house
[124] www.sydneyoperahouse.com/our-story/sydney-opera-house-facts.html
[125] https://eandt.theiet.org/content/articles/2021/07/engineering-places-sydney-opera-house/
[126] http://architectuul.com/architecture/sydney-opera-house
[127] www.theb1m.com/video/sydney-opera-house-building-an-icon
[128] www.businessinsider.com/iconic-structures-eero-sarinen-2016-8?r=US&IR=T - kresge-auditorium-mit-cambridge-massachusetts-7
[129] www.archdaily.com/788012/ad-classics-twa-flight-center-eero-saarinen
[130] www.getty.edu/foundation/pdfs/kim/sydney_final_report.pdf
[131] https://en.wikiarquitectura.com/building/sydney-opera-house/
[132] https://authoritysoccer.com/official-fifa-soccer-field-dimensions/
[133] www.sydneyoperahouse.com/our-story/sydney-opera-house-facts.html
[134] www.sydneyoperahouse.com/our-story/sydney-opera-house-history/utzon-resigns.html
[135] www.sydneyoperahouse.com/our-story/sydney-opera-house-history/the-interiors.html
[136] www.cantorsparadise.com/something-interesting-about-albert-einsteins-ph-d-thesis-a7bf1869059e
[137] Stachel J (ed) 1998 *Einstein's Miraculous Year – Five Papers that Changed the Face of Physics* (Princeton, NJ: Princeton, University Press) and https://history.aip.org/exhibits/einstein/chron-1905.htm
[138] www.nobelprize.org/prizes/physics/1921/summary/

[139] www.theguardian.com/science/across-the-universe/2012/oct/08/einstein-nobel-prize-relativity
[140] www.fourmilab.ch/etexts/einstein/specrel/specrel.pdf
[141] www.nobelprize.org/prizes/physics/1902/summary/
[142] https://arxiv.org/pdf/1205.3904.pdf
[143] https://ethw.org/Maxwell%27s_Equations
[144] https://scienceworld.wolfram.com/physics/Michelson-MorleyExperiment.html
[145] Michelson A A and Morley E W 1886 Influence of motion of the medium on the velocity of light *Am. J. Sci.* **31** 377–86 https://archive.org/details/americanjournal233unkngoog/page/376/mode/2up?q=michelson
[146] Voigt W 1887 On the principle of Doppler *Göttinger Nachrichten* **2** 41–51 https://archive.org/details/nachrichtenvond04gtgoog/page/40/mode/2up?q=voigt (in German) https://en.wikisource.org/wiki/Translation:On_the_Principle_of_Doppler (in English)
[147] https://physicstoday.scitation.org/doi/full/10.1063/PT.3.4429
[148] https://fondationlouisdebroglie.org/AFLB-441/aflb441m900.pdf
[149] https://famousirishscientists.weebly.com/george-francis-fitzgerald.html
[150] http://philsci-archive.pitt.edu/9871/1/On_the_EE_between_SR_and_Lorentz%27s_Ether_Theory_name_included.pdf
[151] https://mathshistory.st-andrews.ac.uk/HistTopics/Time_2/
[152] www.atticusrarebooks.com/pages/books/1062/joseph-larmor/a-dynamical-theory-of-the-electric-and-luminiferous-medium-part-iii-relations-with-material-media
[153] https://mathshistory.st-andrews.ac.uk/HistTopics/Time_2/
[154] www.dutchwatermanagement.com/hendrik-lorentz
[155] Poincaré H 1904 The present and future of mathematical physics http://henripoincare-papers.univ-lorraine.fr/chp/hp-pdf/hp1904bs.pdf (in French, see p 306) www.ams.org/journals/bull/2000-37-01/S0273-0979-99-00801-0/S0273-0979-99-00801-0.pdf (in English, p 27)
[156] http://henripoincarepapers.univ-lorraine.fr/chp/hp-pdf/hp1904bs.pdf (in French, see p 316), www.ams.org/journals/bull/2000-37-01/S0273-0979-99-00801-0/S0273-0979-99-00801-0.pdf (in English, p 34)
[157] Poincaré H 1905 On the dynamics of the electron *C. R. Acad. Sci* **140** 1504–8 www.academie-sciences.fr/pdf/dossiers/Poincare/Poincare_pdf/Poincare_CR1905.pdf (in French) https://en.wikisource.org/wiki/Translation:On_the_Dynamics_of_the_Electron_(June) (in English)
[158] https://sites.pitt.edu/~jdnorton/papers/HumeMach.pdf
[159] https://nautil.us/issue/43/heroes/when-einstein-tilted-at-windmills
[160] Janssen M and Renn J 2015 History: Einstein was no lone genius *Nature* **527** 298–300
[161] www.nobelprize.org/prizes/medicine/1962/award-video/
[162] www.nobelprize.org/prizes/medicine/
[163] www.news-medical.net/life-sciences/History-of-DNA-Research-Scientific-Pioneers-Their-Discoveries.aspx-Frederick Griffith
[164] www.genome.gov/genetics-glossary/Deoxyribonucleic-Acid
[165] www.cif.iastate.edu/services/acide/xrd-tutorial/xrd
[166] www.nobelprize.org/prizes/physics/1914/laue/biographical/
[167] www.nobelprize.org/prizes/physics/1915/summary/
[168] www.nobelprize.org/prizes/physics/1915/wl-bragg/biographical/

[169] www.nobelprize.org/prizes/peace/2014/yousafzai/facts/
[170] Sherwood D and Cooper J 2015 *Crystals, X-rays and Proteins* (Oxford: Oxford University Press) ch 9
[171] www.iucr.org/people/nobel-prize
[172] https://embryo.asu.edu/pages/photograph-51-rosalind-franklin-1952
[173] https://library.cshl.edu/oralhistory/speaker/raymond-gosling/
[174] www.rfi.ac.uk/about/rosalind-franklin/
[175] www.whatisbiotechnology.org/index.php/people/summary/Wilkins
[176] Tobin M J 2003 April 25, 1955: Three papers, three lessons *Am. J. Respir. Med.* **167** 1047–9
[177] Watson J D 1986 *The Double Helix* (London: Penguin) pp 132–3
[178] www.whatisbiotechnology.org/index.php/people/summary/Bell
[179] https://alumni.leeds.ac.uk/news/professor-x
[180] www.leedsinspired.co.uk/events/shaping-course-modern-science-william-henry-bragg-and-his-legacy-university-leeds
[181] www.whatisbiotechnology.org/index.php/people/summary/Astbury
[182] Bell F O 1939 X-ray and related studies of the structure of the proteins and nucleic acids *PhD Thesis* University of Leeds https://explore.library.leeds.ac.uk/special-collections-explore/650413
[183] Astbury W T and Bell F O 1938 X-ray study of thymonucleic acid *Nature* **141** 747–48
[184] http://libgallery.cshl.edu/items/show/55498
[185] Astbury W T and Bell F O 1938 Some recent developments in the x-ray study of proteins and related structures *Cold Spring Harb. Symp. Quant. Biol.* https://citeseerx.ist.psu.edu/viewdoc/download?doi=10.1.1.901.5509&rep=rep1&type=pdf
[186] Manchester K 2003 Discovering the DNA double helix and the secret of life *SA J. Sci.* **99** 460–64
[187] Hall K 2011 William Astbury and the biological significance of nucleic acids, 1938–51 *Stud. Hist. Philos. Biol. Biomed. Sci.* **42** 119–28 https://astbury.leeds.ac.uk/wp-content/uploads/sites/59/2019/11/William-Astbury-biological-significance-of-nucleic-acids-Hall.pdf
[188] https://journals.iucr.org/a/issues/1972/04/00/a08703/a08703.pdf
[189] www.historyofinformation.com/detail.php?entryid=4425
[190] Furberg S 1952 On the structure of nucleic acids *Chem. Scand.* **6** 634–40 http://actachemscand.org/pdf/acta_vol_06_p0634–0640.pdf
[191] https://lpi.oregonstate.edu/about/linus-pauling-biography
[192] https://paulingblog.wordpress.com/2011/03/09/the-alpha-helix/
[193] www.nobelprize.org/prizes/chemistry/1954/summary/
[194] https://nobelprizemuseum.se/en/cultures-of-creativity/
[195] www.nobelprize.org/prizes/medicine/1962/crick/biographical/
[196] Vladimir Vand 1968 Pennsylvania State crystallographer, dies *Phys. Today* **21** 115
[197] https://royalsocietypublishing.org/doi/pdf/10.1098/rsbm.2005.0005
[198] Cochran W, Crick F and Vand V 1952 The structure of synthetic polypeptides. I. The Fourier transform of a helix *Acta Cryst.* **5** 581–86
[199] Sherwood D 1976 *Crystals, X-rays and Proteins* (London: Longman) ch 16
[200] www.nobelprize.org/prizes/medicine/1962/watson/biographical/
[201] www.nobelprize.org/prizes/chemistry/1962/kendrew/biographical/
[202] www2.mrc-lmb.cam.ac.uk/about-lmb/lmb-alumni/alumni/max-perutz-1914–2002/

[203] www.nobelprize.org/prizes/chemistry/1962/summary/
[204] https://player.slideplayer.com/85/13790554/slides/slide_14.jpg and https://courses.physics.illinois.edu/phys102/fa2014/handouts/handout23.pdf slide 23
[205] Pauling L and Corey R B 1953 A proposed structure for the nucleic acids *PNAS* **39** 84–97 www.ncbi.nlm.nih.gov/pmc/articles/PMC1063734/pdf/pnas01587-0028.pdf
[206] Kresge N, Simoni R D and Hill R L 2005 Chargaff's rules: the work of Erwin Chargaff *J. Biol. Chem.* **280** 172–4
[207] Vischer E and Chargaff E 1948 The separation and estimation of purines and pyrimidines in minute amounts *J. Biol. Chem.* **176** 703–14 http://mc142.uib.es:8080/rid=1V35HHS7T-220D4X8-33HH/Article%20Chargaff.pdf
[208] Hargittai I 2009 The tetranucleotide hypothesis: a centennial *Struct. Chem.* **20** 753–6
[209] Vischer E and Chargaff E 1948 The composition of the pentose nucleic acids of yeast and pancreas *J. Biol. Chem.* **176** 715–34 www.jbc.org/article/S0021-9258(19)52687-4/pdf
[210] https://sites.google.com/site/dnareplicationsystem/chargaff-s-rules
[211] www.theguardian.com/news/2002/jul/02/guardianobituaries.obituaries
[212] https://nordicbiosite.com/blog/discovery-of-the-dna-double-helix-65-years-on
[213] https://snaccooperative.org/ark:/99166/w6n89n8b
[214] www.cambridgephysics.org/dna/dna13_1.htm
[215] Watson J D and Crick F H 1953 Molecular structure of nucleic acids: a structure for deoxyribose nucleic acid *Nature* **4356** 737–38 https://exploratorium.edu/origins/coldspring/ideas/printit.html
[216] www.exploratorium.edu/origins/coldspring/ideas/note-furberg.html
[217] https://sfvideo.blob.core.windows.net/sitefinity/docs/default-source/biotech-basics/molecular-facts-and-figures.pdf?sfvrsn=4563407_4
[218] www.historyofinformation.com/detail.php?id=486
[219] www.history.com/topics/inventions/thomas-edison
[220] http://edison.rutgers.edu/inventionfactory.htm
[221] https://artsandculture.google.com/exhibit/edison-s-west-orange-laboratory-thomas-edison-national-historical-park/agJyIGxTlBg6JA?hl=en
[222] www.edisonmuckers.org/wp-content/uploads/2012/06/00090646.pdf
[223] https://edison.rutgers.edu/vote.htm
[224] www.company-histories.com/General-Electric-Company-Company-History.html
[225] https://totallyhistory.com/erwin-chargaff/
[226] www.nobelprize.org/prizes/chemistry/1952/summary/

Part II

How to have great ideas, deliberately

IOP Publishing

Creativity for Scientists and Engineers
A practical guide
Dennis Sherwood

Chapter 6

The 'da Vinci problem'

6.1 Building on Koestler's Law

> *The creative act is not an act of creation in the sense of the Old Testament.*
>
> *It does not create something out of nothing; it uncovers, selects, re-shuffles, combines, synthesises already existing facts, ideas, faculties, skills.*
>
> *The more familiar the parts, the more striking the new whole.*

In my opinion, Koestler's Law is a profoundly insightful statement of what creativity is, and is not.

It is not 'a sudden inspiration'. Nor is 'creativity' the preserve of those 'special people', those who are innately 'creative'. Yes, you, and I, can do it.

What many people find surprising is Koestler's assertion that creativity is not about the 'creation of the new, something out-of-the-blue', but a process of putting together things that *already exist*, but which have not hitherto been assembled into that particular 'pattern'. One implication of this is that creativity is in essence a process of collection and assembly; a second is that 'novelty' is a feature of the resulting assembled pattern, in its appropriate context, for the pattern-as-a-whole can be 'new' even if the components from which that pattern is formed are not new at all, and may have existed, independently, for a very long time indeed. And a third is that it is always possible to deconstruct any current 'pattern' back into its component parts, and to verify that they did indeed pre-exist the new 'pattern'.

The previous chapter gave a host of examples, from music (all music in the Western tradition is formed from the same notes) to the discovery of the structure of DNA (how Watson and Crick assembled the ideas and results of the work of many others to determine the double helix). Those are all specific instances, and it may be that a sceptical reader might think that I chose those examples deliberately, knowing that they 'work', to present a false argument; yes, the sceptic might say, Koestler's Law might be 'interesting', but it isn't a general principle, let alone a fundamental 'truth'.

Scientists in general, and mathematicians in particular, are especially alert to the fragility of special cases, and the rigour required to establish a valid proof. I acknowledge unreservedly that the examples are special cases. But they were selected based on my belief that you will find them illuminating and engaging, rather than as an attempt to pull the wool over your eyes. And I further acknowledge that I do not have a 'proof' of the universality of what I have chosen to call 'Koestler's Law' that any mathematician, for example, would find remotely convincing.

That said, may I leave two thoughts in the sceptical mind?

Firstly, the Koestler Challenge. At the end of chapter 4, I invited readers to identify an idea, concept, product, whatever, that cannot be 'decomposed' into pre-existing components, or traced to appropriate precedents. I have set this challenge to countless people, and no valid counter-example has yet been identified. In any field, from philosophy to physics, from public policy to painting. So if you can find one, please let me know!

Secondly, even if Koestler's Law is not universal, I suggest that it is a very useful working hypothesis. As I trust I demonstrated in the last chapter, it provides a powerful framework for deconstructing an idea, and tracing the story of how that idea came about. This is 'interesting', but necessarily retrospective. Even better would be for Koestler's Law to provide a basis for a process of discovery, so that, looking forward, we can generate ideas, deliberately, whenever we wish to, and wherever a new idea might be of benefit.

Koestler's Law tells us that, once the idea has been proposed, we will be able to identify the component parts form which that idea is formed. Accordingly, before any idea is discovered, the relevant component parts must exist. At any instant, however, there are a truly vast number of components in existence that could be put together in an even vaster number of different ways. To use Koestler's Law as a vehicle for the discovery, the practical problem that needs to be solved is how, before-the-event, to identify precisely which specific 'right' components we need, and how to put them together in just the 'right' way.

This is the practical problem that we all face, and it is indeed a problem – vividly illustrated by the story of DNA. As we saw, in the early months of 1953, the very best minds in the field, all over the world, were trying to solve the structure of DNA: Pauling and Corey at Caltech in Pasadena; Chargaff at Columbia in New York; Wilkins and Franklin at King's College London; Watson and Crick at the Cavendish Laboratory in Cambridge; many others too. We now know, with hindsight, that all the right 'components' were in the public domain in January 1953, and even though Photo 51 had not yet been published, the X-ray diffraction photograph taken by Bell and Astbury, published in 1938, and certainly by Beighton in 1951, had most, if not all, of the key features.

Any of these teams, and individuals, might have made the landmark discovery. Just a few weeks before Watson and Crick published their correct idea, Pauling and Corey had published their incorrect one. Neither Pauling and Corey, nor Watson and Crick, had done the research carried out so carefully by Chargaff which provided that oh-so-important Koestler component, the base pair. Chargaff had been publishing his findings from 1948. All his results were in the public domain. Yet Pauling and Corey missed them; Watson and Crick did not. Why?

I don't know. But I use this example to highlight that the *discovery* of an idea is far, far, harder that analysing the idea after-the-event, decomposing the idea to identify its components. For, in practice, at any time, there are a huge number of potential Koestler's Law fragments 'out there', and an even more huge number of ways in which those fragments might be re-arranged and assembled into new patterns, only one of which will 'work'. How, then, do we identify the 'short list' of likely fragments? And how do we test the resulting patterns?

These are important questions, for it is the answers to these questions that make the process of discovery an intelligent, well-directed, search, rather than a random walk in which we are totally dependent on luck. And yes, as we saw in the Prologue and section 5.14.11, luck does play a role. But there is a real, and pragmatic, difference between a deliberate, well-structured, process in which luck plays but a minor role, and the random walk in which luck plays the dominant, if not the only, role.

The purpose of chapters 6, 7, 8 and 9 is therefore to present a process for the generation of ideas which is by no means a random walk, but which is (certainly) deliberate and systematic, and (I hope!) intelligent too. The process is rooted firmly in Koestler's Law, and looks forwards to discovery rather than backwards to analysis. It is also profoundly pragmatic, and, over the last twenty-or-so years, I have used it, with great success, in all fields, from science to product design, from (hugely) increasing the effectiveness of commercial processes to the invention of new television shows.

But to do this, I need to address the issues identified in section 4.7, entitled 'What Koestler's Law does, and doesn't, do'. Accordingly, in this chapter I explore what I call the 'da Vinci problem', and in the next two chapters how to distinguish a 'good' pattern from a 'poor one' and 'the learning trap'. This then sets the scene for my description, in chapter 9, of my *InnovAction!* process for the deliberate discovery of great ideas.

6.2 The helicopter that couldn't fly

As noted in the Prologue, and also in section 1.6, around 1490, Leonardo da Vinci made a number of drawings for a variety of flying machines, including the 'helicopter' shown in figure 6.1 [1]. The highly imaginative helical aerofoil (ah! helicopter! another, linguistic, example of Koestler's Law: a combination of the Greek words for helix (ἕλιξ, helix) and wing (πτερόν, pteron) = helic -o -pter!) was caused to rotate about the vertical shaft by muscle power, and the key idea – which is in principle correct – is that the rotating aerofoil will 'push' the

Figure 6.1. Leonardo da Vinci's design for a helicopter.
Source: Wikimedia 'Drawings by Leonardo Da Vinci showing a life preserver, the aerodynamics of vertical flight, and a wooden wing operated by a hand crank' https://commons.wikimedia.org/wiki/File: Leonardo_Da_Vinci_-_Lifebuoy_and_Flying_Machines.jpg.

surrounding air downwards, which in turn will cause the machine to lift. I don't know how da Vinci came up with this idea, but there is a Koestler's Law possibility: da Vinci had extensive knowledge of gears, cogs and screws in general, and of the helical Archimedes screw in particular, the rotation of which caused a fluid, water, to flow uphill, against gravity [2]. It is not at all unlikely that da Vinci might have wondered whether a helical screw might also drive the flow of a different fluid, air, so causing his machine to move against gravity. And da Vinci clearly had an intuitive understanding of what, some 200 years later, was to become codified as Newton's Third Law of Motion – that every action has an equal and opposite reaction – an understanding based on his deep understanding of machines, many of which depend on that very principle for successful operation.

Had muscle power been able to lift his device, da Vinci would then have discovered the counter-rotation, which would be a further problem-to-solve, this being another example of the fractal Target Diagram, figure 2.2. Given da Vinci's ingenuity, there is a fair chance he would have solved that too, but we'll never know, for the machine left never left the ground.

We now know that the fundamental problem was that muscle power could never lift a machine of the combined weight of the machine itself and the men rotating the aerofoil, however strong the men. To solve that problem, da Vinci needed some form of engine, such as an internal combustion engine, like the petrol engine designed by the Wright brothers to power the two counter-rotating propellers on their first heavier-than-air machine to make a successful flight in 1903 [3]. Or the superheated steam used by the Italian aeronautical innovator

Enrico Forlanani, who, in 1877, built an (un-manned) machine, very similar in design to da Vinci's (but, as shown in figure 6.2, with two counter-rotating aerofoils), that rose vertically under its own power, reaching a height of 13 metres, and staying aloft for 20 seconds [4].

Figure 6.2. Schematic representation of Enrico Forlanini's un-manned helicopter, powered by a small steam engine, which successfully flew in 1877.
Source: Wikimedia 'Elicottero sperimentale Enrico Forlanini 1877 Museo scienza e tecnologia Milano' CC-BY 4.0 https://upload.wikimedia.org/wikipedia/commons/d/d7/Elicottero_sperimentale_Enrico_Forlanini_1877_Museo_scienza_e_tecnologia_Milano.jpg.

6.3 The problem of the missing component

Had da Vinci had an internal combustion engine available, no doubt he would have used it. And since he didn't, but knew he needed some form of 'engine' more powerful than his team of men, why didn't he invent it? After all, wasn't he the most ingenious of inventors?

Indeed he was. But it wasn't just the petrol-driven internal combustion engine he would need to invent. He'd need to invent much about engineering, and materials science; about mining and manufacturing; about oil refining and…

The list goes on. A list far too long for even an inventive genius of the calibre of Leonardo da Vinci to solve. Any more than Stone Age man was able to build a space shuttle. Yes, all the atoms of the materials used to build the first space shuttle, launch it into space, and achieve its safe return, did exist, and were on planet Earth, during the Stone Age some 10,000 years ago. But they were in the 'wrong' places, such as buried deep within the Earth's minerals, rocks and oil reserves, and in the 'wrong' molecular configurations.

Yes, any idea is indeed a combination of pre-existing components – but those components themselves are likely to be possibly very sophisticated patterns of lower-level components, which themselves are likely to be… And as human development has progressed, more 'sub-assemblies' have been discovered and invented, which themselves have been combined together into increasingly more complex higher-level assemblies, which... Accordingly, at any time, only a given number of 'high-level assemblies' are available. Around 1500, the 'higher level assembly' we recognise as the internal combustion engine was not available to da Vinci, and there were just too many 'levels' missing for him to invent it.

Koestler's Law states that any new idea is formed from patterns of pre-existing components. Which is true. And it suggests that all that needs to be done to discover a new idea is to identify the 'right' components and assemble them in the 'right' configuration. Which is also true. *But for this all to work, all the components that you need must be 'there', waiting to be collected and then assembled.* If the component you need isn't 'there', then you will never be able to find it. The idea therefore remains incomplete, with something important missing – of which da Vinci's helicopter is just one example.

The problem of the 'missing component' is real, and so I refer to it as the 'da Vinci problem'. And unfortunately, the da Vinci problem is more likely to be encountered in academic research than in any other field, for academic research is inevitably on the leading edge. In the DNA story, for example, in 1938, Bell and Astbury were missing two important components: firstly, that the diffraction pattern of a helix is X-shaped, as established in the Cochran, Crick and Vand paper of 1952, and secondly the clue to base pairing given by A = T and C = G, as discovered by Chargaff after 1948. They could therefore go no further than the 'pile of pennies'.

By contrast, in early 1953, Watson and Crick had access to all the right components and put them together in the right way, whereas Pauling and Corey, who in principle had access to the same components, failed to spot one – Chargaff's evidence for base pairs – and so their triple helix idea, published in February 1953, was wrong. And those confident words 'We have now formulated a promising structure for the nucleic acids... The structure that is not a vague one, but is precisely predicted...' that appear in the second paragraph of their paper suggest that they *did not know they were missing a vital component of the right solution.* For a key feature of the da Vinci problem is that *you might not know you are experiencing it,* for you might be convinced that your current answer is right.

So the da Vinci problem is doubly problematic. Firstly, you might not even recognise it. And secondly, if you do, it is quite possible that there is nothing you can do about it – just like da Vinci, who undoubtedly realised his helicopter couldn't fly, and also realised that his big idea would 'never get off the ground'. Literally.

Yes, the da Vinci problem is real, and you might encounter it. So what can you do if that happens? Here are some suggestions...

6.4 You might be a 'victim', now

When Pauling and Corey published their idea that the structure of DNA is a triple helix, they were convinced that they had 'the right answer'. They did not know that they were missing a vital component – the concept of base pairs – so they did not know they were victims of the da Vinci problem. That is a very easy trap to fall into.

To avoid this trap, be alert to the da Vinci possibility, and be very wary of the dangers of convincing yourself that, at last, you have 'the right answer'. Rather, recognise that, although you have 'an answer', there might be a better answer that remains to be discovered.

To help with that, keep asking questions. Questions such as:

- Does my current answer explain all the known data?
- Might there be any other relevant data that I haven't taken into account? If so, where might it be? Who else might I talk to that might have some relevant information?
- What assumptions have I made? What are the bases of those assumptions? How might each of those assumptions be different? What are the consequences of those possibilities?

Had Pauling and Corey asked those first two questions, it would have increased the likelihood that they would have discovered Chargaff's findings; and had Bell and Astbury asked the third one, that might have resulted in something like 'We're assuming a 'pile of pennies'. That makes sense, and is consistent with the diffraction pattern, so that's good. But might something else also be possible – some form of structure rather like a pile of pennies, but not quite the way we've drawn them?'

It might be that the answer to that last question is 'We can't think of anything'. But you don't have to be exceptionally imaginative to have the idea that it might be a 'twisted' pile rather than a 'straight' pile, suggesting a helix, rather than a 'straight' stack.

6.5 Identify the missing component(s) as precisely as you can

If the answers to the questions posed in the last section are, for example, 'yes, there are some known results that my idea does not explain', 'yes, I haven't taken [this] into consideration', 'yes, I've made [this assumption], but [that one] is just as plausible' – or if you are just plain stuck – then that could indicate a da Vinci problem. It might be that it is totally insoluble, as was the case with da Vinci's helicopter, but you won't know that yet. So it's worth taking some trouble to find out. And you do that by identifying, as precisely as you can, what the missing component might 'look like', what question if answered, would crack the problem. The analogy here is a jig-saw puzzle: the more precisely you identify what's missing – it's a piece with three 'bobbles', and probably with a blue colour on one corner and a red stripe close to one side – the easier it is to find it in the box.

Or, in rather more formal language, the more closely you can identify what you don't know, the more precisely you can set out to find it, and the better the questions you will be able to ask.

So, in the case of Bell and Astbury, questioning the assumptions associated with the 'stack of pennies' might, just might, have identified the possibility of a helix. Especially since much of the work carried out in that department was about fibres. Especially since many of those fibres – such as wool – are naturally 'springy'. Especially since many of the springs you see in machines are helical. Especially since, in the early 1930s, Astbury had undertaken some X-ray diffraction studies of keratin, the key protein in wool, and had identified that it could adopt two configurations: an extended form, which he named β-keratin, and a more compact form, which he named α-keratin [5] – which is the origin of Pauling's choice of the term α-helix [6].

If, around 1938, Bell and Astbury might have identified that a helix was a possibility, that might have triggered questions such as:

- How might we prove, or disprove, the hypothesis that DNA is a helix?
- What clues might our X-ray diffraction photographs provide – clues we either haven't noticed yet, or have noticed and either ignored or misinterpreted?
- What other molecules might, possibly, have helical structures, and what do their X-ray diffraction photographs look like?
- Turning things the other way around, if the structure were helical, what would the diffraction pattern look like? Do we know enough mathematics to solve that problem? And if we don't, whom might we approach to seek their collaboration?

I don't wish to criticise Bell and Astbury, for it is of course very easy, with all the hindsight we have now, to pose those questions. My intention, however, is to imagine that it's 1938, and to consider what Bell and Astbury could reasonably have known at that time: are these the kinds of questions that might have been asked? You may judge for yourself. For if they might have been asked at that time, that throws light on the types of question that anyone can ask whenever they might be experiencing a da Vinci problem, especially one of which they are unaware.

That's important, for this is a situation in which many people find themselves, especially academic researchers.

It is possible that, once identified, the problem can be addressed from within your own resources. One example of this is the story of John Rand, the inventor of the paint tube, as described in section 5.4. He wanted to paint outdoors, and he identified that the 'missing component' was a way of being able to carry small quantities of paint, and use it outdoors – a way which was easier to use than a pig bladder tied up with string, or the then-new syringes. It is quite likely that many painters before Rand had recognised the benefit of having easily-used portable paint, but they did nothing about it, and chose either to stay indoors or to fill some of those pig bladders. Which worked, albeit inconveniently. Rand, however, decided to solve the problem, and succeeded.

Another example of someone who solved his own da Vinci problem (or, more likely, got his assistants to) was Thomas Edison in relation to his 1880 patent for the light bulb. In fact, he solved two da Vinci problems. The first related to the filament: Edison recognised that the 'missing component' was a cheap material from which a durable filament, which produced sufficient light, could be manufactured easily. His solution was to test hundreds of materials until he established, after many trials, and that-number-minus-one 'errors', that the best performer was a filament of baked cotton thread [7]. Even then he wasn't satisfied, and in seeking a source of carbon other than cotton, he tested some 6,000 plant-based materials [8] before choosing carbonised bamboo [9].

The second da Vinci problem Edison solved was not directly concerned with the bulb. That's because the light bulb was not Edison's *BIG IDEA*, but only one component within it. Edison's *BIG IDEA* was not just a single light bulb that worked, that was relatively cheap, and that had a usefully long working life of at least several hundred hours before burning out. It was an illuminated city. A city with perhaps thousands of light bulbs all working simultaneously, allowing the city to glow at night like never before in history. And for that to happen, there were many 'missing links' to resolve, many questions to answer. What is the source of the electrical current that would flow through each bulb? Should that source be local to each bulb, or could the required power be generated more centrally, and then distributed to each bulb? If the latter, how? How could everything be made safe? To name just a few.

So, in a sense, getting the light bulb to work, from Edison's point of view, was a solution to just one da Vinci problem amongst many. And over the years from about 1878 to 1882, all were solved. For on 4th September 1882, Edison opened the Pearl Street power station in New York City, which could generate enough power to illuminate around 10,000 bulbs, spread over an area of about one square mile. To provide the power, Edison improved on, and scaled up, the designs of pre-existing steam-driven dynamos to build the 'Jumbo', four times bigger than any previous dynamo. And to connect Pearl Street's six Jumbos to the light bulbs they were powering, Edison (after overcoming much opposition from local politicians) dug up the city streets to lay some 25 km of copper wire [10]. And he designed the sockets for the bulbs too – the screw fitting still known as the 'Edison screw' [11]—this being yet another Koestler's Law story for Edison surely did not invent the concept of the screw thread.

When I told the story of the light bulb, Thomas Edison was running alongside, and perhaps slightly behind, Joseph Swan. But in the context of the 'bigger picture', Edison was far ahead. Swan was content to light a single building, his own house, and that of the industrialist Sir William Armstrong [12]; Edison's ambition was to light a city.

The examples of Rand and Edison are of people who identified their own da Vinci problems and developed their own solutions. That's rare, for it requires that the person either has (Rand), or has access to (Edison, who had a large team of assistants), all the required skills and resources. More generally, the skills required to solve a given da Vinci problem are beyond those of the person with the problem – hence the need for, and value of, collaboration, especially as regards academic research. That itself can be a da Vinci problem – a da Vinci problem in which the missing 'component' is a potential co-worker who can contribute the appropriate knowledge required to solve the da Vinci problem associated with the technical problem. But solving the da Vinci problem of finding the right collaborator can be difficult, for to do so, some tough questions need to be addressed – questions such as: Whom might I approach? How do I make contact? Might I be rejected? Will my ideas get 'stolen'? Will the potential collaborator deliver? These are all associated with the 'structures', 'relationships' and 'You!' segments of the Target Diagram (see figure 3.1), and, as discussed in sections 3.5 and 3.6, the closer the questions are to 'Me', the harder they are to answer… especially when I'm shy…

6.6 Keep your eyes – and ears – open

If you are wrestling with a da Vinci problem that you know about, it may be that someone else, perhaps in a very different location or field of study, is working on exactly what you are looking for. But if you don't spot it, you will remain stuck. And if you don't realise you're facing a da Vinci problem, then noticing something 'else' might identify it and then trigger the solution – which is the heart of Koestler's concept of bisociation.

So the more you know, the more you read, the more you interact with others – even in rather different fields – the more likely it is that you will be able to solve any da Vinci problem you are experiencing, or to identify a solution to a problem you are not explicitly aware of.

An example that I have already used many times is that of the DNA researchers in 1952 who missed Chargaff's evidence for base pairs, in contrast to Watson and Crick who didn't; another is the story of the eminent palaeontologist, Jenny Clack – a story that really is about a 'missing component'.

In the early 1980s, Jenny Clack's key interest was the evolutionary path that linked marine animals to land-based animals: how did fins become fingers? At that time, the fossil *Eusthenopteron,* dating back some 370 million years, was known to have fins showing early signs of limbs, and a land animal, *Eryops*, from some 290 million years ago had four limbs, each with recognisable 'fingers' or 'toes' [13]. But that left an 80 million year gap. There was something missing.

One fossil considered to be the 'missing component' was *Ichthyostega*, discovered in Greenland in the early 1930s. The only known specimens of *Ichthyostega* were at that time in the possession of the Swedish scientist Erik Jarvik, but when Clack approached Jarvik requesting access to the fossil for further study, he refused [14]. Clack's only option was to mount an expedition to Greenland to find her own fossils – but that was quite a risk for there was no certainty that if she were to visit the site where the original fossils had been found that she would be fortunate enough to find any, and a field trip to Greenland was both expensive and complex to organise.

Despite those difficulties, Clack decided to find out more about Greenland. She knew that some people from the near-by geology department had been there, and, as geologists, they might have come across some fossils too. So, in 1985, in a conversation with the geologist Peter Friend, she was told that 15 years previously, in 1970, there had been an expedition to Greenland, close to where the *Ichthyostega* fossils had been found [15]. And whilst trawling through the expedition's note books, she discovered that a student, John Nicholson, had indeed found some fossils, identified as *Ichthyostega* [16]. But being a geologist rather than a palaeontologist, other than noting the find, Nicholson hadn't given them much attention.

Clack was shocked. Might Nicholson have brought the fossils back from Greenland? And, although Nicholson had left the geology department some ten years previously, might any specimens still be in a box somewhere? If so, she needn't have worried about Jarvik's non-co-operation. She needn't incur the expense of an expedition to Greenland. The specimens she was seeking might be right there, in a neighbouring building. And had been there for 15 years. Within 100 metres or so. If only she had found out sooner!

And yes, the fossils were there, in the basement.

But when she saw them, she had a surprise. They weren't *Ichthyostega*.

But she recognised them as three nearly complete skulls of another, somewhat similar, species, *Acanthostega*, only fragments of which had hitherto been found. And Nicholson's notes had recorded the precise location from which the fossils had been collected. So, soon thereafter, in 1987, Clack joined a Danish geological expedition to Greenland which, by chance, happened to be going to the region visited by Nicholson in 1970. Where she found more than enough *Acanthostega* fossils to keep her and her colleagues, Michael Coates and Sarah Finney, very busy after her return to her laboratory.

And *Acanthostega* did indeed prove to be the 'missing component', filling the gap in the evolutionary between *Ichthyostega* and *Eryops*.

Jenny Clack had solved her da Vinci problem, truly discovering the 'missing component'. She knew that there was a gap to be filled, she knew she would not have access to Jarvik's fossils, she knew that she had to find her own specimens in the only known place on the planet where these fossils were known to be present, Greenland. But she didn't know that the critical clue was, quite literally, right beside her, in an adjacent building, and had been there for 15 years. And she only found out by talking to Peter Friend, who didn't know that something of value to Jenny Clack was right there, in the basement. It was only the conversation between them that unlocked all this.

The 'take-home' message is self-evident. Talk to people. Find out what they know, and (more importantly) what they know *about* – and let them know what you know, and know about. At any time, that conversation might lead nowhere in particular. But conversations don't have to be 'instrumental' or 'transactional', with 'purpose'. Sometimes; often; it's just good to talk.

And as we shall see in chapter 17, one of the characteristics of a physical environment that helps innovation flourish is to provide spaces in which people might meet, ideally randomly, and just chat.

6.7 Be patient

In section 4.5, I told the story of the bar code, and described how, in 1949, Joe Woodland was sitting on a beach in Florida, running his fingers through the sand, and spotted how the resulting stripey pattern might form the basis for a way to identify products at a supermarket check-out. Woodland had recognised that the stripey pattern was in essence a pictorial version of Morse code, and so could carry information. Woodland and his friend Bernard Silver then developed the idea further, filing for a patent later that year [17].

After some experimentation, Silver and Woodland discovered that it was proving impossible to print labels, suitable for supermarket products such as foods, with sharp enough edges to the stripes, and impossible to develop a reliable reader using the only source of illumination then available – a light bulb. So they let the matter drop [18].

About 10 years later, in 1959, David Collins, then working on a project for the Pennsylvania Railroad, designed a system using blue and red reflective stripey patterns painted on the side of rail wagons to enable them to be recognised and tracked as they rolled past a scanner. But not very reliably: the mud and dirt that inevitably accumulating on the wagons made the patterns progressively unreadable [19].

Something else was happening in the 1950s and 1960s too, in a very distant context. In 1954, Charles Townes demonstrated the first maser, using microwaves [20]; in 1958, he extended this concept to visible wavelengths, patenting the idea in March 1960 [21]. Then, a few months later, Theodore Maiman, a physicist working on a defence contract at Hughes Research Laboratory in California, built the first operational ruby laser [22]. The laser, of course, is a light source very different from an incandescent bulb...

...which David Collins noticed, for in 1968 he founded his own company to develop bar code technology, incorporating a laser in his reader. And in 1969, his device was successfully used to track car axles, coded with two digits, in a General Motors factory in Michigan [23].

Whilst bar codes were beginning to be used in factories, American retailers became increasingly interested in developing a standard for product labelling and coding, resulting in an open competition for the best design for a bar code. The winning entry, announced in April 1973, was submitted by IBM, who, as well as designing the bar code we use today [24], also manufactured all the necessary equipment – including a laser scanner – first launched in October 1973 [25].

The team that submitted IBM's winning proposal included someone who has appeared in this story before: Joe Woodland, who had joined IBM in 1951 [26], and who, 14 years previously in January 1949, almost to the day, had run his fingers through the sand.

Woodland therefore had 14 years to wait until his da Vinci problem was solved, in this case, by the improvement of printing technologies, and – more importantly – by the invention of the laser, and its subsequent development into a relatively low cost, mass-produced, component that could be incorporated in all those supermarket checkout scanners.

It is quite rare for the same person to be the originator of an idea – and an idea that had to remain incomplete because of a significant da Vinci problem – and also be present at the end, when the idea becomes reality, with the da Vinci problem having been solved by something 'totally over there' – in this case, the laser – and subsequently incorporated as a key Koestler's Law component.

The 'take-home message' from this story is be patient, don't throw any 'partly formed Koestler patterns' away, and keep looking around, as widely as you can. Yes, your idea might be blocked now because of a da Vinci problem you can't solve – but someone else, somewhere else, might have the answer at some time. In da Vinci's case, he would have had to wait around 400 years for a suitable internal combustion engine, so his judgement was right in leaving his helicopter on his most exquisite drawing board. Woodland had to wait 'only' 14 years, but it was worth it in the end.

6.8 In conclusion

The da Vinci problem – the problem that a component that you need to complete a 'Koestler's Law assembly' is missing – is real. You may be (probably painfully) aware of the component's absence ('Agh! I'm stuck!'), but perhaps not (as were Pauling and Corey in relation to their proposed triple helix structure for DNA) – which might, at the time, feel rather better, but is probably more dangerous.

So bear in mind the value, and importance, of:

- identifying all your assumptions, and asking 'how might [this] be different?'
- checking that all your conclusions recognise, and incorporate, all the evidence you have
- identifying, as specifically as you can, 'what you don't know', so that you *do* know what you're looking for, and are in a position to recognise it when it turns up
- talking to as many, and as many different, people as you can
- working with people from other fields
- scanning, scanning, scanning to keep track of what's 'out there'
- being patient
- not throwing anything away, just in case it's a 'fragment', the value of which only becomes apparent when something else happens, somewhere else, later
- not being scared to 'down tools' and do something else, as da Vinci did with his helicopter, rather than struggling on too long...
- ...and when you do 'down tools', keep on looking around, just in case that internal combustion engine gets invented...

Leonardo da Vinci

Leonardo da Vinci, self-portrait, 1505
Source: Wikimedia 'Self Portrait c.1505' www.wikiart.org/en/leonardo-da-vinci/self-portrait-1505.

The term 'Renaissance Man' surely is embodied by Leonardo da Vinci – artist, engineer, architect, anatomist, musician, scientist, inventor, adviser to dukes, kings and popes. And, for those days, long-lived, for he died in 1519, aged 67, in France [27]. The church in which he was buried was destroyed in the upheavals following the French Revolution, and so – like Kepler's (see page P-10) – da Vinci's tomb has been lost [28].

Relatively little is known of da Vinci's early life; the record becomes clearer in his teens, when he served as an apprentice in the studio of Andrea del Verrochio, a Florentine sculptor, goldsmith and painter, which is where he learnt and refined many of his technical skills.

Politically, Italy was a turbulent place during da Vinci's life-time, with many independent dukedoms, republics and city states, not forgetting the papacy and foreign intervention too. So over the next few decades, da Vinci accepted commissions from many individuals and institutions (such as the Augustinian monks of San Donato in Florence, for whom he painted *The Adoration of the Magi*), as well as from (among many others) the Medici family in Florence, the Sforzas in Milan (where he designed his helicopter), the Republic of Venice (where he devised defences against naval attack), Cesare Borgia, the son of Pope Alexander VI (once again as a military architect), and for the last three years of his life, King Francis I of France, whose service he entered in 1516.

According to the 1550 account by art historian Giorgio Vasari [29], and as poignantly depicted by Jean-Auguste Ingres in a dramatic painting of 1818 [30], the King held the head of the dying da Vinci in his arms. Francis I held onto something else too – the *Mona Lisa*, which da Vinci had started around 1503 (when he was back in Florence), and was still working on in 1517, after he had moved to France. da Vinci's masterpiece has remained in France ever since – subject to a gap between 21st August 1911 and 4th January 1914, whilst it had been 'borrowed' by the Italian art thief, Vincenzo Peruggia [31]!

References

[1] https://theconversation.com/leonardo-da-vincis-helicopter-15th-century-flight-of-fancy-led-to-modern-aeronautics-116241
[2] www.sciencephoto.com/media/892725/view/leonardo-da-vinci-s-water-lifting-devices
[3] www.nps.gov/wrbr/index.htm
[4] www.aviastar.org/helicopters_eng/forlanini.php
[5] Astbury W T and Street A 1931 Studies of the structure of hair, wool and related fibres. General *Phil. Trans. Roy. Soc.* A **230** 75–101
[6] www.bionity.com/en/encyclopedia/Alpha_helix.html
[7] www.signify.com/global/our-company/blog/innovation/edison-perfects-the-light-bulb
[8] www.fi.edu/history-resources/edisons-lightbulb
[9] www.amusingplanet.com/2019/05/how-japanese-bamboo-helped-edison-make.html
[10] https://ethw.org/Milestones:Pearl_Street_Station,_1882
[11] https://worldwide.espacenet.com/patent/search/family/002320853/publication/US251554A?q=pn=US251554A
[12] https://williamarmstrong.info/science
[13] www.corzakinteractive.com/earth-life-history/416_pennsylvanian.htm
[14] https://carlzimmer.com/coming-onto-the-land/
[15] www.pbs.org/wgbh/nova/link/clack.html
[16] www.theclacks.org.uk/jac/Expeditions.html
[17] www.channel4.com/news/barcode-inventor-who-transformed-commerce-has-died
[18] www.smithsonianmag.com/innovation/history-bar-code-180956704/
[19] https://barcode-labels.com/early-barcodes-brief-history-how-barcodes-changed-our-lives/
[20] www.nobelprize.org/prizes/physics/1964/townes/biographical/
[21] https://patentyogi.com/this-day-in-patent-history/this-day-in-patent-history-on-march-22-1960-schawlow-and-townes-were-issued-a-patent-for-the-laser/
[22] https://ethw.org/Theodore_Maiman_and_the_Laser
[23] https://datacaptureinstitute.com/about-david-collins/
[24] www.relegen.com/blog/bullseye-barcode-history/
[25] www.ibm.com/ibm/history/ibm100/us/en/icons/upc/
[26] www.theguardian.com/technology/2012/dec/14/barcode-inventor-norman-woodland-dies
[27] www.leonardodavinci.net/
[28] www.leonardodavinci.net/last-wishes-and-gravesite.jsp
[29] https://salemcc.instructure.com/courses/451/pages/giorgio-vasari-the-life-of-leonardo-da-vinci-1550
[30] www.petitpalais.paris.fr/en/oeuvre/francis-i-receives-last-breaths-leonardo-da-vinci
[31] www.picturesfromitaly.com/florence/stealing-the-mona-lisa

IOP Publishing

Creativity for Scientists and Engineers
A practical guide
Dennis Sherwood

Chapter 7

Emergence – why some patterns are better than others

7.1 Emergence

In section 4.7, you will find these words: 'There's something else, too, that Koestler's Law doesn't do, and which is really important. Koestler's Law identifies creativity as the formation of a new pattern from existing elements. Which is true. But in his definition, Koestler makes no statement about the *quality* of the resulting pattern. And some patterns are much, much, better than others.'

Yes, some patterns are indeed better than others, and so the purpose of this section is to gain a deeper understanding of 'quality', enriching our appreciation of Koestler's Law, and putting us in a stronger position to use it insightfully to discover great new ideas from scratch.

My starting point is the simple sentence 'I went to the bank'.

Everyone reading this book will know what that sentence means; everyone will be able to form a mental picture of what is being described. For this sentence has a property, within the English language, that we can call 'meaning'.

But where, precisely, is that meaning located?

Suppose, for example, that I study the single word 'went'. I could do that for a lifetime, but never discover the meaning of 'I went to the bank'. The meaning of the sentence is not 'located' within any single word, or any small group of words: the meaning of the sentence is a property of 'I went to the bank' as a whole, as a single entity, and is not a property of any of the individual words from which that sentence is formed.

In Koestler terms, the sentence 'I went to the bank' is a 'pattern' of individual 'components', the words from which that sentence is formed. And it is a pattern that shows the special quality of having meaning, a meaning that is intrinsic to the pattern-as-a-whole, and that cannot be inferred from any component by itself.

This is an example of an important concept known as 'emergence' – the existence of a property at the level of a pattern, a property that is not associated with any of the individual components from which that pattern is formed [1]. And, as I describe in the following paragraphs, it is emergence that unlocks the problem of why some patterns are better than others.

7.2 Same components, different patterns

Consider now the 'sentence' 'bank I to went the'. In the context of Koestler's Law, this too is a pattern of individual components, but it is somewhat problematic. It doesn't make any sense. It has very little, if any, meaning: hence the use of the inverted commas in 'sentence', for it isn't really a sentence at all.

An important characteristic of the non-sentence 'bank I to went the' is that it contains exactly the same components as the meaningful sentence 'I went to the bank', but in a different sequence. In fact, given these five words, there are 120 different possible patterns that can be formed. Only one of those patterns is 'I went to the bank', which conveys meaning directly, and there are 119 that do not, although there are a few patterns – such as 'To the bank I went' – which can be understood, just about, for this sequence is most unusual in the normal, every-day way in which the English language is currently used, but might be found, for example, in poetry, or in a text written in an earlier century.

If the 'goodness' of a Koestler pattern of words in the English language is assessed by the clarity of its meaning, then the pattern 'I went to the bank' is 'very good'; 'To the bank I went' is 'OK–ish'; and most of the remaining possibilities such as 'bank I to went the' are 'not good at all'.

The quality of any pattern is therefore determined by the emergent properties of that pattern, as judged against whatever criteria of 'goodness' are appropriate.

We'll return to that, important, point shortly; for the moment, there are two other features of emergent patterns to which I wish to draw attention.

7.3 Not too little, not too much

The first feature is 'not too little, not too much'.

So, for example, 'I went to the' has some meaning, but only very little, for the pattern is incomplete – where had 'I' been going to? That said, 'I went to the' – or rather 'I went to the...' could work in a detective novel as the last words of someone who is about to tell the detective the key clue, but whose sentence is cut short by the villain, who wishes to stop the speaker from 'spilling the beans'. And 'spilling the beans' is itself an 'interesting' emergent Koestler's Law pattern which means something very different when the words are interpreted literally, as compared to the use of that phrase in the context of a detective novel.

'Not too much' is the other possibility – 'I went to the bank trousers' is also devoid of any meaning, for the addition of that extra word destroys the meaning of the preceding five.

So the simplicity of that sentence, 'I went to the bank', and the fact that we construct such sentences instantaneously whenever we talk, let alone whenever we write, masks a bewildering complexity. In our minds, we have a vocabulary of tens of thousands of words. And whenever we speak, we select just the right words, assembling them in just the right order, with nothing missing, nothing superfluous. All without conscious thought. Wow.

Although that statement was made in the context of language, the general principle that an emergent pattern has just the right components in just the right order, with nothing missing and nothing superfluous, holds in all fields. In music, for example, the emergent properties of a pattern of notes relate to the subjective appeal of the melody or harmony, often associated with its recognition as somehow familiar, but different. For a commercial product, the emergent properties relate to the appeal of the product to the customer, so resulting in high sales. For a scientist, the emergent properties of the findings of some research relate to the extent to which the results, for example:

- unify hitherto apparently unrelated observations (Mendeleev's Periodic Table made sense of all those individual elements [2])
- identify some fundamental principles (Einstein's theories of Special and General Relativity)
- develop a technique that becomes of widespread use (partition chromatography, for which John Martin and Richard Synge won the 1952 Nobel Prize in Chemistry [3])
- create a framework for much more reliable, or unexpected, predictions (Newton's Laws of Motion), or
- provide an insight or outcome that others find helpful (as measured, for example, by the number of citations).

7.4 Patterns within patterns

The second feature to note is that emergence can be present at many levels simultaneously. In the pattern 'I went to the bank', for example, there are many lower-level patterns, such as the word 'went' which, in its own right, is an emergent pattern of the four components, the individual letters 'w', 'e', 'n' and 't'. A different sequence such as 'tnew' has no meaning (at least in the English language), whilst another pattern, 'newt', does have meaning, but a totally different one, referring to the amphibian. 'ent' has something missing; 'wents' has too much. So the Koestler pattern 'went' complies with all the characteristics of emergence.

At a lower level still, the single letter 'w' is the emergent pattern of four straight line segments in a particular configuration, a configuration that we understand as a letter with a particular sound. A very similar four-line pattern is another letter, 'M'; a third, letter 'E', also has four lines, as does the Greek letter Σ; moving away from letters, we have, for example, >>, the mathematical symbol for 'much greater than', or just a plain square, □, which is used by mathematicians to denote the end of a proof.

And we can go to higher levels too. 'I went to the bank and watched the swans glide past' embeds the original five words within a more complex pattern, giving more context, and also adding some important information: the longer sentence suggests that the 'bank' is not the building where I carry out financial transactions but rather the bank of a river. And that sentence itself can be incorporated within a paragraph, a chapter, a novel, a trilogy... each of which has emergent properties at the appropriate level. But not randomly, for those higher-level patterns must also 'make sense' in their own right: the chapter must tell a 'mini-story', with linkages back to previous chapters and a context that will be revealed in future chapters; the novel must form a satisfying whole, but not necessarily a complete one, for there are still two more novels in the trilogy...

7.5 Emergence is often subjective

I've just made the point that the longer sentence 'I went to the bank and watched the swans glide past' suggests that the bank is the bank of a river. But that's an inference, rather than a definitive reading of the longer sentence – suppose, for example, that the building in which I carry out financial transactions just happens to overlook a river, so that, from inside the building, I can see the swans... The meaning we infer from the words – the emergence we perceive – is therefore not totally an 'arms-length' transaction; rather we are bringing our experience, expectations and perhaps beliefs and opinions to bear too, and so the 'emergence' we perceive is the objective experience of whatever the Koestler's Law pattern happens to be, blended with the subjective experience of 'me'.

This is especially the case for Koestler's Law patterns related to music and art. Two people can hear exactly the same music, but have very different judgements of that pattern's 'goodness' – to some, Jimi Hendrix is aural torture, to others aural nirvana. Neither person is 'right' or 'wrong'; the judgements are just different. Likewise, are Jackson Pollock's paintings merely the results of a grown-up child throwing paint around, or intricately inter-twined fractals [4]?

The importance of 'me' as being integral to the interpretation of patterns is illustrated in figure 7.1, which reproduces the masterpiece *La Seine à la Grande Jatte*, painted by Georges Seurat in 1888, and also a close-up of a small section within it.

La Seine à la Grande Jatte depicts a tranquil scene looking across from one bank of the river Seine to the other, featuring a sail boat and a rowing boat. The detail in the lower part of figure 7.1 corresponds to the edge of the bank just to the left of the tree, and highlights Seurat's painting technique, known as 'pointillism' [5], in which individual dabs of colour are closely juxtaposed. Although some colours occasionally overlap, there is no attempt to blend the colours or deliberately to blur the edges. Each dab is small, on the scale of a few mm, about the width of a painter's brush – it's as if Seurat were colouring in a screen, pixel by pixel. And in the language of Koestler's Law, the detail visible in the lower part of figure 7.1 vividly illustrates the 'components' which collectively form the 'pattern' shown in the complete picture.

When we look at the full painting, we can immediately see what it depicts – the tree on the right, the river bank, the river itself, the boats. But when we look at the detail, yes, we see some green and some blue, but those pink and yellow blobs

Figure 7.1. *La Seine à la Grande Jatte*, Georges Seurat, 1888, and a close-up detail.
Source: Wikimedia 'The river Seine at La Grande-Jatte' www.wikiart.org/en/georges-seurat/the-river-seine-at-la-grande-jatte-1888.

in the upper left are unexpected in a river, and what is that red blob (over which there are two creamy blobs) on the lower left? And those prominent darker blue 'stripes', in the centre sloping upward to the right, and on the far right? If we just saw the lower part of figure 7.1 'cold', with no prior knowledge, how would we interpret it? As an abstract pattern? Might the blue be an evening sky, and the green perhaps the top of a tree? Is it a close-up of a fabric, such as a coarsely knitted sweater, mainly green and pale blue, but with a darker blue diagonal stripe, now somewhat stained with the remnants of the wearer's breakfast – perhaps a bowl of muesli followed by a boiled egg? It's hard to tell...

But when you see the whole picture, the overall emergent property is evident. 'The whole' is undoubtedly greater than the 'sum of its parts' – for that commonly-used phrase captures the essence of emergence.

There's another, deeper, emergent property in Seurat's art too – a property that is totally dependent on the interaction between the physical object – the painting – and me. Seurat's technique was to place individual, separate, blobs of differently coloured pigments next to one another. But these combinations were not chosen arbitrarily, for Seurat had a deep understanding of the work of the chemist Michel Chevreul [6], who had shown that our perception of a particular pigment is not 'absolute', but rather *dependent on the neighbouring colours*. So how we see a particular colour, say, a specific tone of green, is the emergent result of that particular tone in a particular context [7].

Figure 7.2. Plate IV-3a from The Interaction of Colour, Josef Albers, 1963. Reproduced courtesy of Yale University Press.

This was understood, and deployed, intuitively by Seurat, and was investigated systematically and thoroughly by the artist Josef Albers [8], as detailed in his book *The Interaction of Colour*, from which figure 7.2 is reproduced.

The upper grid appears to be a lighter shade of green than the lower grid. Not so. The two grids are the *same hue of green*. Really. And the difference we perceive in the greens is not attributable to 'transparency', by which some of an underlying colour might be visible through the green. It's about the juxtapositions: the green in the context of blue appears lighter than the same green in the context of yellow.

That's emergence: a property associated only with the pattern-as-a-whole, and not with any component from which that pattern is formed.

7.6 An enriched definition of creativity

According to Koestler's Law, any new pattern is an 'act of creation', whatever that pattern's quality. Intellectually, that's fine; pragmatically, it doesn't recognise that, if we are deliberately seeking to generate ideas, then our time is better devoted to searching for ideas that have a reasonable chance of being 'good' rather than 'poor'. But as we saw in section 1.9, the actual 'goodness' or 'badness' of an idea cannot be assessed, with certainty, from the idea; it can only be determined later, after the idea has become real. But what we can do at the early stages of innovation, whilst we are actively generating ideas, is to seek those ideas that we believe have the potential to be good, and that seem to have 'promising' signals. Yes, that is subjective, and we might be wrong. But it might help direct our energies.

With that in mind, here is my extension of Koestler's Law, which acknowledges the importance of emergence:

> *Creativity is the process of forming new patterns from pre-existing component parts.*
>
> *The more the resulting pattern shows emergent properties, such as perceived beauty, the likelihood of enhanced understanding, the prospect of utility, or the possibility of creating value, the better the corresponding idea is likely to be.*

The first sentence is my paraphrase of Koestler; the second, my recognition of emergence, using words such as 'perceived', emphasising the subjectivity of judging the emergent 'beauty' of music, art or literature, and 'likelihood of', 'prospect of' and 'possibility of', emphasising that the extent to which an idea is intellectually insightful, commercially useful or economically valuable can be determined only later, when the idea has become real.

When an idea is first formulated, it is usually 'raw', incomplete, deficient, and so it is inappropriate to attempt to identify any emergent properties at that stage. Over what is often a relatively short time – sometime a few days, sometimes a few weeks – the idea becomes more complete, sharper, richer, especially if the idea is discussed with colleagues. Once the idea has become fully 'incubated', emergent properties then become more apparent, and soon the idea is robust enough for those emergent properties to be assessed: even though the idea is still an idea, we can begin to imagine enough about what the 'world might look like' once the idea has become real, so envisioning whether those emergent properties might actually be beautiful, useful, or valuable. We won't know for certain until the idea has indeed become real, but we understand the idea well enough to be in a position to assess the likelihood that those emergent properties will become realised in the future. The process of making this assessment is what I call 'wise evaluation', the second step of the four-step innovation process represented in the Target Diagram of figure 2.1, as was discussed in section 2.4, and as will be discussed in more detail in chapters 14 and 15.

So, for now, back to creativity, for I still have yet to describe my process for generating great ideas deliberately, 'on demand'. But before I can do that, there is one more puzzle to solve: yes, we know from Koestler's Law that the outcome of creativity is the formation of a new pattern from pre-existing components. But if I don't have the idea yet, if I'm seeking to generate an idea, just where are the 'components' that I might need? Where do I look? And how do I put them together in just the right way so that the resulting pattern is truly emergent?

References

[1] Sherwood D 2002 *Seeing the Forest for the Trees – A Manager's Guide to Applying Systems Thinking* (London: Nicholas Brealey) p 14
[2] www.chemistryworld.com/features/the-father-of-the-periodic-table/3009828.article
[3] www.nobelprize.org/prizes/chemistry/1952/summary/
[4] Taylor R P, Spehar B, van Donkelaar P and Hagerhall C M 2011 *Front. Hum. Neurosci.* https://doi.org/10.3389/fnhum.2011.00060
[5] www.nationalgallery.org.uk/paintings/glossary/pointillist
[6] www.lib.uchicago.edu/collex/exhibits/originsof-color/color-theory/chevreul/
[7] www.sothebys.com/en/articles/pointillism-7-things-you-need-to-know
[8] www.themarginalian.org/2013/08/16/interaction-of-color-josef-albers-50th-anniversary/

IOP Publishing

Creativity for Scientists and Engineers
A practical guide
Dennis Sherwood

Chapter 8

Knowledge, experience, learning, and unlearning

8.1 Where are the Koestler's Law 'components'?

If you're a musician, the components you need for a new Koestler's Law pattern – your new symphony, pop song or blues riff – are to be found on a piano keyboard, or on any other musical instrument. The notes are there, ready to be combined into any pattern you wish.

If you're a painter, the components are the pigments in your paint box.

If you're an author, the components are the words in whatever language you are writing in, to be found in the appropriate dictionary, held within the vocabulary in your mind, in a book, or on-line.

In each of these cases, the fundamental components you need are freely available, in their elemental form, for anyone and everyone, ready to be assembled into whatever pattern you might like.

But if you're a product designer, where are your 'components'? Where is your 'piano', your 'paint box', your 'dictionary'? You might have a box full of 'bits', gathered together over the years and not thrown away, and some of those bits might help. But are enough 'bits' there, and are they the right ones?

And if you're an academic researcher, where are the 'components' you need? You're in an even more problematic position than the product designer, for you don't have that box.

At this point, the absence of the 'components' in these – and indeed in many other – contexts might lead to a conclusion something like, 'Well, for musicians, artists and writers, Koestler's Law seems to work. But for everyone else, the 'components' just aren't there, so Koestler's Law just can't be valid. And even for musicians, artists and writers, there are so many 'components' – all those notes, pigments and words – that the number of possible combinations will be staggeringly huge. Most of those combinations will be rubbish, so Koestler's

Law doesn't seem to be of any use even when it does work. Overall, Koestler's Law might be 'interesting', but nothing more.'

This conclusion is understandable, but it's missing something important. Yes, for musicians, artists and writers, the fundamental components that they need – notes, pigments, words – are indeed immediately available in their most 'elemental' forms. But the fact that, in other contexts, the components are not immediately available in their 'elemental' form does not imply that the components do not exist, nor that Koestler's Law is consequently invalidated.

For it might be that the components are indeed there, *but not freely available in their 'elemental' form* – they might exist, but in a *different* form.

What might that form be?

In principle, in *already-existing patterns*; patterns that need to be 'deconstructed' to 'reveal' the fundamental components which can then be re-combined into new patterns of creativity.

As an example, consider the metal iron. As a chemical element, symbolised as Fe, iron exists. But in nature, iron does not exist in its elemental form: iron within the Earth's crust is present predominantly in the form of an oxide, such as ferric oxide Fe_2O_3, which itself is usually mixed alongside other components in an iron ore mineral such as hematite. And it's not until that ore has been mined, crushed, separated and smelted that the iron metal, pure Fe, can be made available and used – used, for example, to combine with carbon (and perhaps some other components such as chromium) to form steel.

In the context of our current discussion, iron ore is the 'already-existing pattern' in which the component we need, elemental iron, is contained; the process of smelting is the 'deconstruction' of the ore, the 'already-existing pattern', to 'reveal' the elemental iron; the manufacture of steel – a product that does not exist naturally – is the 're-combination' of the elemental iron with carbon and chromium to form the new Koestler's Law pattern we know as steel.

Fine. But I'm a scientist. I don't do mining and smelting. Where are my 'components'?

The answer to this important question is that, yes, 'intellectual components' are indeed 'there', but bundled together in pre-existing patterns. Patterns that we label as *learning, knowledge* and *experience* – where those terms refer not only to the learning, knowledge and experience of the individual, but also to the collective learning, knowledge and experience of a team, or larger group.

And just as the forging of steel requires that the source material, the ore, must be 'deconstructed' to release the elemental iron before that iron can be re-combined with carbon into the new pattern of steel, then to be intellectually creative, you have to do something very similar: you have to deconstruct your learning, knowledge and experience to reveal the 'elemental components'; you then have to 're-combine' those together with the 'elemental components' of others; and then there is every possibility that you will discover your 'steel'.

Central to this is the need to deconstruct your learning, knowledge and experience – a process I refer to as 'unlearning'. And it so happens that many people find 'unlearning' extremely hard to do. Our learning is deep within us; we've relied on it, and trusted it, for years; it's part of our 'identity'. And so tearing it all up really hurts. I'll return to unlearning shortly, for unlearning is better appreciated in the context of having a richer understanding of the opposite, and in many ways more fundamental, process of learning. What, then, is 'learning'?

8.2 Donald Hebb's Theory of Learning

In 1949, the Canadian physiologist (and nominee for the 1965 Nobel Prize in Physiology or Medicine [1]) Donald Hebb published a landmark book, *The Organisation of Behaviour* [2], which included his ideas about the neurophysiological basis of learning.

As we saw in section 1.2, creativity is an activity that takes place within the human brain, and it was Hebb who was the first to propose an explanation of how learning takes place. In the 1940s, when Hebb was formulating his theory, the knowledge of the brain was, in current-day terms, rudimentary at best. But there were four key features of the brain that were known.

The first is that the brain of a new-born infant is about one-quarter of the size of the corresponding adult brain, and grows very quickly, approaching its full size when the child is aged about five [3].

It might be thought that the increase in size results from an increase in the number of brain cells, the neurons, but this is not the case: the adult human brain has about the same number of neurons as the corresponding infant [4], as was known to Hebb. That's the second feature: the size difference is attributable not to the increase in number of neurons, but to an increase in each neuron's size.

The third key feature that Hebb knew was that by far the most significant difference between the brain of a new-born infant and the corresponding adult concerns not the neurons themselves, but the connectivity between them: the adult brain is far more richly connected than the corresponding infant, and it is this increase in connectivity that explains the increase in the size of each neuron – to make the required connections within the brain, neurons grow extensions that interact with other neurons by exchanging chemicals, known as 'neurotransmitters', across a small gap, the 'synapse' [5].

Hebb did not know the relevant numbers, but today we do – the current estimates are of order 10^{11} neurons [6] in both the infant and adult brain, with each neuron in the infant brain having on average about 2,500 synapse connections, increasing to about 15,000 synapses per neuron at around age 5, and falling to about half that number in the adult brain [7]. Although the average number of synapses for each neuron is only a factor of about three greater in the adult than the infant, the impact on the number of possible neural pathways is colossal.

The additional 'connecting-together' of the adult brain must take place as the infant grows and develops. And Hebb knew that something else happens as the infant grows. The infant learns. An infant can't speak, get dressed, ride a bicycle, solve a differential equation. But by the time the infant becomes an adult – with the appropriate guidance – all of these can be done. That's the fourth feature: as the same time as the brain becomes progressively more interconnected, the infant is learning. Is that just a coincidence, or might the one be associated with the other?

In the late 1940s, Hebb put those four Koestler's Law fragments together to form his *BIG IDEA*, now known as Hebb's Theory of Learning: that learning is the manifestation of the formation, and progressive strengthening, of neural networks resulting from the repetition of the corresponding activity [8].

At birth, an infant can't do very much at all, but over the next several months, a number of abilities begin to be developed, such as turning the head towards a voice, opening and closing the hand, reaching towards a moving toy. Take, for example, moving the head towards a voice. This requires the infant to be aware of a sound, to recognise that the sound comes from a specific direction, and then to turn her head towards her mother. To turn her head, the right muscles in the neck need to be contracted in the right way, so a different set of muscles are required depending on which direction her head is turned.

Muscles, however, do not contract of their own accord: they contract only in response to a signal transmitted along the appropriate nerve, a signal that must have originated by the firing of a corresponding neural circuit in the infant's brain. And that circuit is very specific: it's the circuit that says (metaphorically!) 'contract precisely those muscles that will move my head in the direction from which I think the sound is coming from', rather than 'contract precisely those muscles that move my leg' or 'contract precisely those muscles that close my hand'. Furthermore, the neural circuits that recognise sound will be firing too, and those circuits will be linked to the circuits that control movement.

Hebb recognised that every action that we observe – such as the infant's movement of her head – is associated with the firing of just the right neural circuits in the brain. But he took matters further by hypothesising that the *more frequently an action is repeated, the stronger the connections within the corresponding neural circuit, until, after a sufficiently large number of repetitions, the connections are sufficiently strong that the circuit can be invoked without directly conscious thought.* That progressively strong connectedness, proposed Hebb, is the basis of learning.

We have all experienced this. Take, for example, any everyday activity such as getting dressed, crossing the road, riding a bicycle, driving a car. These are all activities that we have learnt to do at different times during our development from infant to adult. Crossing the road in particular is a very complex cognitive skill: we need to know where to stand in relation to the pavement and the roadway; we need to know where to look and how to listen; we need to recognise if there is a corner which might be concealing a car; we need to assess how long it will be before that bus we see coming towards us will actually arrive; we need to take account of the immediate conditions, such as whether or not the surface of the road might be slippery; we need to estimate how long it will take for us to cross the road safely. And we all know how much care we devote to ensuring our children learn to cross the road safely, and how long this can take.

But once this skill has been repeated and learnt, we can invoke it, whenever we wish – invoke it safely, and without consciously thinking about it.

Similarly, everyone who now routinely drives a car will remember when they first sat behind the wheel: they gripped the wheel so tightly that their knuckles glowed white; they concentrated so hard on what they are supposed to be doing with their feet that they did not hear the instructor's advice; and – WATCH OUT – that other car is approaching SO FAST! It was all so strange, requiring a degree of co-ordination between eyes, ears, hands and feet that seemed to be quite impossible. But, with practice, we all learn, and can safely drive from 'here' to 'there' whilst listening to the radio, talking to a colleague, noticing things by the side of the road, and without 'consciously' thinking of, and deliberately 'directing', the physical actions. Before learning to drive, many of the required neural circuits were not present; the process of learning formed the right inter-connections; the repeated practice strengthened the network; the result is the ability to invoke those newly-formed circuits almost 'automatically'.

Another example is a story told by Dame Darcey Bussell, a distinguished ballerina and dancer, and Principal at the British Royal Ballet from 1989 to 2007. 'When you're learning to dance,' she said, 'you place your foot 'there', look down, and then, in your head, ask 'Is my foot in the right place?' If the answer is 'no', you move your foot to the right place; if the answer is 'yes', you take the next step.'

Yes, we all recognise that.

She then continued, 'But when you're on stage, dancing Odette in *Swan Lake*, you haven't got time for all that. You have to have confidence that your foot is in the right place, confidence gained by having done it so many times. You just know it's in the right place, and you don't even have to think about it.'

Or, as Hebb might have said, Dame Darcey knew her foot was in the right place because the steps had been so well practised that the corresponding neural circuits had become 'hard wired' [9].

The more you repeat something, the deeper the learning, and the more skilled you become – and the stronger the synapses between the associated neurons. This is the explanation behind the '10,000 hour rule', featured in Malcolm Gladwell's book *Outliers* [10], which states that if you spend about 10,000 hours doing something, you become quite good at it. So professional sports stars will have devoted 10,000 hours to honing their skills when they were younger, as did musicians. And although 10,000 hours sounds like a very long time indeed, in fact, it isn't: someone working conscientiously for 5 hours a day, 5 days a week, will clock up 25 hours a week, or 100 hours every four weeks – say, about 1,200 hours every year. For someone dedicated, that's not too arduous a commitment, so the 10,000 hours goal is reached after about 8 years; perhaps 5 or 6 if the work rate is higher. So, during one's teens that's quite feasible. And a really committed Mozart could have put in his 10,000 hours between the ages of about 4 and 9.

Many of the specific claims of Gladwell's book have been disputed, as well as the black-and-whiteness of corollaries such as 'the only way to become really good at something is to devote 10,000 hours to it' and 'if you spend 10,000 hours doing something you will inevitably become a world champion'. The number 10,000 is of course approximate, but the basic message that 'the more you practice something, the better you become at it', is, I believe, true. And to me, the interpretation of this in terms of the strengthening of neural synapses makes sense too.

When Hebb put his idea forward in 1949, there was little direct evidence that neural networks were becoming strengthened by learning. But over the subsequent decades, much research has been carried out into nerve function, the structure of synapses and the neurophysiology of the brain, and the devices that can be used to assist such studies – such as imaging devices and of course computers – have become increasingly sophisticated. So much so that Hebb's hypothesis has now been proven: synapses really do become stronger with repeated use, as recognised by the award of the 2016 Brain Prize – neurophysiology's Nobel – to Graham Collingridge, Richard Morris and Timothy Bliss who 'independently and collectively have shown how the connections – the synapses – between brains cells can be strengthened by repeated stimulation' [11].

And once those strong neural networks have been formed, they can be invoked quickly and efficiently, without conscious thought – and become even stronger as a result.

A metaphor for this process is the formation of a physical landscape. Aeons ago, some land might have been relatively flat. But over time, rain, falling on that landscape, might carve some shallow depressions, along which small rivers flow towards the sea. And over more time, those depressions attract more rain, causing the carving of ever deeper valleys, and the flow of ever wider rivers... until we end up with the richly carved landscape of alpine Europe, the Rocky Mountains, the Himalayas. These landscapes were not 'designed' according to some 'grand plan'; rather, they evolved naturally over time, influenced by meteorological and geological events.

In the context of this metaphor, the relatively sparsely-connected infant brain corresponds to a gently undulating landscape; the much more densely-connected adult brain, to the 'Alps'. And the detailed 'topography' of the adult landscape has been forged over the individual's life, according to that person's experiences, interests, motivations, influences, and learning over time.

That 'shaping' of the 'mental landscape' is dramatically fast during the development of an infant, for so much learning takes place over such a short time. In adulthood, the rate of 'shaping' slows down, but it never stops. The brain does not reach a 'final static state' in which all the neural circuits are 'done'. Rather, throughout life, the brain maintains its ability to form new circuits and re-form older ones, as referred to by the term 'brain plasticity'; neural circuits disintegrate too, as associated with forgetting, especially if the circuit is only rarely invoked (as is the case for many of the connections formed by infants, connections that were 'experimental', and are subsequently found to have no use), or damaged by, for example, 'amyloid proteins', as associated with Alzheimer's Disease.

8.3 The learning trap

Learning is, of course, immensely valuable, for it enables us to deal with a multitude of situations easily and quickly. To take two simple examples: if I hadn't learnt how to get myself dressed, I'd still be in my bathroom, wondering what to do next (think, for example, how long it takes to manoeuvre an infant into a 'Babygro', especially before the infant has at least learnt to co-operate!); if I hadn't learnt how to cross a road safely, I'd spend all day standing bewildered and frightened on the kerb. But because I've learnt how to do these tasks, I don't have to solve the problem afresh each time I encounter it; I can use my learning to invoke 'the right answer' at once – I can get dressed in minutes and look presentable as a result; I can cross the road in seconds and safely.

As adults, the neurons within our brains are immensely interconnected, for everything we know, everything we've ever learnt, has its corresponding neural circuit. Using the landscape metaphor, our brains have many, deep 'valleys', valleys of learning that, like a falling raindrop, we are 'swept down' whenever we encounter the corresponding situation. One valley is labelled 'getting dressed', another 'crossing the road', another 'how to chair a meeting', another 'how to do research'...

Which is all fine whenever doing things in the same way as they have always been done before works. If I want to run today's meeting as I have done many times in the past, then flowing down the 'how to run meetings' valley will deliver the outcome I seek. But at the last meeting, I noticed that something didn't go so well – perhaps because there was someone new around the table. That person will be present at this afternoon's meeting too, and might cause difficulties. If I just do what I always do, then that difficulty might be aggravated. I want to avoid that, so I need to run the meeting differently. But how? I'm so used to running meetings in 'the usual way', I just don't know how to do things differently...

That's just one example of what I call 'the learning trap'. Yes, as I have just stated, learning is indeed immensely valuable – but only for addressing familiar situations in the same way, and with the objective of achieving a familiar outcome. If the situation is unfamiliar; if the situation is familiar but needs to be approached in an unfamiliar way; if the 'same old' outcome is no longer fit-for-purpose and a different outcome is required, then much, if not all, of my learning might be of no use at all. Or perhaps worse. My previous learning might be a *barrier* to the discovery of the new, if I can't 'let go'...

To use the landscape metaphor, if I want to do something – say, chair a meeting – differently, then I need to take my 'raindrop' from deep within the 'how to chair a meeting' valley, lift it up the slope to a neighbouring ridge, look around for a different, more appropriate, 'valley', and flow down that one. In the language of Hebb, I need to 'unwire' the neural circuits normally associated with 'how to chair a meeting', and reconnect them, perhaps to different neurons, to form a different circuit.

But those mental valleys are deep, and it is hard to lift a raindrop out; the synapses are strong, and 'disconnecting' them is difficult. And even if I can do that, the state in which the 'raindrop' is on a ridge looking for another, better, 'valley', in which the synapses have been broken and the neurons are 'waving about in the air', is unstable, uncomfortable. How do I know that there is a 'better valley' somewhere else? That there is a more suitable neural circuit? I don't. So that's risky. Maybe it's better to leave the raindrop where it is, to leave that already-working circuit intact. Even if the outcome of the meeting might not be as good as I might wish...

8.4 Unlearning

The depth of those mental valleys, the difficulty of 'lifting the raindrop', and the precarious nature of the intermediate state 'wobbling on a ridge', collectively conspire to entice you to stay at the 'bottom of the valley', doing things in the familiar way. But if you wish to do something differently, if you want to generate a new idea, you must 'climb the valley'. You must escape from the learning trap. You must be willing to 'unlearn'.

And unlearning can be difficult; very difficult; especially if you are senior; especially if your very identity is bound up with the *status quo.*

So, at this point, let me return to first principles. Fundamentally, Koestler's Law tells us that, after-the-event, we will recognise creativity by identifying a new pattern of pre-existing components. Before-the-event – and assuming that the da Vinci problem is not, in the current context, stopping things – it therefore follows that the pre-existing components we need are somewhere, now. It may be that they are 'laid bare', directly available for immediate use, like the notes on a piano. But they may not. Rather, the components we need are there, but 'bundled' together in existing 'patterns', patterns of our knowledge, experience and learning. In order to access those components, so enabling them to be combined with others to form the idea we are seeking, we need to examine or knowledge, scrutinise our experience, deconstruct our learning. Overall, we must unlearn.

To illustrate the difficulty that we all have in unlearning, in my workshops, I carry out an exercise based on an example given in one of Edward de Bono's books, *The Mechanism of Mind* [12]. The exercise involves two participants, sequentially, such that the second participant does not have sight of the first participant's activity.

A participant in the group is invited forward, and a small square (say, 1 unit × 1 unit) is placed on a table. The participant is then given a rectangle, 3 units × 1 unit, and asked to place the rectangle wherever they choose, such that the combination of the two pieces forms the simplest possible overall shape. There is no prompt as to the criteria by which 'simplest possible overall shape' might be determined; the participant may use his or her own judgement. Most participants place the 3 × 1 rectangle against one side of the 1 × 1 square, so forming a longer, 4 × 1 rectangle. The participant is then given a third piece, a 2 × 1 rectangle, and asked to do the same thing: to place it so that the now three pieces collectively form the simplest possible overall shape. The most frequent result is an even longer, 6 × 1, rectangle.

The participant is now given a fourth piece, a 2 × 2 square from which a 1 × 1 corner has been removed, resulting in an 'L'. Once again, the participant is invited to form the simplest overall shape. There is usually some hesitation, since the 'L' doesn't obviously 'fit', but what frequently happens is the formation of the overall shape shown in figure 8.1:

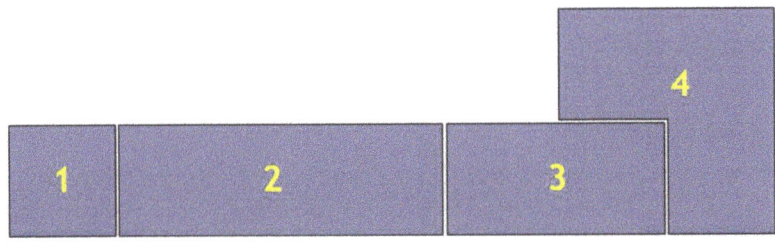

Figure 8.1. The result of positioning four pieces, offered separately in the sequence shown.

The participant is thanked, and the four pieces cleared away.

A second participant, who did not witness what happened with the first participant, is then invited forward. As before, the second participant is asked to assemble four pieces, one-at-a-time, so that, at each stage, the overall result is the simplest possible shape. The second participant, however, is given the pieces in a different sequence: the 'L' first, then the 1 × 1 square, followed by the 2 × 1 rectangle, and finally the 3 × 1 rectangle. The result is almost always as shown in figure 8.2, a 3 × 3 square.

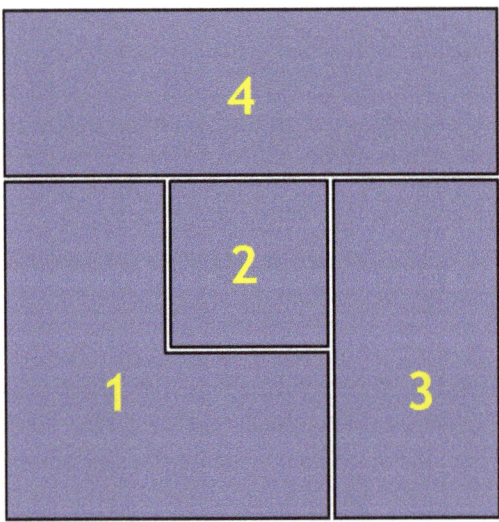

Figure 8.2. The result of positioning four pieces, offered separately in the sequence shown - a sequence different from that shown in figure 8.1.

As can be seen, the outcomes are different. But the four pieces are the same. The only difference was the sequence in which the pieces were presented to each participant.

Even such a simple exercise gives a powerful insight into learning, and the formation of neural networks. Every individual experiences the world as a sequence of events, from the moment of birth, if not before. As those experiences take place, we remember, we learn, we form and re-form our neural networks. And the outcome at any time, representing that accumulation, is whatever it is. But one thing is sure. It is different for every human being. No two individuals have exactly the same experiences in exactly the same sequence in exactly the same context. Not even genetically identical twins. The 'wiring' of each human brain is unique, and different from the wiring of all other brains.

An important question is why the first participant does not end up with a square. In fact, that does happen on rare occasions, and I'll explain why shortly: for the moment, I'll explore the majority of cases in which the first participant ends up with a four-piece pattern as shown in figure 8.1, and not the square as shown in figure 8.2. The instructions as each successive piece is given to each participant were always the same: 'combine the pieces to form the simplest overall shape'. Although there is no definition of 'simple', and each participant can form their own subjective judgement, few would disagree that, for the four pieces overall, the 'simplest possible overall shape' is the square, as shown in figure 8.2.

So why does the first participant usually fail to form it?

To explain this, I need to go back to the situation in which the first participant is placed after positioning the third piece, as depicted in figure 8.3:

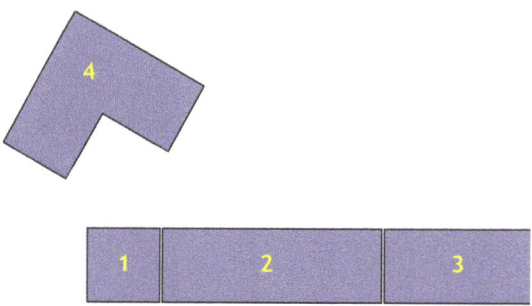

Figure 8.3. The pattern formed by participant 1 after having been given three pieces; the fourth piece is waiting to be positioned.

Having positioned the 1 × 1 square, the 1 × 3 rectangle and the 1 × 2 rectangle alongside one another to form a 1 × 6 rectangle, participant 1 has indeed formed, to that point, 'the simplest possible shape' – or perhaps more correctly, one of two 'simplest possible shapes', the other being two 'layers', the lower layer of the single

1 × 3 rectangle, with the 1 × 1 square and 1 × 2 rectangle side-by-side on top, so forming an overall 2 × 3 rectangle.

But on being presented with the fourth piece, the 'L', the participant has a problem. Given either the 1 × 6 rectangle as shown in figure 8.3, or the alternative 2 × 3 rectangle, the 'L' doesn't readily fit. Usually, the participant moves the 'L' to a number of different positions to see if anything works, and then places it at one end or the other – perhaps hoping that the next piece is a 1 × 5 rectangle to complete a more symmetrical shape, a 2 × 7 rectangle. There is, however, no further piece – but the participant does not know that.

Yet, given the 1 × 6 rectangle shown in figure 8.3, it is very easy to combine this with the fourth piece to form the 3 × 3 square. *But to do that, you have to tear the* 1 × 6 *rectangle apart first.*

Or, before you can be creative, before you can form a new pattern from pre-existing components, you have to deconstruct the existing patterns to release the underlying components, so allowing them to be re-combined to form the new pattern of creativity. A pattern which, in this particular case, has the emergent property of 'simplicity'. Before you can be creative, you must deconstruct, de-compose, everything you know. You must unlearn.

8.5 Why is unlearning so difficult?

Yes, to be creative, you must unlearn. But, for many of us, especially if we are senior and experienced, unlearning can be very difficult.

Why so?

Well, by reference directly to the four-piece puzzle, and indirectly to much wider experience, here are four reasons, each associated with a fundamental human behaviour or emotion.

8.5.1 Love

The first reason concerns love. If you are 'in love with your ideas', then disassembling them, deconstructing them, can be so painful that it's almost impossible to do. That 'three-piece linear combination' is 'mine'. I crafted it. It's my intellectual property. I made my reputation on it! It took me years to perfect. There's absolutely nothing wrong with it. It works. I just can't bear to tear it apart…

8.5.2 Laziness

The second reason is laziness. Deconstructing that pattern, fiddling around with that extra piece, seeing if there is something better, all take time and effort. I just can't be bothered…

8.5.3 Fear

The third reason is fear. That 'three-piece linear combination' is good, and complies with 'the simplest possible shape'. It's a pity that the 'L' doesn't obviously 'fit', but maybe that's only temporary, and that there is another, fifth piece that will make everything work better. My fear is that in deconstructing the three-piece pattern, I will be transforming something that's quite good into a fragmented mess. And even with the 'L', I can't envisage what a better pattern might look like with all four pieces – if indeed there is a better pattern there at all. So I might waste a lot of time, trying all sorts of combinations that don't work, and are even worse than just putting the 'L' on the end, and hoping for the best. Worse still – suppose the boss comes in whilst I'm randomly pfaffing around with all those separate pieces, and asks, 'What are you doing?'. What do I reply? 'Pfaffing around looking for a better pattern'. And then the boss will say, 'You're doing WHAT?!?' Oh dear. Can't risk that. Better to play safe, and put the 'L' on the end.

8.5.4 Arrogance

And the fourth is arrogance, often associated with power and authority. Suppose, for example, that 'the boss' is preparing a presentation from work done by his team, a presentation represented by the 'three piece linear combination'. The 'boss' hears a knock on the door, and a junior member of the team comes in, and says, 'Hi. I've just been doing some more work, and have just come across some new information', and holds up the 'L'.

'How can you be so stupid?!?' says the boss, angrily. 'You know we're looking for 'long, thin, bits'. How dare you come to me with a wrongly-shaped piece? Go away, and find some right bits!'. Meaning, of course, 'pieces that already comply with my mental model, which, by definition, is right'.

It might be, however, that the boss had attended a 'how to treat your junior staff with dignity and respect' course the previous week. In which case, the boss might then say, albeit through tightly-clenched teeth, 'Thank you!'. The junior then leaves the office, at which point the boss throws the 'L' into the litter bin, and continues to work on the three-piece presentation...

8.5.5 Unlearning is indeed difficult...

Yes, for those understandable reasons of love, laziness, fear and arrogance, unlearning is difficult. And that last one can be especially pernicious as it is most usually associated with the powerful. This has been recognised and acknowledged for some time: for example, in 1959, Arthur Koestler wrote in *The Sleepwalkers*,

> *The inertia of the human mind and its resistance to innovation are most clearly demonstrated not, as one might expect, by the ignorant mass – which is easily swayed once its imagination is caught – but by professionals with a vested interest in tradition, and in the monopoly of learning.* [13]

And nearly 450 years earlier, in Book VI of Machiavelli's *The Prince*, which dates from 1513, we read

The innovator makes enemies of all those who prospered under the old order... and whenever those who oppose the changes can do so, they attack vigorously... [14]

Indeed so. Unlearning is difficult.

But it is essential for creativity.

8.6 Hegel, and genetics

The principle underpinning unlearning is that before you can create the new, you must deconstruct the old. This too is the principle underpinning the 'thesis, antithesis, synthesis' philosophical method often associated with Georg Hegel: an idea is proposed (the 'thesis'), then negated, contradicted or otherwise deconstructed ('antithesis'), from which a better idea emerges ('synthesis'). Despite the attribution of these words to Hegel, he never actually used them – his terms were 'abstract/subjective/being', 'negative/objective/nothing', and 'concrete/absolute/becoming' [15], but they don't quite have the same 'ring'. The first use of 'thesis, antithesis, synthesis' is to be found in the work of Johann Fichte, who had preceded Hegel as Professor of Philosophy at the University Berlin [16].

And a totally different field in which an act of creation is preceded by an act of destruction is genetics.

In sexual reproduction – which is widespread across the living world – the child inherits genetic information from each of two parents. The physical manifestation of this is the set of 'chromosomes' in each individual, these being biological structures, primarily of DNA, in which genes are embedded. Chromosomes are associated in pairs, with one chromosome in each pair coming from each parent: humans, for example, have 23 pairs of chromosomes; potatoes, 24; camels, 35; red king crabs, 104 [17].

Very briefly, during 'meiosis' [18], the process by which a male forms a sperm cell, and a female, an egg, each of the chromosomes in a pair firstly replicates, with the two identical versions remaining attached together at the 'centromere'. One maternal and one paternal chromosome then 'cross over', and subsequently physically break, forming two different, composite, chromosomes, as represented schematically in figure 8.4.

The two pairs of chromosomes then separate, resulting in four chromosomes, two of which are identical to a chromosome of the individual's mother or father, the other two, a combination. Each of these four chromosomes can then become one of the chromosomes within a sperm cell or an egg, such that each sperm or egg cell contains the appropriate number of single chromosomes, this being 23 for a human, 104 for a red king crab. Taking the human as an example, for each of the 23 chromosomes, there is a probability of 1 in 4 that the single chromosome in any sperm or egg cell is from a single parent, and so the probability that all 23 chromosomes are from that same parent is $(0.25)^{23} \sim 1.4 \times 10^{-14}$. That is a very small number indeed! So the genes to be passed to the next generation will be a mixture from the previous generation, and the generation before that...

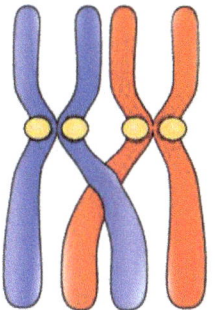
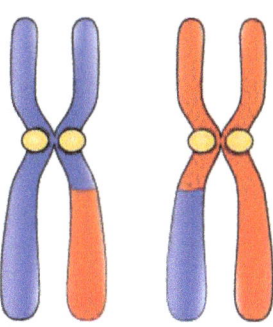

Figure 8.4. Schematic representation of an intermediate state in meiosis, in which a replicated maternal chromosome (blue) 'crosses over' a replicated paternal chromosome (red), resulting in recombination. Source: Wikimedia 'Figure 17 02 01' https://commons.wikimedia.org/wiki/File:Figure_17_02_01.jpg.

If cross-over and recombination did not happen, the same pair of chromosomes would be transmitted generation after generation, and the result would be generations of clones.

But in the presence of cross-over and recombination, each generation is different, a true creation. But for that to happen, the previous generation's chromosomes had to be broken apart, and then recombined.

In the language of this chapter, each chromosome is a pre-existing Koestler's Law pattern of genes, the overall result being a given individual. To create a new individual, the 'genetic patterns' of both the mother and the father have to be 'deconstructed' to 'reveal the components', which are then 're-assembled' to form the 'creation' of the next generation.

To create, you must first deconstruct, to destroy. And in an intellectual context, that means that to create the new, you must first unlearn the old.

8.7 A brief pause...

Let's pause for a moment to draw together the four key threads I have developed so far.

The fundamental 'first principle' is Koestler's Law, the profound insight that all creativity is the formation of new patterns from pre-existing components.

For this to work in practice, the appropriate pre-existing components must indeed pre-exist. If they don't, that's an instance of the da Vinci problem, a possibility that we must always bear in mind.

The next important concept is emergence: the existence of properties at the level of the pattern, and not at the level of any of the components from which that pattern is formed. The nature of the emergence depends on the context: emergence in art is the necessarily subjective assessment of 'beauty'; emergence in science is a rather more rigorous test, such as the explanation of previously fragmented observations or the reliability of predictions.

Koestler's Law does not distinguish between [this] pattern and [that] one: as long as both patterns have never been formed before, in their appropriate contexts, then both, according to Koestler, are 'acts of creation'. But we don't want to generate just 'any old idea', we want to generate ideas that have the promise of being useful, of 'making the world a better place'. So our objective is to generate ideas associated with patterns that have a good likelihood of having beneficial emergent properties. But we need to remember that emergence is a property of the pattern-as-a-whole, so the emergence associated with any idea can be assessed only after the idea has been generated, and – recognising that all ideas are very fragile and incomplete when first suggested – given some time to 'mature'.

The fourth thread is unlearning. The components that we are likely to need to recombine into the new pattern of creativity are not 'laid bare' as are the notes on a piano. They are 'bundled together' in existing patterns of knowledge, experience and learning, patterns that must be deconstructed to reveal the underlying components, so enabling them to be re-combined into the new patterns of creativity. To form the new, we must unlearn the old. And the knowledge, experience and learning that needs to be deconstructed and then recombined is not just 'my' knowledge, experience and learning, but 'ours' too – and drawing in the knowledge, experience and learning of many others as well, as documented, for example, in academic papers.

So – at last! – the stage is set to present my process for the deliberate discovery of ideas. I call it ***InnovAction!***

Donald Hebb

Photograph of Donald Hebb reproduced courtesy of Richard Brown, Dalhousie University.
www.dal.ca/faculty/science/psychology_neuroscience/news&events/celebrating-70-years/HebbLecture.html

Hebb's life-long interest in learning and education probably started very early: until aged 8, his mother home-schooled him, using Montessori methods. Following a degree in English, in 1925, he became a teacher in his home town of Chester, Nova Scotia. Between 1928 and 1932, he studied for a Master's degree at Montreal's McGill University whilst also being a full-time headmaster at a nearby school. Frustrated by the limitations of the school curriculum [19], Hebb left school-teaching for full-time study, firstly at the University of Chicago, and then at Harvard, taking his PhD in 1936.

Returning to Montreal, Hebb then researched the effect of brain surgery on brain function, causing him to note, in particular, that the brain's frontal lobes were particularly important in the ability of a young child to learn – an important clue in linking the development of brain connectivity to learning. To study learning further, in 1939, he and a colleague, Kenneth Williams, designed the variable-path Hebb-Williams maze [20], used extensively thereafter by many researchers around the world to examine the ability of small animals to solve problems.

Between 1942 and 1947, Hebb studied primates at a research centre in Florida. Despite finding chimpanzees hard to teach (!), he had the time to think, and to draft his major work, *The Organisation of Behaviour: A Neuropsychological Theory*, which was published in 1949, after his appointment in 1947 as Professor of Psychology at McGill, where he remained until his retirement in 1972.

Hebb's theories and insights have been profoundly influential, and the key message of 'Hebbian learning' is often paraphrased in what is now known as 'Hebb's Law': 'Neurons that fire together wire together'. He also recognised the importance to learning of a rich and stimulating environment, so laying the foundations of the 'Head Start' programmes [21] that various governments around the world sponsor for early years education and development [22].

References

[1] www.ant-tnsjournal.com/Mag_Files/15-2/15-2_p127.pdf
[2] Hebb D O 1949 *The Organisation of Behaviour: A Neuropsychological Theory* (New York: Wiley)
[3] www.firstthingsfirst.org/early-childhood-matters/brain-development/
[4] https://extension.umaine.edu/publications/4356e/
[5] https://knowablemagazine.org/article/health-disease/2020/what-does-a-synapse-do
[6] www.kcl.ac.uk/research/bibs-brain-imaging-in-babies
[7] https://theconversation.com/what-is-brain-plasticity-and-why-is-it-so-important-55967
[8] www.historyofinformation.com/detail.php?id=3902
[9] The quotation I've attributed to Dame Darcey Bussell is my memory of what she said on a television programme – but I've been unable to verify my memory by reference to an available source. Much the same sentiment, however, will be found on the section Ballet Evolution on this web page coloursofdance.com/articles-on-dance-learning, and also on page 60 of Dodge N 2008 *The Brain that Changes Itself* (London: Penguin)
[10] Gladwell M 2008 *Outliers* (New York: Little Brown)
[11] www.fens.org/news-activities/news/the-brain-prize-winners-2016-announced
[12] Bono E de 1969 *The Mechanism of Mind* (London: Jonathan Cape) ch 27
[13] Koestler A 1931 *The Sleepwalkers* (London: Hutchinson) p 427
[14] Machiavelli N 2003 *The Prince* (London: Penguin) (Engl. transl.) George Bull p 21
[15] https://egyankosh.ac.in/bitstream/123456789/38410/1/Unit-3.pdf
[16] http://scihi.org/johann-gottlieb-fichte-german-idealism/
[17] www.lcps.org/cms/lib/VA01000195/Centricity/Domain/4726/2N-chart.pdf
[18] https://openoregon.pressbooks.pub/mhccmajorsbio/chapter/meiosis-i/
[19] https://can-acn.org/donald-olding-hebb/
[20] Brown R E 2020 Donald O Hebb and the organization of behaviour: 17 years in the writing *Mol. Brain* **13** 55
[21] https://medium.com/dish/75-years-of-innovation-head-start-program-f6a83d3993ad
[22] Britto P R and Gilliam W S 2008 Crossing borders with head start *Infants Young Child.* **21** 82–91

IOP Publishing

Creativity for Scientists and Engineers
A practical guide
Dennis Sherwood

Chapter 9

How to have great ideas 'on demand'

9.1 *InnovAction!*

If you want to have great ideas, this box defines how to do it. Its six simple steps encompass – implicitly, rather than explicitly – Koestler's Law, unlearning, and emergence, as well as fundamental concepts such as the importance of observation, curiosity, permission, and the value of teamwork. I call the process *InnovAction!*

InnovAction!

1. Define the 'focus of attention'.

2. Individually and in silence, write down everything you know about the agreed focus of attention.

3. Share.

4. Then choose one feature, and ask 'How might this be different?'

5. Let it be …

6. …and then, when that discussion runs out of steam, choose another feature and repeat Steps 4 and 5.

As I'll describe in chapter 11, it's not the only process. But it's a very practical one, and it's one I have used, with success, on countless occasions.

It is possible for an individual to be creative alone, and the ***InnovAction!*** process works for a solo individual, save for step 3, Share, for there is no one else to share with.

I strongly believe, however, that creativity is far more effective in small groups. So almost all my creativity experience has been gained in workshops, usually of between 12 and 24 people, where most of the activity takes place in groups of about 6, with each group working independently. I'll talk more about the organisation of these workshops, the composition of the teams, and the way they are conducted, in the next chapter; in this chapter, I explore each of the steps.

9.2 Step 1: Define the 'focus of attention'

The 'focus of attention' is a statement, or a question, defining the context in which a new idea might be useful. It might be quite specific ('How can I improve this poster?') or wide-ranging ('How can we make our next conference a much richer experience for everyone?'), and it can relate to any of the six segments of the Target Diagram shown in Figure 3.1, for example, 'How can we manufacture a vaccine in much less time, but without compromising on quality and safety?' (that's primarily about 'process', and was asked, and answered, to all our benefit in relation to Covid-19), 'How might we improve the way collaborations work?' (primarily that's about 'relationships', and also 'You!' too), and 'Suppose [this material], which is central to much of our work, is no longer available?' (primarily 'content', but perhaps 'process' and 'strategy' too).

Very importantly, the 'focus of attention' *does not* have to be a 'problem that needs to be fixed'. If there is a problem – for example, if a particular product is attracting complaints – then, certainly, the ***InnovAction!*** process will prove very useful in discovering solutions. But even if that particular product is getting top reviews, ***InnovAction!*** could still be very valuable in making it even better – not only identifying possible improvements, but also helping keep ahead of the competition. Furthermore, the focus of attention 'How might we improve our conference?' is not implying that 'right now, our conference is a disaster'. Rather, it accepts that the conference is as it is, and instead of attributing blame and prompting questions such as 'why is our conference not as good as it might be?' or 'why did we take that bad decision last year?', it asks 'what can we do differently that would make our next conference even more successful?'. That is positive and constructive, looking forwards not backwards, about 'building a better world'.

At the beginning of my workshops, people often ask 'What is the 'problem statement'?', or – if the community manufactures a product – 'What is the problem we're trying to fix?' and 'What are customers complaining about?', often associated with 'What do the customers want? If only they'd tell us, then we'll do it!'

Certainly, if a customer is complaining, then it makes sense to take a look and fix things, and so the customer is defining the 'focus of attention'. And the question 'What do the customers want?' presupposes that the customers actually know what they want, what is available, and what might be possible. Sometimes they do, but sometimes they don't: as Henry Ford is alleged to have quipped, 'If I'd have asked people what they wanted, they would have said faster horses' [1].

So when I open the workshop by saying, 'There are no complaints, and the customer hasn't asked for anything different, so we're just going to explore how [this product] might be different, and hopefully better...', I often then face a barrage of questions like, 'That's going to be a total waste of time – if it works, why bother?' and 'How do we know that whatever we come up with will be accepted by the customer?'.

To which the answers are 'We're bothering because we might discover something we've never noticed before, and that might be very valuable. And even if we don't, we've verified that what we are doing now is a good as it can possibly be, which is also useful information', and 'We don't know what this workshop will come up with, and we don't know the value that whatever those things might be will have to an actual or potential customer. But if we discover something, we will then be in a strong position to determine that at the right time. And far better for us to discover such things, rather than having to react once our competitors have done it.' Indeed, the complacency resulting from an attitude of 'if it ain't broke, don't fix it' can make an organisation vulnerable to being swept aside by what the business school community refers to as 'disruptive innovation' [2]. So a wise organisation doesn't wait until there is a 'burning platform': even when things are going really well, they continue to search for ideas to make things even better.

At the start of the workshop, it's therefore important to give a clear statement, defining the focus of attention, and checking with everyone that they both understand it and are happy with it. That ensures everyone is 'looking in the same direction'.

Also, the focus of attention must be grounded in the reality of 'now', anchored in something that we can see, touch, feel, and experience, so that everyone can draw on their existing knowledge. For the search is not for 'something out-of-the-blue'. It's a search for something that is *different* – different from what is happening *now* – and hopefully better.

9.3 Step 2: Individually and in silence, write down everything you know about the agreed focus of attention

At most meetings, there is a lot of talking, with some people talking more loudly than others. 'Brainstorming' workshops, conventionally, are no exception, with the facilitator, after introducing the session, getting things going by saying, 'OK. Let's start... Alex, what ideas have you got?'

So what happens at my workshops, after the introduction, is found by many to be quite startling, positively unusual, and perhaps distinctly odd.

My workshops start in utter silence; silence that rarely lasts less than about 30 minutes, and that can last as long as an hour, or even more, depending on the focus of attention.

And during that silence, the community, working as individuals, are writing notes on everything they know about the focus of attention, in as great detail as they can. The notes can take different forms: perhaps a narrative or a list, perhaps a mind-map, perhaps a diagram or rich picture. Whatever works for the individual.

The purpose of this activity is two-fold: to encourage observation, and to begin the process of unlearning.

With reference to the 'landscape' metaphor of learning as discussed in sections 8.2 and 8.3, the knowledge we have accumulated over each of our lives has resulted in each of our brains being a very richly 'carved landscape of mountains and valleys'. Everything we know corresponds to a 'valley' with the appropriate 'label', and so, in this part of the process, each participant is invited to go deep into his or her 'valley' corresponding to the appropriate 'focus of attention', to settle down in the bottom of that 'valley', and then to take a good 'look around': what do you 'see'?

And the act of writing everything down, in whatever format is appropriate – a list, a drawing, a process chart – encourages the participant to begin 'deconstructing' a complex concept into its underlying component parts, so initiating the process of unlearning.

Suppose, for example, an academic community gets together to explore how to improve team working. Everyone has worked with other people in a variety of contexts, such as on a project team, on a social activity, or on the sports field, so everyone can draw on their own personal experience. The starting point is for everyone to write down their own observations, as exemplified by the items shown in this box:

Some representative descriptions of team working, as compiled by four different people

Participant 1

- Some teams work really well; others much less so.
- The good ones have a continuous sense of energy…
- …with everyone highly motivated…
- …producing good results…
- …with much personal satisfaction.
- Having clear roles and responsibilities helps.
- And the best teams I have worked on communicated really effectively…
- …both formally at meetings and seminars…
- …and informally too…
- …when people just happened to be together.
- …

Participant 2

- Being on a team is much better than working by myself…
- …because I really enjoy learning from others…
- …and I like the opportunities for social stuff too.
- On one team I worked with recently, I was junior…
- …and felt that I was always being judged…
- …especially at team meetings…
- …so my 'safety play' was to keep as quiet as I could…
- …so at least I didn't say anything wrong…
- …but maybe my silence was judged – adversely – too…
- …

Participant 3

- Usually, everything starts optimistically…
- …but then things can fall apart…
- …so one of the big challenges is how to keep the momentum going.
- As well as being an academic, I'm a musician too…
- …and play the trumpet in an orchestra…
- …which is also a 'team'.
- In a good orchestra, everyone does what they're supposed to do…
- …at exactly the right time…
- …in the right way…
- …because we are all listening very attentively to each other…

> **Participant 4**
> - The most recent team I was on I led...
> - ...which was OK when things went well...
> - ...but one of the problems I had was with a particular individual who failed to deliver the results that had been promised...
> - ...and was also either late, or absent, from some important meetings.
> - I am by nature quiet...
> - ...and I don't like confrontation...
> - ...especially with people that – at least from my standpoint – are belligerent...
> - ...so I avoided the issue...
> - ...and did nothing...
> - ...

Those are just very brief extracts: for any real topic, the lists will be very much longer. And in compiling these descriptions, it's important for those present:

- To note everything – no detail is too trivial.
- To draw on their own, personal, experience, describing the 'world' as they see it themselves: there are no 'right answers', and the objective is *not* 'to give the answer that the boss wants'.
- To describe only features of 'today's world' that are there, now, or that have been present in the recent past: they might be verifiable facts, they might be opinions, they might be feelings. Importantly, this excludes any thoughts about the future, about what would be better, about ideas. Yes, those are important, but we'll get there later: this step is grounded in the reality of 'now'.
- To keep only to descriptions and to avoid any interpretation or explanation of why 'now' is as it is. And – very importantly – to avoid blame. So a statement such as 'some teams work better than others' is an observation, a description: it happens, and can be seen to happen – in contrast to '[Sam] is not a team player...', which, if said by someone other than [Sam], might not be the whole story, and might be throwing a brick.
- And, very importantly, to be as honest as possible.

In any group, some will be more knowledgeable or have more experience than others, and of course some people write more quickly than others. As I've already mentioned, it usually takes about half-an-hour to compile an insightful and complete description of any meaningful focus of attention, sometimes longer for more complex topics. To set expectations, I usually suggest that participants should spend about 30 minutes thinking and writing – certainly, those who are 'finished' within, say, 10 minutes are likely to have only just skimmed the surface.

That said, some people will finish sooner than others. That person should then be patient, and allow others to continue. It is discourteous to get up and leave, to rustle papers, to play with your mobile, or otherwise indicate to everyone else how irritated you are becoming.

Likewise, if you are still writing when you sense that most others have stopped, write rather more quickly, or signal that you are ready to have a group discussion even if you haven't quite finished – you can always write down some more points later.

9.4 Step 3: Share

9.4.1 What the 'share' is about

Step 2 has provided an opportunity for each participant to reflect upon their own personal experience of the focus of attention; Step 3, 'Share', draws these individual perspectives together to compile a collective description, which is inevitably richer, more multi-faceted, more complete, than any one individual's view.

When the focus of attention is something 'impersonal' like 'roof tile' (an important focus for a manufacturer of roofing products), you might think that everybody would write down the same list. Certainly, when the topic is more subjective, sensitive or emotional – such as 'meetings', or 'teamwork' – then, yes, different people will have different experiences and perceptions, and so we would expect different people would write different things. But 'roof tile'? Surely everyone will have the same list...

But no. I've run creativity events with people that make, or use, roof tiles, and even for something as apparently 'dull' as a roof tile (although, if you manufacture them, or climb a ladder to position them, they are not 'dull' at all!), everyone writes down something different. Of course, there is much in common (such as 'they are used to make the roof on a structure'), but there is also much that is different too.

Once again referring to the 'map' metaphor, every individual has a different, unique, mental 'landscape', and so when anyone is invited to 'stand in a particular valley', as 'labelled' by the appropriate focus of attention, each individual will 'see' different things. So when a group shares their individual observations, the resulting aggregate description is far richer than any single individual's.

9.4.2 Managing the group dynamics

The group dynamics of this sharing process are important.

Quite often, when a group convenes to discuss a given topic, a number of behaviours might emerge, for example:

- Dominant, energetic people might be dominant and energetic, drowning everyone else out.
- Two people engage in a detailed discussion about a specific point, so taking the group down an endless 'rabbit-hole'. Or, in a more virulent form...

- ... two 'alpha males' get into a dispute, if not fight, whilst the rest of the group watch, firstly in amusement, subsequently in disbelief, finally in horror.
- One person might read through their entire list, or draw their complete diagram, whilst the rest of the group get bored, read text messages, send emails...
- ...and when that person (at last!) finishes, he or she sits down and pays no attention to anything anyone else says.
- The chair, wishing to maintain the impartiality of the role, feels reluctant to express his or her own views.
- Younger people might feel scared to contribute, and just say 'yes, I agree'.
- Quieter people just keep quiet, or check out. Which such people are used to doing, so they don't mind; and which others acquiesce to as they know, from previous meetings, that these people usually just keep quiet.
- As a sequence of points are made, and perhaps written on a flip-chart, someone seeks to structure or group the points in some way, which in principle is helpful, but can often, in practice, be a distraction whilst the individual points are being collected.

We all recognise these.

But they are all totally counter-productive to achieving the objective of aggregating our individual 'mental landscapes' to obtain the most complete set of observations of the focus of attention that we can.

With that objective in mind, here are some thoughts as to how the 'share' can be organised so that everyone contributes, in as lively a way as possible, whilst everyone else pays attention and listens. The setting I have in mind is a group of, say, 6 to 8 people, all of whom have, individually and in silence, been writing notes on the appropriate focus of attention.

9.4.3 Self-facilitate

Don't expect the session facilitator to say 'pens down, you may now talk'. It's not a school exam. So, within the group, be alert to one another, notice who still has their head down writing, who is looking around. Make eye contact with one another, and, when everyone is naturally ready, someone can say something like 'It looks like everyone has finished – is it OK to start the share?'.

This process does not require a 'chair', so there are no 'special' roles – everyone is 'in the same place'.

When people are ready, the share can then start. The method I advocate is for the group to stand, and gather in a loose circle around a flip-chart, with everyone able to see the flip-chart clearly, and to make eye contact with one another.

Standing injects energy into the group, and moving around signals a change in pace.

One person should take the pen, and then one person makes a single point, such as 'Team meetings are of different types'. The essence of what was said is written on the flip-chart. A second person then makes a different point, say, 'Progress meetings are scheduled to start at 4 pm every Thursday', which is also captured. And then a third person…

9.4.4 One point at a time

Each person makes a single, brief, point at any one time, which is captured on the flip-chart. Once this is done, the next person contributes, and the conversation then goes around and around the group, until all of each individual's observations have been written down. This process is dynamic, with everyone involved in succession, so avoiding any one person dominating the conversation; each point is brief, so no one individual is talking for any length of time; it also provides a 'level playing field' enabling everyone, even those who are usually quiet to contribute; furthermore, since everyone has just spent perhaps half-an-hour thinking about the focus of attention, and writing things down, everyone has something relevant to say.

9.4.5 Writing on the flip-chart

The person who is writing on the flip-chart is part of this process, and so when it is that person's 'turn', he or she contributes alongside everyone else.

The person doing the writing has a duty to write down the essence of what is said. Not what the writer thinks was said. Not what the writer wanted to hear. The writer will rarely write down all the exact words used by any speaker – some paraphrasing, or simplification, will inevitably happen. So the writer should, as appropriate, check with the speaker by saying something like 'I had to shorten that a bit – is what I've written down a fair representation of what you wanted to say?'; likewise, the speaker can always say 'I'm not sure what you've written is quite what I had in mind - how about [whatever]?'

Adopt the 'rule' that when a flip-chart is filled, a new person takes the role of writer. And attach completed flip-charts to a nearby surface, such as a wall, a door, a window, a white-board, so that all the completed flip-charts remain in view, and so can be read, rather than being obscured.

9.4.6 Be succinct

Whoever is speaking should keep points brief. Any one person's 'turn' will come again in a few minutes' time, so more can be said then.

A consequence of this is that any one person's points are separated, and embedded alongside others. That's fine.

9.4.7 Don't worry about structure

Another consequence is that the sequence of points, as recorded on any one flip-chart page, is unstructured, disjointed, and doesn't tell 'a coherent story'. That's fine too, for everything can be organised and structured later, when all the information has been captured. As the aggregate list is being compiled, however, it won't be structured, and this might cause someone to day 'shouldn't we get a new flip-chart going where can put similar things together?'.

This person does have a valid point: [this] probably does go with [that], and yes [these things] do indeed go together. My experience, however, is that this re-structuring and re-organisation, though useful, not only takes time but more importantly diverts attention towards structural perfection and away from the primary objective of the 'share' – to provide an opportunity for everyone to contribute and to listen, attentively, to one another. So although it can seem somewhat anarchic, my preference is to live with the absence of structure for the moment – grouping things together, and linking one concept with another, are good and valuable things to do, but later; not around the flip-chart, now.

9.4.8 Don't argue…

Don't argue. When asked to describe the same thing – as everyone was invited to do whilst working individually and in silence – different people will notice different things. So during the share, those differences will become apparent – indeed, one of the main objectives of the activity is to gain as rich, and hence varied, a description as possible.

This can, however, cause some tension, as illustrated by an event that took place during one of my workshops when the focus of attention was 'the game of chess'. During the share, one participant stated 'the pieces are either black or white', which was duly written on the flip-chart. At which point, another participant, in a loud voice, said, 'No they are not. They are red and white'. These two participants happened both to be extreme 'alpha males', who – to the increasing incredulity of everyone else – engaged in a progressively more intensely heated argument.

The reality, of course, is that some chess sets are composed of black and white pieces, whilst others are red and white, and – quite possibly – yet others are some other combination of colours. So don't argue. If one person sees [this], that's what they see. Accept it. And if someone else sees [that], accept that too. It's not a question of 'right' and 'wrong'; rather, it's all about 'difference'. This is especially important for the person writing on the flip-chart: the role is to record, faithfully, what was said. Not to act as a censor. And be alert – and constructive – too: something like 'each of the pieces in either player's set are of the same colour, a colour different from that of the opponent's pieces' should resolve the argument.

9.4.9 ...but do ask questions

But although arguing is a 'bad thing', asking questions is a positively 'good thing': questions such as 'Can you tell me more about that?', 'That's interesting – I hadn't noticed that. Can you explain further?', 'Do you have an example?'. Questions of this type are seeking further information, inviting the speaker to enrich their point, and stimulating a lively conversation.

9.4.10 Don't duplicate points already captured

As the conversation goes around and around the group, people realise when a point they have written down has already been made, so there is no need to repeat the same point exactly. 'Exactly' is important – if what someone has written down isn't *exactly* the same as has already been captured, then the point should be made: the (slight) difference might make all the difference.

9.4.11 Don't start generating ideas (yet!)

During the conversation, it's very easy, very natural – and hugely tempting – for people to start generating ideas. Try to resist that temptation! This part of the process needs to keep to what exists, not what might be – it's the next part of the process that will get there. So be patient! Even if it's a bit frustrating!

9.4.12 And eventually...

Eventually, everyone has said everything they wish, and there could be many flip-charts, full of observations, full of knowledge, all relating to the focus of attention. And if there is just one point on the flip-charts that was not on the notes of any individual, then the process has been successful: the collective observations of the group are richer than the observations of any one person.

The process as just described is appropriate to a focus of attention described by most people most naturally as a list, or as the equivalent of a list, such as a mind-map. If the focus of attention is a process, then people might more naturally draw a process flow chart; if it is a system, perhaps a systems thinking 'causal loop diagram' [3]. In these cases, a rather different 'share' process might be more appropriate, such as a series of presentations in which each participant describes their own diagram; subsequently, all the diagrams can be suitably displayed, allowing everyone to walk around, look at each other's diagrams, and ask questions. After that, the group can gather together and compile a single, composite, diagram that combines and synthesises everyone's observations into a single picture.

My experience is that the discussions which naturally happen during the 'share' energise the group, making everyone feel more comfortable with each other, helping to build the team. As is crucially important for helping create the conditions in which creativity flourishes... as is about to happen...

9.5 Step 4: Then choose one feature, and ask 'How might this be different?'

Step 2, 'Individually and in silence, write down everything you know about the agreed focus of attention' and Step 3, 'Share', are totally focused on observation, and on a rich and complete description of the focus of attention. Everything on the flip-charts, or in the diagrams, is firmly rooted in the reality of 'today', everything must be objectively verifiable as a fact ('Having clear roles and responsibilities really helps a team work well'), or as a feeling that at least one person in the group is willing to express as a truth ('I don't like confrontation').

Given that the workshop is all about being creative and generating ideas, to have spent so much time *not* doing that might appear counter-productive. But not so. This is the all-important groundwork to enable creativity to happen: gathering together as many observations, from as many different perspectives, as possible, and – very importantly – helping participants unlearn. By writing lists or drawing diagrams, the 'bundled concept' corresponding to the focus of attention is gently deconstructed into the Koestler's Law components from which it is formed.

The scene is now set to be creative. And for that to happen, we draw on the fundamental principle that we're not seeking something new. We're seeking something *different. Different from what is happening now.* And, hopefully, better too – but in the first instance, it's difference that counts.

And the flip-charts and diagrams that now surround the group are full of observations, of details, of what 'now' looks like. So if just one of those might be different, then the overall pattern will be different, so complying with Koestler's Law.

To continue with the example of working together, consider the feature *'I felt that I was always being judged'*. That statement refers to that (probably more junior) individual's past experience, but it may still be happening, or be a potential problem in the future.

How might that be different?

What needs to happen to ensure that more junior team members don't fear being judged by those more senior? How might more junior people be encouraged to contribute to meetings, and not remain silent for fear of 'saying something wrong'?

Those are powerful questions, and they lead to real, innovative, answers.

The concept of creating a team environment in which team members feel 'safe' is of course not new, innovative or creative in a general sense. But for this particular community, and for some of the individuals within it, being sensitised to this, and behaving differently accordingly, could well be hugely beneficial.

In many organisations, however, matters such as 'safety' are 'elephants in the room' – matters that we all know are present but which are not directly addressed, for they relate to 'difficult' topics such as personal behaviours, courtesy, mutual respect, the relationships and interactions between those who are more junior and those more senior, when it is appropriate to 'judge' people, and when not. Such themes are rarely examined in the normal course of the day; indeed, as hinted at by the points noted by 'Participant 4', they might be deliberately avoided. But the associated problems won't just disappear simply because they are not discussed. So a well-conducted creativity workshop provides a 'safe space' to explore them.

In terms of the Target Diagram of Figure 3.1, the discussion is partly in the 'relationships' domain, partly in the 'You!' domain, and, as noted in sections 3.5 and 3.6, these domains are, in general, more 'difficult'. But they can also be highly important.

In practice, the lists resulting from Step 3, 'Share', can be long, with many tens of items, perhaps even a hundred. In principle, all are potential triggers for asking 'How might this be different?', but some are more 'interesting' than others, in that they identify issues that the group senses to be more important, more perceptive, or more likely to lead to valuable ideas. A brief group discussion will help identify which item to choose first, but don't spend too much time agonising over that: if the resulting exploration turns out to be rather dull, no matter, for there is a wealth of material from which to choose another. Also, as will be examined in more detail shortly, that central question 'How might this be different?' can take many forms.

But once an initial feature for exploration has been chosen, don't force the conversation, but…

9.6 Step 5: Let it be…

…which is what Step 5, 'Let it be…' is all about: to follow whatever avenues open up, whatever themes emerge, as the discussion evolves. Importantly, the only requirement is for there to be a free-flowing, imaginative exploration. There is no obligation to 'generate 8 good ideas by 4 pm'. Don't force it. Let it be…

For it is during this conversation that much unlearning is taking place. Continuing the example of the previous section about more junior people feeling that they are being judged, and therefore remaining silent in meetings, a participant who is more senior might, during this discussion, start thinking 'What might I be doing in meetings that – inadvertently – is making the more junior people feel unsafe? Am I perhaps too forceful? Do I ask questions too aggressively? Do I allow enough time for more junior people to think?' Those are tough questions, really tough – and perhaps questions that have never been posed before. And there are of course any number of reasons why that senior person behaves like that: from 'the cut and thrust of academic challenge' to 'the need for rigour' to 'that's how more senior people behaved to me when I was more junior, and so that's how I learnt to behave'. So unlearning those behaviours is hard, very hard. And takes time.

To make matters even more complex, just because the question 'how might this be different?' is asked about any particular issue does not imply that whatever-it-is *should* be different – so perhaps some of the more senior person's behaviours are OK, and don't need to change: maybe it's the more junior person who is just too sensitive, too quiet, and should be more thick-skinned, braver. So that's the unlearning the more junior person is going through – which, once again is hard, very hard.

The end-game is quite likely to be one in which everyone has come to terms with a very challenging issue. Those more junior steel themselves to contribute more at meetings; those more senior are sensitised to behave less judgementally; everyone agrees that having a 'safe environment' in which a junior person does not feel penalised for saying 'I'm sorry I don't know' is a 'good thing', and there is general agreement to try to make that happen.

Or, to take a very different example, perhaps a factory manager is participating in a workshop to design a new product. The manager has no particular interest in what the actual product is, but when people are discussing 'how might this be different?' and exploring possibilities to improve the design, the manager can't help thinking that the particular idea the designers are discussing right now has implications on the factory. Some staff will probably have to be retrained, and some new equipment commissioned. Those are headaches the factory manager just does not need right now, so the natural inclination is to be rather downbeat about the idea and to contribute to the discussion with comments like 'that idea will cause major disruption to our manufacturing capability'. But maybe the factory manager has to unlearn all that too…

As discussed in section 8.5, unlearning is difficult, very difficult. So don't force the discussion, and when someone ways 'that will be difficult because…' respond not with 'no, it isn't' or 'stop moaning'. Respond with a question of the 'how might this be different?' style, such as 'Yes, that's a valid point. So what has to be different to make things work? Different, perhaps, about the design? Or perhaps different about the skills of the team in the factory?'

9.7 Step 6: …and then, when that discussion runs out of steam, choose another feature and repeat Steps 4 and 5

Sooner or later, the discussion of any specific feature will naturally come to an end. That's fine. But one of the benefits of Step 3, 'Share', is the compilation of an often very long list of features, each of which is a possible trigger for generating ideas. So when any one discussion finishes, go back to the original lists or diagrams, choose another feature, and repeat the process: ask 'how might this be different?' and let it be…

To take the team working context: another feature is that '*I don't like confrontation*'. How might that be different? Perhaps it doesn't need to be different, after all, very few people 'like' confrontation, let alone 'being confronted'. But that statement was made in a broader context, concerning the role of a leader in a situation in which a team member is failing to deliver their obligations, letting others in the team down, and possibly jeopardising the team's collective outcomes. So the issue is not about 'confrontation' in the abstract; it's about some very pragmatic, and real, matters indeed – the roles of a team leader and a team member; the obligations on everyone to 'pull their weight'; how everyone in the team can be fairly held to account; how the team is best 'managed'; and how the essentially 'bureaucratic' concept of 'management' relates to something much cherished by many - 'academic freedom'. Yes, tough matters indeed. For all these are firmly in the 'relationships' and 'You!' domains of the Target Diagram (see sections 3.5 and 3.6), which, as discussed, are much more difficult to address than those in the 'Content' and 'Process' domains. Designing a new product is much easier than changing your mind.

9.8 The nine dots puzzle revisited

Section 1.8 discussed the nine dots puzzle – how to join all nine dots by drawing a single straight line – and described how three people solved it. The first person had a flash of inspiration, which is fine. Sometimes that happens, and we all rejoice when it does. But that's unreliable – sometimes it happens, but more often, not. The second person is lucky, and that's good too, but also unreliable.

The third person, however, solved the puzzle using a deliberate process based on careful observation, which provided the platform for curiosity, within a climate of permission. In fact, the process that person used, as described in section 1.8, is a 'solo' version of **InnovAction!** The focus of attention was 'how to solve the nine dots puzzle'; the description of what we know included, for example, 'the paper is still, and not moving', 'the paper is a continuous surface', 'the paper is flat', 'the pencil lead is a millimeter or two wide'. Since there were no other participants, there was no 'share' and so the process moved directly to the first 'how might this be different?' question, 'What might happen if the paper could move?' After 'letting it be' for a short while, the solution of rotating the paper emerges, after which 'choose another feature' triggers 'if the paper were not a continuous surface, what would it look like?'.

Yes, the process works, and really is reliable, as the examples in the next chapter will illustrate…

References

[1] www.forbes.com/sites/forbesfinancecouncil/2017/10/19/on-building-a-faster-horse-design-thinking-fordisruption/?sh=2ec613ee49f9
[2] Christensen C 2016 *The Innovator's Dilemma: When New Technologies Cause Great Firms to Fail* (Cambridge, MA: Harvard Business Review Press)
[3] Sherwood D 2022 *Strategic Thinking Illustrated: Strategy Made Visual Using Systems Thinking* (New York: Productivity Press)

IOP Publishing

CREATIVITY for Scientists and Engineers
A practical guide
Dennis Sherwood

Chapter 10

InnovAction! in action

10.1 Ideas for games based on chess

Here is an example of how *InnovAction!* works for real, in which the focus of attention is to invent a new game. In fact, 'a new game' is too broad, so the scope is narrowed as 'a new game based on chess'.

I often use this example in my training programmes as a first experience of putting *InnovAction!* into practice, and it works very well – even for those who say, 'but I don't know anything about chess!' Certainly, some people know a lot about chess, and might be keen, if not expert, players. But everyone knows something – even if only 'it's a board game' or 'it's played by two players' or 'it's boring'. So everyone can participate, and, in response to Step 2, 'Individually and in silence, describe…', some people write relatively short lists, and others, longer lists, list which are then aggregated in Step 3, 'Share'.

10.2 Some things we know about chess

Here are the results of a typical 'share':

- It is played on a flat...
- ...chequered board...
- ...of 64 squares...
- ...in the overall form of an 8 x 8 square...
- ...usually coloured black and white.
- The game can be portable.
- The game is usually played by two opposing players...
- ...each controlling a total of 16 pieces...
- ...distinguished by being of two colours...
- ...usually black and white.
- The pieces on each side are of 6 different types...
- ...placed, at the start of the game, in 2 rows...
- ...at opposite ends of the board...
- ...close to each payer...
- ...in specific, pre-defined, locations...
- ...with one piece on its appropriate square.
- There is only one king...
- ...and one queen on each side...
- ...and also two castles...
- ...two bishops...
- ...two knights...
- ...and eight pawns.
- The front-line pawns are 'expendable'...
- ...the back-line pieces are more 'valuable'.
- All the pieces are on the board from the start.
- Each piece has its own rules for movement...
- ...pawns: one or two squares first; then one;
- ...castles: straight lines, any number of squares;
- ...knights: two squares straight, one to the side;
- ...bishops: diagonals, any number of squares;
- ...queen: all directions, any number of squares;
- ...king: all directions, only one square.
- Pieces can move only across vacant squares...
- ...except the knight, that can jump over other pieces...
- ...and is the only piece that does not move linearly.
- Each piece has a value.
- It is an ancient, traditional, game...
- ...with a medieval theme...
- ...and a 'combat' style of play...
- ...but without being a contact sport.
- White takes the first move...
- ...and players take one move alternately...
- ...moving just a single piece...
- ...of their own colour.
- Players must take a move when it is their turn.
- There are a huge number of possible moves.
- To win the game, a player must achieve 'check mate'...
- ...the imminent capture of the opponent's king.

> - The game can also end in 'stalemate', a draw.
> - There is no time limit...
> - ...unless as agreed at the start.
> - The game is one-on-one...
> - ...and each player can see their opponent's moves.
> - The game is played by players of all ages.
> - Good players think strategically, many moves ahead.
> - 'Speed chess' is played very quickly, with very little time for each player to think before each move.
> - One player can be a computer.
> - The game can be played on-line.
> - The game is usually played in silence.
> - All squares can be inhabited by any piece.
> - A square can be vacant...
> - ...or occupied by only a single piece.
> - A captured piece is removed from the board.
> - If player B moves a piece onto a space currently occupied by a piece of player A, player A's piece is removed, and replaced on that square by player B's piece.
> - Player A cannot move onto a square already occupied by another of player A's pieces.
> - In principle, any of player A's pieces can displace any of player B's pieces.
> - A pawn may capture another piece one square diagonally to either left or right.
> - If a pawn reaches the opposite side of the board, it may 'transform' into a piece of the player's choice.
> - ...

That list is not exhaustive, for there are many other features of chess that can be added – indeed, you might think of some. Such a list, however, has more than enough content to stimulate creativity, as we will shortly see.

This list is all about observation, noticing as many aspects of the focus of attention as possible. Importantly, every item on the list is a verifiable fact of chess-as-we-know-it; furthermore, the list allows for known variety, for example, 'the board is usually coloured black and white' allows for other colour combinations. Also, identifying each feature as a separate bullet point helps with unlearning, as we deconstruct the concept 'chess' into its constituent elements.

The share is a collective activity, drawing together the observations of everyone in the group. When this is complete, I always ask two questions. The first is 'Is there any item on anyone's list that has not been captured here?' – just to check everything has been recorded; and then 'Is there any item on the group's collective list, as shown on these flip charts, that was not explicitly identified on your own personal list?'. In response to which everyone in the group nods. There are always more features on the flip charts than on any one participant's list. Collective observation is always richer, wider and more complete than any individual. Always.

Depending on how much any individual knows about chess, some of these might be familiar, some not. And some might be in the category 'ah... I knew that, but have never noticed it' – for example, the very fundamental rules that any square can be either vacant or occupied by a single piece, and that a player cannot 'self-capture' (as described by 'Player A cannot move onto a square already occupied by another of player A's pieces').

10.3 Ideas, ideas, ideas...

10.3.1 Self-capture

And as soon as that is noticed, that immediately triggers an intriguing 'how might this be different?' question – suppose two players mutually agree that for the next game, self-capture is allowed...

Certainly, players would not wish to self-capture wantonly. But a situation might arise during a game in which sacrificing a piece is the only way to avoid something even worse. Furthermore, the possibility of self-capture will surely affect the way both players deploy their pieces, so the entire strategy of the game will be changed as a direct result of this very simple modification of the conventional rules.

That's just one possibility... here are some others...

10.3.2 'Hobbled' bishops

Conventionally, a bishop can move any number of squares along the appropriately coloured diagonal. How might that be different? Suppose the number of squares were limited to, say, no more than three? Or perhaps a die might be rolled to determine the maximum number for any one move? Which might include zero, implying that the piece moves zero squares, and so has to stay where it is, but counting as that player's 'move'... and the same could apply to the castles too...

10.3.3 Taking turns

Another fundamental rule is that players take a single move, in turn. How might that be different? Mmm... suppose that players could take two moves at the same time... Or suppose the player rolls a die – and if it shows 1, 2 or 3, the player is allowed a single move; if 4, 5, or 6, two moves... or, once again, rolling a die, if 1 or 2, one move; if 3 or 4, two moves... or 5 or 6, no move...

10.3.4 Square occupancy

The rule that any square can either be vacant or occupied by a single piece might be different by imagining that, say, a square might be occupied by two pieces – which might apply to a 'special magic square', rather than all squares in general. What might happen when two pieces are on that 'magic square'? Perhaps nothing at all... but perhaps they could each be immune from capture... in which case, how long might any one piece remain in this 'safe haven'?

And do the two pieces have to be of the same colour, or of different colours, or do the colours not matter?

And what might happen if the two pieces are the king and the queen of the same colour? Mmm.... maybe they have a 'prince' or a 'princess'? So, for example, if the player throws a die, a 1 or 2 results in nothing, a 3 or 4 a prince (maybe that's another knight), and a 5 or 6 a princess.

But there is no 'princess' piece. So why not invent one… a princess is a 'little' queen… so if a queen can move any number of squares in any direction, perhaps a princess can move in any direction but only by, say, two squares… and if the king is in checkmate, but there is a prince, perhaps the prince succeeds as king so that the game can continue…

10.3.5 The initial layout

At the start of every chess game, both players know the locations of all of the pieces: in two rows, with each piece in a specific position – the pawns in front, the castles in the corners, then the knights, then the bishops, and the king and queen in the middle, with the black queen on a black square and the white queen on a white square, so that equivalent pieces on each side are opposite one another. And because the starting positions are always the same, good players learn sequences of initial moves, known as 'opening gambits'.

Suppose, that, at the start of the game, pieces could start in other places. So, for example, each player might put their own pieces on the board in accordance with some shared rule – for example, still in two rows as usual, with the pawns in the front, but with a choice as to where the other pieces can be positioned on the back row – implying that the castles don't have to be in the corners, but might be somewhere else.

How does this actually happen? Does each player position their own pieces 'in secret', only revealing the positions when the pieces are all on the board? Or are the pieces placed on the board one-at-a-time, in full view, so each player can choose what to do next in full knowledge of the evolving layout?

Or does an 'umpire' lay out both sets, which the players then have to play as given? And of course the umpire can deliberately break the convention that similar pieces are opposite one another, so introducing a spatial asymmetry.

Another possibility is that each player lays out not their own pieces, but their opponent's, either 'in secret' or one piece at a time in full sight…

10.3.6 Combinations

And there's a game that combines 'hobbled' bishops, and a board with a 'magic square' that can hold two pieces…

…and perhaps the pieces might be laid out differently…

…and…

10.4 It really is as simple as that!

Those are just a few examples – there are many others, for (almost) every feature of chess is susceptible to 'how this might be different?' – and the abundance of ideas resulting from the discussions of 'multiple occupancy' and 'the initial layout', as just described, illustrates what can happen when you invoke Step 5, 'Let it be…'.

Some the resulting games work better than others, and not all of the ideas are particularly good. Those statements, however, are evaluative, and so are not applicable during idea generation, the purpose of which is to generate as many ideas as possible. Once there is a (possibly quite large) number of ideas on the table, they can each be reviewed to identify the most promising, as we shall see in chapters 14 and 15.

As I've already noted, I often use this as a training exercise, for it always works well, and quickly too – the whole activity, from 'individually in silence, describe...' to the generation of a host of ideas can usually be accomplished within an hour. And once people get used to 'how might this be different?', and realise that objections such as 'that won't work because' have no validity, the ideas flow and flow, often so quickly that the person trying to keep a record on the flip-chart just can't keep up. Even those who 'know nothing about chess' and who self-believe 'I am not creative' are wide-eyed in amazement at the number and variety of the ideas generated – generated by using a very simple, but none the less highly powerful, process: start grounded in the reality of what you know, describe it in detail, share with others and listen to others, then ask 'how might this be different?' and let it be...

It really is as simple as that. But that simple process is based on some deep, profound, principles. On Koestler's Law. On the importance of searching for difference, not novelty. On making it as easy as possible to unlearn. Of working together.

And talking of novelty, it turns out that few of these ideas are 'new' in the sense of 'never in human history has this idea ever been discovered'. Chess experts are often familiar with what are known as 'chess variants' – these being chess-like games, but different in some way from 'conventional' chess. Indeed, there is a book, *The Encyclopedia of Chess Variants*, that lists about 1,400! [1] So it's quite likely that the games identified in my workshops can be found in this book. But that doesn't matter. If the game is new to the community that 'discovered' it, then that proves that the process works, and that, yes, you can do it.

10.5 The central step – Step 4: 'How might this be different?'

10.5.1 Three key rules

The central step of *InnovAction!*, the step during which ideas are generated, is Step 4, 'How might this be different?' Very often, this question immediately triggers any number of possibilities as people in the team say, 'Suppose [this] were like [that]...', 'What if [this] were different [this] way?' 'Suppose [this] was not [here] but [there]...', 'What would happen if we added [that]?' and 'What would happen if we eliminated [that] altogether?'.

There are three important matters that should be born in mind at this stage.

10.5.2 Don't be negative

Firstly, don't be negative – so any response along the lines of 'that won't work', and its more benign, and indeed harsher, variants are banned. The effect of these statements is to close the discussion down, which is not the objective here – the requirement is to open the discussion out, and explore. It could well be that 'eliminating [that]' will indeed not work, for [that] is probably present for a reason. But by imagining what might happen if [that] were eliminated, the team can explore, for example, how the function currently performed by [that] might satisfactorily be achieved using other methods, as well as testing that the functionality is indeed still required. The question 'What would happen if we eliminated [that] altogether?' is therefore not a statement, still less a demand, that [that] should indeed be eliminated; rather, it is a question opening an enquiry into an area that perhaps has not been explored before, and that might lead to an 'interesting' place. But it might not, we don't know. That's why Step 5, 'Let it be…', allows for exploration, and Step 6, 'Choose another feature', keeps things going when any one exploration naturally peters out.

10.5.3 Don't lose the key questions…

Secondly, don't lose all those 'How might this be different?' questions – which is very easy to do in a lively workshop with an energetic group. Make sure they are all noted, and deal with each one in turn – they could well lead to different places.

10.5.4 …and keep a record of all the ideas

And thirdly, during the discussions, capture as many ideas as possible, perhaps on flip-charts, perhaps on screen shots, in small drawings, in people's notes, or on coloured cards or post-it notes that can be stuck on the walls for others to see and read. I call these 'meteor cards', for they capture flashes of thought going through your mind, just as a meteor is a flash across the sky. Remember Koestler's Law – that every idea is a pattern of components. Each of these notes and drawings might contain a vital Koestler's Law fragment which, when combined with another – that may emerge later, or had emerged earlier but has perhaps been forgotten – might form a powerful idea indeed. At the end of the workshop, all these notes, drawings, screen shots, flip-charts and meteor cards need to be collected, sorted and written up.

10.6 Different ways of being different

As already noted, the question 'How might this be different?' is general, and usually sparks the discussion at once – that said, here are some variations, each highlighting a particular type of difference that might be helpful in appropriate contexts.

10.6.1 Size and scale

All physical objects have dimensions and all processes take time. Physical dimensions and time can be measured, and these measurements can form the basis for 'How might this be different?' questions such as:

- How can [this] be smaller?
- How can [this] be faster?
- How can [this] be lighter?
- How can [this] be made from fewer parts?

In my experience, the discussion is richer, and the resulting ideas better, when the question sets a target that is both specific and startling – say, 100 or even 1,000 times smaller; 100 or 1,000 times faster; just 10 parts. The intention, of course, is not to achieve those particular numbers, but rather to stimulate a wide-ranging exploration, a mind-game, an imaginary 'world', and to keep going, searching for ever more beneficial ideas. And in so doing, you might discover a realistic way of achieving a 10-fold difference, which will be of considerable value. But if the original challenge had been to achieve a 10-fold difference, that could well have been regarded as 'unachievable', and the discussion would stop with only a 2-fold improvement.

10.6.2 Sequence, flow and configuration

Almost every process comprises a series of steps, taking place in a specific order; almost every product, machine or instrument comprises a number of parts assembled in a specific way. Those processes and machines are quite likely to work well, but perhaps they could be even better... and to help discover new ideas, here are some questions that throw a spotlight on sequence, flow and configuration:

- Suppose [this step] precedes [that one]...
- Suppose [this step] happens at the same time as [that one]...
- Suppose [this] were on top of [that]...

10.6.3 Function and scope

These questions examine methods, objectives, purposes and outcomes, for example:

- Why do we do [this] in [this way]?
- What other ways can be discovered of doing [this]?
- How else might [this] function be successfully delivered?
- Why do we do [this]?
- What would be the consequences of not doing [this]?

Although these are quite general, they prise open why [whatever] is done in [this] way, often surfacing underlying assumptions that might hitherto have been taken as 'givens'. And if the answer to 'why do we do [this]?' is 'because we have always done it', be alert to the possibility that there might be a very big idea indeed waiting to be discovered...

10.6.4 Roles and responsibilities

The focus so far has been on quantities (measurement), processes (sequence and flow) and purpose (function and scope); these are primarily about people:

- Suppose that roles were organised differently...
- Suppose that [this role] were eliminated...

The second is particularly powerful. Would the absence of the role be noticed? If something important were not to happen, how might that gap be filled, but without re-creating the role? These are good questions to ask of any organisation... and to keep asking...

These suggestions are summarised in figure 10.1.

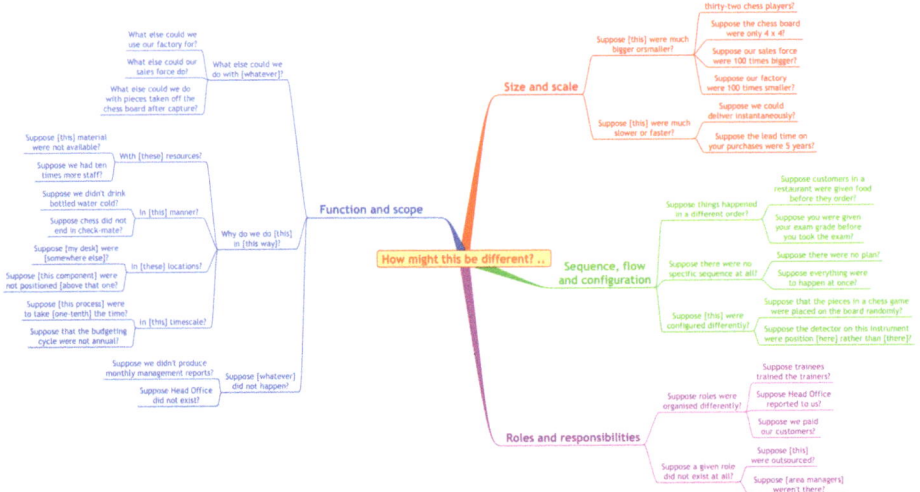

Figure 10.1. Different ways of being different – some examples.

10.7 Some examples

The next several pages contain just a few, real, examples of applying the *InnovAction!* process at my workshops...

InnovAction! in action – Finding collaborators

Bigger projects, and interdisciplinary research, require bigger teams – which in turn requires that people agree to collaborate. But how can willing, and appropriately skilled, collaborators be identified and approached?

For a senior researcher with an established reputation, that might not be an issue. For younger researchers, though, it's not so easy. But the more people you know, and who know you, the more likely it is that, at any time, you will know whom you might approach – and the greater the likelihood that you might be approached by others. Which is all about 'networking'. 'Networking', though, might be something 'I' feel uncomfortable doing, for 'I' am rather shy…

So here are some ideas generated by an academic team about how to make 'networking' just a natural thing to happen:

* The more seminars and conferences a team can host, inviting outside speakers, the greater the number of people you come into contact with…

* …and the more opportunity to go to seminars and conferences organised by others…

* At conferences, do everything you can to meet new people, like deliberately sitting next to people you don't know at meals and conference sessions. And to help 'break the ice' safely, before you sit down, ask the person who is already seated something like 'is it OK if I sit here?' or 'is this seat free?', and as you sit down, just say 'Hi, I'm….'.

* More senior people have more extensive networks. So introduce your more junior colleagues into your networks.

Here are some ideas about enhancing each individual's visibility:

* Everyone is encouraged to produce a well-executed video – say, on YouTube – about their own work, which can then be posted on the web.

* When you collaborate with others, do a good job, and build a reputation as someone who is reliable and trustworthy… word can get around.

* Don't lose contact over time. Networking relationships can last for decades…

And one further idea…

* How about compiling a departmental 'network map' of who knows whom?

> ## *InnovAction!* in action – Managing knowledge
>
> Here are two statements often seen on the lists compiled by workshop participants discussing how knowledge is acquired:
>
> - I read journals in my field.
> - I attend relevant conferences.
>
> How might these be different?
>
> The key words here are 'my field' and 'relevant', which limit the scope of the knowledge acquired. It is of course quite understandable that anyone would only read journals 'in my field' and only attend 'relevant' conferences – after all, why would you waste time reading any-old journal, or attending 'irrelevant' conferences? Especially when there isn't enough time to read all the journals that should be read anyway, and funds for attending conferences are limited too.
>
> All true. But there is a paradox. Koestler's Law tells us that the more 'fragments' of knowledge we have, the more opportunity there is to form new 'patterns', and maybe the 'fragment' we need to solve a particular da Vinci problem might be 'somewhere else'. Those are strong incentives to read journals from 'other' fields, and to attend (superficially) 'irrelevant' conferences.
>
> Theoretically, that might be valid. But in practice…
>
> So here are some practical ideas, ideas that work well in research groups:
>
> Identify fields adjacent to, or linked in some way, with 'our' field. Then compile a list of journals and conferences in these other areas. Choose as many journals and conferences as there are members of the team, and allocate one journal and one conference to each. That person is then asked to keep an eye on that journal, and alert other team members to any articles that might be of interest – which requires that each team member knows what all the other team members are interested in. That person also attends one 'external' conference a year, with a duty to find out as much as possible, and to report back to the team afterwards.
>
> Implementing those ideas does not make the individual, or the team, 'omniscient'. But it does extend the boundaries…

> ## *InnovAction!* in action – Which team?
>
> You hear the distinctive ping announcing the receipt of a text message.
> 'Hi! Great news! I've been picked for the school team for the match next Wednesday afternoon!'
>
> And of course you want to go to watch. But then you remember that there is an important research group meeting scheduled for that day. Oh dear. You can't be in two places at once…
>
> Yes, we all want to be good team players. But a reality is that, at any one time, everyone is a member of more than one team – in this case 'team research' and 'team family'. If the objectives of both teams are in harmony, and if the priorities of each do not conflict, then everything is fine. But when they clash, the result can be stress, guilt, frustration.
>
> How might this be different?
>
> What needs to be different here is the management of the consequences of multiple team membership, so that stress, guilt and frustration are minimised, and people aren't, and don't feel, let down. So, for example, some of the things that might be different are:
>
> **Acknowledge this possibility at the start of all projects**
>
> At the start of a project, when a team gets together, it's important that everyone knows what other teams each participant is also a member of, and the nature of the corresponding commitment – from 'I need to do the school run every afternoon at 4 pm' to 'I attend [this meeting] every fourth Tuesday'. That sets realistic expectations. It's also sensible to recognise that some of these other commitments are not regular, so the team needs to agree how they will keep each other informed.
>
> **Be honest, with others…**
>
> There will inevitably be occasions on which one team member has committed to do something, but is unable to fulfil that commitment, which can let others down. The team should therefore agree to be as honest as possible in informing others, at the earliest opportunity, that this might happen.
>
> **…and with yourself too…**
>
> …which is all about not committing to one team that you will deliver something without taking full account of all the commitments you've already made elsewhere.

> ### *InnovAction!* in action – Detecting water leaks
>
> Water is a utility that we rarely think about – at least in the developed world: when we turn on the tap, water of drinkable quality instantly flows, as much as we need. But for that to happen, there is an extensive network of underground pipes, from a water treatment plant, where that pharmaceutically pure product is made, to your home. In many cities, those pipes have been there for perhaps a century or more. And from time to time they leak. When that happens, much water can be wasted, so water companies take considerable care to detect leaks as soon as possible after they happen, and to fix them as quickly as they can.
>
> Some leaks are very obvious – the water spouts from a road. But many leaks are much less visible, but still lose much water by dripping steadily into the soil.
>
> Detecting these leaks is not so easy, and one of the main ways of doing this is by what the industry calls 'acoustic leak detection' – which means that an engineer listens to the flow of water along a pipe, seeking to identify a characteristic gurgling which is the tell-take sign of a nearby leak. And to listen to the flow, the engineer – the 'listener' – places his or her ear to one end of a 'listening stick', the other end of which is pressed against the pipe, or a valve.
>
> #### 'Listeners' can hear
>
> Of course they can. They wouldn't be 'listeners' if they were deaf! Indeed so, and when this topic was discussed at a workshop, the word 'deaf' triggered a participant to think of the word 'blind'. Which in turn triggered the thought that sight-impaired people often have especially acute hearing, and so would make very good 'listeners' – especially since the 'listeners' usually work in small teams, at night, when the noise pollution is least. Which is an idea about recruitment…
>
> #### The pumps are turned down at night
>
> The pressure of the water within the network is often controlled by pumps, and in many urban areas, the pressure is raised just before dawn in anticipation of the demand when people wake up, and reduced in the early hours of the morning, when demand is at its lowest. But it is early in the morning, when it is still dark, that the 'listeners' examine the network to discover leaks. A consequence of the lower pressure, however, is that any leak will flow more slowly, making it harder to detect. The policy of reducing the pressure when demand is low therefore (inadvertently) cuts across the objective of the 'listeners' to discover leaks. In many urban areas, served by large networks, it is possible to raise the pressure in one area, whilst keeping it lower in others. So wouldn't it be sensible to raise the pressure in the area where the 'listeners' are working, so giving them the best opportunity to discover the leaks?
>
> Yes, it is. But that idea wasn't discovered until some of the 'listeners', and some of those who control the network pressure, happened to be at the same workshop. For they are in different parts of the organisation structure, and rarely meet…

Reference

[1] Pritchard D B 1994 *The Encyclopedia of Chess Variants* (Godalming: Games & Puzzles Publications)

Chapter 11

Springboards and retro-fits

11.1 *InnovAction!* is not the only way to have idea 'on demand'

11.1.1 The 'mountains and valleys' metaphor

Chapters 9 and 10 described my process, ***InnovAction!***, for having great ideas, 'on demand', whenever and wherever a great idea might be useful to the organisation. ***InnovAction!*** is certainly very effective, and my method of choice, but, as was mentioned in section 9.1, it isn't the only technique. Indeed, if you go into a bookshop or browse the web, you will find any number of books [1] with titles such as *The Ideas Book: 60 ways to generate ideas more effectively* [2] or *How to Think like da Vinci: Seven steps to genius every day* [3], as well as packs of cards (such as the *Creative Whack Pack* by Roger von Oech [4]) and the 'creativity box', *Spark Creativity: 50 ways to Ignite Bright Ideas* [5] – all of which list many different methods to stimulate creativity. And for scientists in general, and mathematicians in particular, George Pólya's 1945 classic *How to Solve It* [6] is a treasure trove.

To me, however, all the methods fall into two categories that I refer to as 'springboards' and 'retrofits' – with ***InnovAction!*** being a 'springboard'.

My distinction is best understood in the context of the 'mental landscape' metaphor I introduced in connection with Hebb's Theory of Learning, as discussed in section 8.2.

As you will recall, Donald Hebb suggested that learning was associated with strengthening neural networks, so, by metaphor, 'carving' deep valleys in our 'mental landscapes' – 'valleys' that are valuable in enabling us to repeat what we have already learnt, but act as traps if we want to think differently, if we want to have exciting ideas. Escaping from those valleys is difficult – hence the discussion of unlearning in sections 8.4 and 8.5.

With this metaphor in mind:

- **Springboards** are techniques that start within the 'valley' associated with the focus of attention, and seek to 'escape' to discover a new, distant, 'valley', the valley of the exciting idea.
- **Retrofits** are techniques that start in a distant, apparently unrelated 'valley', and seek to 'find a way back' to the valley of the focus of attention, discovering an idea along the way.

This is best illustrated by some examples...

11.1.2 *InnovAction!* – a springboard

One of the key features – and indeed benefits – of ***InnovAction!*** is that the process is grounded in the reality of what happens now, a reality that is well-known and familiar, and can be experienced, touched, seen, described. So once the 'focus of attention' has been defined, Step 2, 'Individually and in silence, write down everything you know about the agreed focus of attention', places the participant squarely in the bottom of the 'valley' corresponding to the chosen focus of attention, inviting the participant to 'look around' and to describe in detail what can be seen. This 'maps out the contours of the valley', and is instrumental in the process of unlearning. Then, asking 'how might this be different?' encourages the participant to 'climb of the valley' and to stand on a 'ridge', from which a 'distant valley' might be seen, the 'distant valley' of an idea.

The ***InnovAction!*** process therefore acts as a metaphorical springboard, 'projecting' the participant from the 'valley of the familiar' to the previously undetected 'valley of the new'.

11.2 Some other springboards

11.2.1 Challenge assumptions

In this method, the team is invited to identify all the assumptions associated with the focus of attention, the starting points that are the 'givens' we don't usually notice, let alone think about, for example, '[this process] obeys [this equation]', 'the most senior person leads the project', 'for [this purpose] we use [this material]', 'that unexpected event is an error'.

Having noticed any assumption, this then sets up questions such as 'what are the conditions for which [this equation] does not apply?', 'what might be the consequences if the project were led by someone other than the most senior person?', 'what other materials might we use?' and 'what might this unexpected event be telling us?', from which ideas flow.

Since fundamental assumptions, such as 'the earth is at the centre of the universe', 'time is invariant', and 'light is always a wave', are just that – fundamental – then challenging them can lead to truly *BIG IDEAs*: Dame Jocelyn Bell Burnell's discovery of pulsars [7] was not 'expected', nor, as we saw in section 2.1, was Fleming's identification of penicillin.

This process is, in fact, a variant of ***InnovAction!***, for all assumptions are features of the relevant focus of attention, and should be identified during Step 2, 'Individually and in silence, describe'... A complete description of the focus of attention, however, contains many features in addition to the fundamental assumptions, so by focusing on the assumptions, the resulting list contains fewer items, and – usually – items of rather more significance; items which, if changed, are likely to lead to 'bigger' ideas.

11.2.2 Decomposing and recombining

One of the most influential books on mathematical problem-solving, and a cracking good read too, is *How to Solve It*, written by the Hungarian mathematician George Pólya, and first published in 1945. Pólya describes 67 different methods, including what he calls 'decomposing and recombining', in which 'you decompose the whole into its parts, and you recombine the parts into a more or less different whole'. This is, in principle, a re-statement of Koestler's Law, and maps very closely onto ***InnovAction!***: Pólya's 'decomposition', in my terms, is the identification of the component parts within the focus of attention, so facilitating unlearning, whilst one answer to 'how might this be different?' is 'recombine in a different pattern'.

11.2.3 Edward de Bono's 'PO'

'Wow! That's a tough one. We really need to think laterally about that...'

You've probably heard, or indeed used, the phrase 'lateral thinking' any number of times whenever there's a difficult problem to solve, whenever a good idea would really help, whenever you've needed to be creative. This now-ubiquitous term may be traced back to a book entitled *The Use of Lateral Thinking* [8], published in 1967, and written by Edward de Bono, one of the most ardent advocates of creativity, and the author of many books, training courses and other helpful materials [9].

De Bono introduced 'lateral thinking' as a way of breaking free from what he calls 'vertical thinking', this being the usual thought patterns that are constrained by the tram-lines of familiarity – what I call the 'learning trap'. 'Lateral thinking', by contrast, is, to quote de Bono directly, about 'the search for different ways of looking at things'. And as a by-the-by, those words 'lateral thinking' are, perhaps, a Koestler's Law story in that, as mentioned in section 4.5, Koestler referred to 'thinking aside' in *The Act of Creation*, published in 1964, quoting Paul Souriau's 1881 insight that to be creative, 'il faut penser à côté' – 'you must think aside'.

In Chapter 34 of his book *The Mechanism of Mind* [10], published in 1969, de Bono introduces the term 'PO' as a means of stimulating lateral thinking, later developing the concept in *PO: A Device for Successful Thinking* [11], published in the US in 1972, and under the title *PO: Beyond Yes and No* [12] in the UK in 1973. 'PO' is described in the first chapter:

"PO is a new word... in 20 years' time (that would be 1992), when computers have taken over most of human thinking, no other word will be more important... PO is a magic word. It will do all you want it to do is you believe in it... PO is the first word to arise from the mathematics of the mind... The world is divided into two sorts of people: the PO people and the NOPO people..."

Bold claims indeed.

As de Bono then explains, PO, stands for 'Provocative Operation', this being a 'dramatic question' or 'startling assertion' that so disrupts 'normal' thought patterns as to cause new ideas to emerge.

Here is an example from de Bono's *Serious Creativity* [13]: 'PO: planes should land upside down'. This form of statement is referred to as a 'PO-1'; somewhat different forms of statement are referred to as 'PO-2' and 'PO-3', as will be described shortly.

The possibility that planes might land upside-down is indeed 'provocative'! But once you get over the shock, you might notice that, as the plane comes in to land, the pilot (who is presumably sitting up-side down, but let's ignore that for the moment) has a good view of the runway, especially of the ground immediately beneath the plane, ground that becomes progressively closer, and very quickly too, as the plane descends. And although the 'normal' position of the cockpit, on the top of the plane, gives the pilot a very good view of the way ahead, the view directly downwards is blocked.

We would all agree that the more of the ground the pilot can see, the safer the landing is likely to be, and so although landing the plane upside down is itself not a particularly appealing idea, the 'provocation' does indeed trigger some more feasible ideas such as having a 'landing cockpit' underneath the nose of the plane into which the pilot might go for landing, or, as enabled by currently available technology, mounting a camera underneath the nose of the plane to display an image of the ground underneath on a screen in the (normal) cockpit.

The result of 'PO: planes should land upside down' is therefore the discovery of these ideas.

What de Bono doesn't describe is how to create the 'POs' in the first place. His examples pop up out-of-the-blue, and are then followed by often convoluted stories that end with (often very sensible) ideas.

In fact, PO-1 is a 'springboard', very much like **InnovAction!**, but with some (rather important) components missing – in particular, Step 2, 'Individually and in silence, describe...' and Step 3, 'Share' – and leaping directly to something akin to Step 4, 'How might this be different?'. So, in the case of 'PO: planes should land upside down', Step 1, 'Define the focus of attention' might be, for example, 'how planes are designed' or perhaps 'how to make plane landings more safe'. Step 2 would be to describe everything we know about these, so that might include:

- Planes are conventionally designed with the cockpit towards the front of the plane, looking ahead, above the nose.

or

- As a plane comes into land, the pilot has a good view ahead, but only a restricted view downwards.

There will be many other items too, but focusing on these two in particular, some relevant 'how might this be different?' questions are:

- Where, other than in the conventional cockpit, might the pilot sit as the place comes in to land?
- Where might the cockpit be, other than on the top of the nose?
- What might happen if the pilot sat underneath the nose rather than on top?
- Where might a pilot get the best view for landing?

All of which lead to the *BIG IDEA* – but in my view by a rather easier and more practical process than by suddenly saying 'PO: planes land up-side down'.

That's what de Bono calls PO-1; PO-3 is in essence the same as 'challenge assumptions' as discussed in the previous sub-section; PO-2 will be described in section 11.4.1.

11.3 Random words – a retrofit

All organisations seek to attract talented recruits – new trainees in an industrial firm, next year's PhD student intake. To support its recruitment activity, suppose that an organisation's website is being re-designed, and that someone suggests that this presents an opportunity to incorporate a strap-line or slogan to go alongside the organisation's name – 'The place to be' – that sort of thing.

The identification of slogans is very much the type of creativity associated with advertising campaigns: 'Diamonds are forever' (the de Beers diamond mining company), 'The real thing' (Coca Cola), 'Because you're worth it' (L'Oréal cosmetics) are very well known, and can have great longevity – the 'diamonds' slogan was first used in 1947 [14]. All are 'obvious' with hindsight, but not so easy to discover, which is one of the reasons why advertising agents are so well paid!

There is, however, a way to stimulate creativity that is especially useful for slogans – the 'random word' method, which follows this process:

1. Identify a single, truly random, word – perhaps from an arbitrary page in a book, or a newspaper.
2. As a team, generate a large number (say, 100) of further random words.
3. Then, once again as a team, explore the list of around 100 words to see if there might be some 'interesting' combinations or associations.

The easiest and most effective way of carrying out step 2 is to go around the team, with each person saying a single word. These words should be random in that no pair of words – and certainly no successive words as suggested as the list is compiled – should be related, especially as regards their meaning. In reality, this is very hard to achieve, for most of us find it difficult to escape from the mental valley labelled 'sequences of words must make sense'. But if everyone in the group knows that the rule-for-the-time-being is 'these words don't have to make any collective sense', then a list which is 'random enough' can be compiled, such as this list of 100 words starting from 'envelope':

Envelope	Ghost	Brazil	Ration
Elephant	Tremendous	Tea	Family
Gorgeous	Lively	Ship	Rubber
Pistol	Spark	Flat	Course
Fast	Mexico	Tap	Bicycle
Ladder	Vampire	Trap	Sofa
Automobile	Catastrophe	Snow	Mariner
Run	Real	White	Draw
Gravity	Meal	Horse	Romance
Freedom	Mile	Vase	Request
Money	Energy	Bright	Match
Childlike	Footprint	Planet	Book
Fun	Why	Train	Trade
Loudspeaker	Merit	Splendour	Bleak
Football	Ceiling	Architect	Spaghetti
Banana	Heaven	Beer	Awe
Bandana	Grit	Darkness	Lampshade
Cream	Grass	Mountain	Goat
Dream	Red	Politics	Share
Spin	Redeem	Golf	Exact
Laughter	Castle	School	Many
Soon	Round	Stream	Chant
Giving	Goose	Image	Peaceful
Today	Phone	Subject	Many
Cook	Law	Public	Split

Even though these words were meant to be random, there are some patterns – childlike, fun; snow, white; golf, course – and there are a few rhymes and word-plays too (cream, dream; banana, bandana; tap, trap), but those are quite helpful for although the words are related in a particular way, their meanings are not.

Having compiled the list, each participant is invited to scan the list, dwelling on whichever words attract their attention, and then to examine what a particular word means; what its associations are; if a noun, what the object is used for; if a verb, what actions are implied – as many aspects of the word, and its associations, as possible. And all the time, the participant also has 'recruitment' in mind too ... and it might happen that a link is made, and a slogan emerges.

Looking through the list just presented, here are a number of possibilities:

- 'The ladder for your career' (of general applicability)
- 'We climb intellectual mountains. Do you?' (general)
- 'The sparkiest research' (electrical engineering, perhaps?)
- 'Fulfil your dreams with us' (a psychology department?)
- 'Building your future' (as triggered by 'Architect')
- 'Matchless' (renewable energy?)
- 'Tap into your talent' (mechanical engineering?)
- 'Cast iron success' (metallurgy? as triggered by 'golf' ... club ... iron)
- 'Mariners of the stars' (astronomy?)
- 'Your career won't slip here' (rheology? triggered by 'banana')

And each might, perhaps, be accompanied by a suitable picture.

Perhaps you can think of some more – and don't feel limited by the words on the list, for their purpose is simply to project your mind into a number of 'locations' that you would not normally be thinking about, for that is the essence of a 'retro-fit' – to take your mind to a very different place.

Suppose, for example, that a university physics department is seeking to formulate a slogan. The normal starting point is the physics, so the high energy team might come up with 'we accelerate your career', those studying magnetism, 'we attract the best', and the quantum physicists, 'unentangled excellence'. What the random word process does is to project your thinking into a mental space far from physics, into, in the terms of Hebb's metaphor, a 'distant valley' – a valley associated with 'mountain' or 'goat' or 'football'. That's not to imply that you never think about mountains, goats or football – of course you do when you're in the Alps, visiting a farm, or watching a match. So those words, and the related concepts, are not in any sense 'remote' in general; the point is that they are not usually associated with 'physics'.

By leaping to 'football', your mind starts thinking about things associated with that game, and noticing that 'red' is in the list too, how about 'Kick start a red-hot career with us'? Might that work for a group researching nuclear fusion?

Retro-fits project your mind to a 'distant' location, and the process is then one of 'looking around' from that 'distant' location to seek to find a relationship between that location and the 'focus of attention'. In the language of Koestler, retro-fits set up potential bisociations (see the Prologue and section 4.5) between the focus of attention and, in the case of random words, any one of a number of random concepts, leaving it to the ingenuity of your mind to discover a relationship.

In fact, one of the best examples of this process emerged in a workshop that I ran for an insurance company. One of the random words was 'tennis'; the slogan, 'You can't fault our service'!

It is of course always possible to discover any of these slogans by starting from the 'right' end: the renewable energy scientists, for example, are totally aware that their work is about discovering ways to generate energy without combustion, without the need to light a fire, without using matches. So 'matchless' is, in a sense, 'under our noses'. But whether or not 'matchless' might be spotted is another matter. The word 'match', however, is in the list of random words, making it more likely that 'matchless' will be noticed, with the 'twist' that the word 'match' has several meanings, such as 'game' (as in cricket match), 'igniter of fire', and 'equal' (as in [this] matches [that]), with the slogan straddling the latter two (we don't use matches; we are unique). It is indeed quite accidental that the word 'match' happened to be on the random word list – had it not been there, it's most unlikely that 'matchless' would have been identified as a possible slogan. So the appearance of 'match' in the list is lucky, and as has been noted elsewhere, luck has a role in creativity. But as has also been noted, luck can be 'encouraged' to happen, and the random word method is one way of 'encouraging'.

11.4 Some other retro-fits

The random word technique is just one example of a retro-fit. Here are some others...

11.4.1 PO-2

As mentioned in section 11.2.3, de Bono refers to another form of 'provocative operation' PO that he calls PO-2. This is a 'random juxtaposition' in which the focus of attention is directly juxtaposed to a random word or proposition, for example 'recruitment PO spark', so triggering the possibility of discovering the slogan 'The sparkiest research'. PO-2 is therefore identical to the random word method just discussed.

11.4.2 Simile, metaphor and analogy

These closely-related methods also start from a list of random words, which, once compiled, are used in the context of specific questions, for example (as in section 11.3, in the context of seeking to formulate a recruitment slogan):

- How many ways can I think of in which my organisation is like spaghetti? (simile)
- How can my organisation be a castle? (metaphor)
- What does my organisation and a ship have in common? (analogy)

At first sight, questions of this type can often appear to be totally non-sensical. How on earth can I compare my organisation to spaghetti? How stupid is that? But pause a moment; sometimes asking what appears to be a very strange question can lead to 'interesting' places – so, for example, spaghetti is often tangled or chewy, suggesting 'untangling the knottiest problems' (topology research?) or 'research you can get your teeth into' (food science?); castles have strong, high walls, triggering 'where your imagination is not walled in' (general); once built, ships are launched – 'the ideal launching pad for your career' (general, with particular applicability to space science).

In many cases, the focus of attention is a noun (in this case, 'organisation'), in which case similes and analogies require nouns; metaphors can work with both nouns and adjectives. If these methods are used, then they are more productive if the list of random words contains mainly nouns, with perhaps a few adjectives; any verbs on the list will need to be expressed as a corresponding noun (so, for example, 'redeem' becomes 'redemption'), adverbs as a corresponding adjective ('soon' becomes 'prompt', 'quick', 'immediate'...), whilst other parts of speech might require some ingenuity ('why' might become 'reason' or 'cause').

11.4.3 Other people's shoes

The objective here is to try to see the focus of attention – a slogan to accompany a scientific recruitment campaign – through the eyes of someone very different from yourself, someone with a deliberately dissimilar perspective, such as a particular category of person (such as, say, a journalist), or a specific, named individual.

Journalists write stories about the news, and use brief, catchy, headlines – which gets the imagination looking for what might be newsworthy about the research the team is doing and what a headline might be. And while it is in general true that news stories and headlines have increasingly less impact over time, that might not matter, for the slogan can be changed whenever its 'newsworthiness' has diminished.

As an example of using a named individual, what sort of recruitment slogan might, say, Abraham Lincoln come up with?

Abraham Lincoln is associated with honesty and integrity, with the Declaration of Emancipation, with a stove-pipe hat, with being President of the United States. Mmm... how about 'Where great ideas are set free' (triggered by 'emancipation'); 'The coolest team' (from 'stove-pipe', to 'stove', to 'hot', to 'cold'), and perhaps especially well-suited to low temperature physics; 'We attract the brightest stars' (the flag of the United States is often referred to as the 'stars and stripes'), and maybe just right for an astrophysics department.

Abraham Lincoln was not a scientist, and so there are no direct associations with science or engineering. Even so, 'standing in Abraham Lincoln's shoes' can stimulate some ideas that are relevant. Also, as the last two examples demonstrate, the ideas do not have to emerge from the direct associations with Abraham Lincoln, but result from thinking about Abraham Lincoln and 'letting it be...' (stove-pipe hat, stove, hot, cold, cool, and United States of America, stars and stripes, stars). Rather, the process is one of putting you in a 'valley' in your mind deliberately distant from the focus of attention, and inviting you to 'look around'...

11.4.4 Journeying

You're travelling from [here] to [there]. Who is your travelling companion? What do you talk about? What do you see? Where do you visit along the journey? What is the weather like? What methods of transport do you use? What adventures happen?

You get the idea – just a series of questions to take your mind to a very different mental space, a series of far-distant valleys. And as you ask those questions, and explore them with the others in the team, the original focus of attention is always there... and maybe a connection emerges, a connection that leads to another idea...

So maybe part of the journey is along a river ('Where ideas flow' – fluid mechanics?); you climb a high mountain ('A PhD with us – the summit of achievement'); you meet a wizard ('Help make magical materials' – materials science).

11.4.5 Visioning

Suppose that a company wishes to design a new product. The 'visioning' process invites the team to imagine that the new product has been invented, and to describe in as much detail as possible what it looks like. Which might elicit responses such as 'it must be lighter than our existing product', 'it must be made from fewer components', 'it must incorporate sensors so as to take advantage of the internet-of-things'... Each of these in turn leads to questions such as 'what do we need to do to make it lighter?', 'how can components be eliminated?', 'how would we use the internet-of-things?', 'what features of the product need to be monitored, and so what sensors are needed?'. And in discussing these questions, ideas will be generated.

This, of course, is *InnovAction!* 'backwards'. As described in section 9.3, for any given focus of attention (such as 'we need a new product'), the starting point of *InnovAction!* is a very detailed description of the 'world-as-it-is-now', each feature of which is then the trigger for asking 'How might this be different?', so acting as a 'springboard' into the desired future. So, in this example, this would result in a list of features, including 'the product weighs [3 kg]', 'the product is made from [37] individual components', 'the product was designed in 1987, before the internet was widely used', triggering 'How might this be different?' questions such as 'Suppose the product were lighter?', 'How can we reduce the number of parts?', 'What opportunities does the internet open up?'

Visioning goes in the opposite direction – from as detailed a description of the ideal future as can be imagined, the question 'if the world looks like [this] in the future, how did we get [there] from [here]?' starts in the imagined future and seeks to retrofit a path back to the reality of the present – a path that could, ideally, be followed from the present to the future.

11.4.6 Working backwards

One long-established format for questions in mathematics exams is 'prove that [this algebraic expression] is the same as [another form of algebraic expression]', and I remember being taught that there were, in general, two ways of doing this: either to start with [this], and then, carefully, to go step-by-step towards [the other form], or to start with [the other form] and then, carefully, go 'backwards' step-by-step to [this]. Sometimes, it turned out that tackling the problem 'backwards' was easier, and indeed 'working backwards' is one of the methods explicitly described in Pólya's *How to Solve It*.

This is another example of a retrofit, and in exams, both 'ends' of the problem are made explicit, with the task being to discover the path between them. In a real situation, the starting point is the known current reality – the product we have now, the way the team is actually working – and the desired future state is perhaps an aspiration or envisaged ideal. If that end state can be imagined, then it might be possible to 'work backwards' towards the current reality, and so discover ideas which, if implemented and followed 'forwards', will lead from the current state to the desired outcome. In practice, this is in essence identical to 'visioning'.

11.5 Springboards and retrofits – which to use?

The last few pages have examined only a small selection of the 'creativity techniques' that can be found in various books, and I trust that the classification under the two headings of 'springboards' and 'retrofits' is helpful in making some sense out of what could be perceived as a rag-bag of apparently different approaches.

The method used in any particular circumstance depends on both personal preferences and context. Some people are more familiar with, and more comfortable using, [this] method rather than [that] one, and if [this] method works, that's fine. As regards context, if the objective is to discover a metaphor – as is often the case in discovering names for products or advertising slogans – then the random word approach is very likely to be the method-of-choice, for it 'forces' comparisons of the focus of attention with a large number of random concepts, so increasing the probability that a metaphor will be discovered. But if the key requirement is the discovery of something that is different from, and better than, whatever is happening now, then in my experience springboards are the more effective.

Why so? Because springboards are grounded in the reality of what happens now – the reality that can be seen, heard, touched, experienced, depicted, described. That is very helpful in that it enables everyone with any experience of that reality to participate, and do so actively from the outset. Retrofits, however, require some initial imagination, which some people revel in (so they really enjoy using retrofits, and do so productively), whilst some others freeze, finding it genuinely difficult to come up with another random word, or to imagine how Alexander the Great might redesign a pump. Certainly, some imagination is required when asking 'How might this be different?' as in Step 4 of *InnovAction!*, but, as discussed in section 10.6, that can be well-focused – for example, 'What would happen if [this component] were removed?' requires participants to envisage a well-defined and specific outcome, which is not too demanding.

Overall, therefore, my 'method of choice' is the springboard, ***InnovAction!***, which has worked time after time to great effect – provided, of course, that there is an existing 'reality' which can form the platform for discovering 'differences'. This is almost always the case, but in those (very rare) situations in which there is no existing 'reality', then a retrofit must be used – in which case, visioning can be very effective, for it is about imaging the 'reality' being sought, which should be possible to describe; alternatively, random words can be quite fast and productive, for example, when seeking, as already discussed, some type of slogan, or a name for a new product.

References

[1] There are very many books that describe techniques and methods to stimulate creativity, and some of the old ones remain among the best – let me mention in particular van Gundy A 1988 *Techniques of Structured Problem Solving*, 2nd edn (New York: Van Nostrand Reinhold) (the first edition was published in 1980), and also Osborn A F 1983 *Applied Imagination: Principles and procedures of creative problem-solving* 3rd rev edn (Buffalo, NY: The Creative Education Foundation Press) (the first edition dates back to 1953!). Arthur van Gundy was an academic; Alex Osborn was the archetypal 'creative', being the 'O' in the leading Manhattan Avenue advertising agency BBDO, and also the originator of 'brainstorming'.

[2] Duncan K 2019 *The Ideas Book: 60 Ways to Generate Ideas More Effectively* (London: LID Publishing)

[3] Gelb, M J 2009 *How to Think Like Leonardo Da Vinci: 7 Easy Steps to Boosting Your Everyday Genius* (London: Harper Collins)

[4] von Oech R 2002 *Creative Whack Pack* (Stamford, CT: US Games Systems)

[5] 2018 *Spark Creativity: 50 Ways to Ignite Bright Ideas* San Francisco, CA Chronicle Books

[6] Pòlya G 1957 *How to Solve It: A New Aspect of Mathematical Method* 2nd edn (New York: Doubleday)

[7] www.cam.ac.uk/stories/journeysofdiscovery-pulsars

[8] de Bono E 1967 *The Use of Lateral Thinking* (London: Penguin Books)

[9] www.theguardian.com/books/2021/jun/10/edward-de-bono-obituary

[10] de Bono E 1969 *The Mechanism of Mind* (London: Jonathan Cape)

[11] de Bono E 1972 *Po: A Device for Successful Thinking* (New York: Simon and Schuster)

[12] de Bono E 1973 *Po: Beyond Yes and No* (London: Penguin Books)

[13] de Bono E 1995 *Serious Creativity* (London: Harper Collins) pp 156–7

[14] https://theeyeofjewelry.com/de-beers/de-beers-jewelry/de-beers-most-famous-ad-campaign-marked-the-entire-diamond-industry/

IOP Publishing

Creativity for Scientists and Engineers
A practical guide
Dennis Sherwood

Chapter 12

Creativity workshops

12.1 Observation, curiosity and permission made real

Chapters 9 and 10 included a number of suggestions as to how the *InnovAction!* process can be conducted, but now that the process has been explored from beginning-to-end, let's step back for a moment and look in more detail, as well as from a wider perspective, as to what is happening.

Firstly, the process embodies the three key principles underpinning all creativity: observation, curiosity and permission.

Observation is, of course, the essence of Step 2, 'Describe what happens now, in as much detail as you can', and Step 3, 'Share' your observations with everyone else, so as to compile a richer picture. And to ensure that the discussion is well-focused, everyone is looking at the same focus of attention, as determined in Step 1.

Curiosity is Step 4, 'How might [this] be different?'. Importantly, this question is open, and carries no 'baggage'. So, for example, there is no implication that [this] is good, bad, right, wrong, functional, dysfunctional, 'the best thing since sliced bread', 'an urgent problem to fix'. It might be any one of those, but in this context, that doesn't matter. It just 'is'. Why it's the way it is might be 'interesting'. But it isn't important. Because that's all in the past. What *is* important is the future: and to make the future *different* from the past. And the easiest way to explore that is to take a feature of 'today', and ask 'how might this be different?'. This is forward-looking and constructive, not backward-looking and blaming. So people don't feel obliged to defend their past actions, their past decisions; rather, everyone can work together to discover something that is different from the way things currently are, and – we all hope – better.

Permission is more subtle, for it is not an attribute of any particular step in the process, but about the context in which the process takes place. In this regard, Step 2, 'Individually and in silence, write down...', and Step 3, 'Share', play an important role.

By inviting everyone to write down what they know about the focus of attention, two things are happening. Firstly, it ensures that the subsequent discussion is well-structured and focused – focused, indeed, on the focus of attention! That's because everyone has been thinking before speaking – a benefit at any meeting. Secondly, it is a tacit recognition that everyone has the right to contribute, and that everyone's knowledge and experience is recognised and respected.

And orchestrating the 'share' such that each person makes a single statement in sequence, going around everyone in the group until everyone's points have been captured, ensures that everyone has as much 'air time' as they need, and that everyone makes as many contributions as they wish. This avoids the situation, experienced in many meetings, in which the loud people make speeches, whilst the quiet ones become progressively distracted and feel increasingly disenfranchised. Furthermore, it creates the conditions in which those who are not speaking are actively paying attention and listening, for they know it will be their 'turn' to speak again quite soon, and no one wants to be told 'we've just had that one!'.

12.2 The workshop themes

Yes, you can invoke the *InnovAction!* process by yourself, but as I have discussed on many occasions in this book, generating ideas is far more effective when done in small groups, with everyone contributing their knowledge, experience, and curiosity, so as to maximise the opportunity to combine all those Koestler's Law fragments in different ways. From which creativity springs. Much of my work is organising and running creativity workshops, attended by from six people to sometimes 24 of more (which corresponds to three groups of eight, or four groups of six, working simultaneously), and lasting from about three hours to five days (with a preference for two days).

If a group of people are assembling, perhaps from distant locations, it's important that the time is well-structured. My workshops are therefore very well thought-through, with an agenda that does not prescribe what is going to happen at 10-minute intervals, but rather in half-day sessions. This allows plenty of freedom to explore, but with guidance as to what that exploration should be about, as expressed by one or more appropriate themes – each of which, for idea generation, is a particular focus of attention, described by a one-page brief, as discussed in more detail shortly. That ensures that all members of the group are 'looking in the same direction', ideally from rather different perspectives.

Any focus of attention should be an issue relevant to the appropriate community. Sometimes it relates to a well-defined problem-to-solve, such as 'how to reduce the costs of [this process]', 'how to increase our market share', 'how to improve the durability of [this product]'. Sometimes, however, the focus of attention is more exploratory ('How might we be more effective in recruitment?'), or simply a topic that is not associated with any problems at all, but might benefit from some fresh thinking – as we saw, for example, for the theme 'new ideas based on chess', discussed in sections 10.2 and 10.3: there is nothing at all 'wrong' with chess-as-we-know-it, yet many ideas (and some very good ones) can be generated by asking 'how might this be different?' In fact, it is my experience that some of the very best ideas have emerged when discussing a focus of attention that was open-ended, and not a problem-to-solve. It's very easy to be complacent about something that apparently works well, and in a competitive market, knowing how something that works well can be made to work even better is a very strong position to be in.

12.3 Who should participate?

That's an important question, for those who gather together at a creativity workshop are giving their time, which would otherwise be spent doing other things, and that time should not be wasted. Furthermore, it's important that the workshop is as productive as possible, and so all the 'right' people should be present.

Much depends, of course, on the specific focus of attention, but as a general rule, the broader the range of knowledge and experience that can be assembled, the better. Since everything is based on observation of the 'current state', the richer those observations can be, the longer and more extensive the basic 'list' of features compiled as a result of Step 3, 'Share'. This in turn provides a wealth of material for Step 4, 'How might this be different?', and the more varied the experience of the participants, the more likely it is that the subsequent exploration will discover 'interesting' Koestler's Law patterns.

There is a strong temptation to involve only the 'experts', those with deep knowledge and appropriate expertise, those who 'own' the problem. Accordingly, the product design team would convene to generate ideas for a new product, the sales team to explore how to improve customer relations, the senior academics to discuss research opportunities. Yes, all these experts do have much to contribute. But so do many others.

So, for example, the factory team can often make very valuable contributions to discussions about new products, for they will know much about the existing products – especially as regards how they are manufactured – as well as anticipating some of the manufacturing implications of new possibilities; service engineers bring the perspective of the product in use; and customers notice many aspects of any product that the designers have never thought of. Discussions on customer relationships are much enhanced by inviting the customers to participate too, for there are many features of the customer-supplier relationship that only the customer experiences. And academic questions benefit enormously from a rich mixture of less, and more, experienced people, for the less experienced are less 'trapped' in their learning, and can (if the permission is right!) ask 'the emperor's new clothes' questions (in Hans Christian Andersen's 1837 story [1], it is a child that speaks up; nor is Andersen's version 'original' – its antecedents date back to

14th century Spain and even earlier stories told in South Asia [2], so yet another example of Koestler's Law).

Also, bear in mind that academic discussions can sometimes be badly hampered by the 'expert' syndrome, which manifests itself in three ways: firstly, if the level of permission is low, as it often is, then only the 'experts' speak; secondly, those who do not consider themselves 'experts' self-censor, and don't consider themselves 'qualified' to contribute; and thirdly, those 'not in the club' are either not invited to participate in the first place, or, if invited, self-exclude on the grounds that 'this is nothing to do with me'. This despite overwhelming evidence that much of the greatest science has happened on 'boundaries', the edges between conventional disciplines, where skills can cross-fertilise – just like happens naturally in genetics, just as insightfully described by Koestler's Law.

As I've already mentioned, idea generation is most effective in small groups of six, seven or eight people. This represents the minimum number of workshop participants, but workshops can, very effectively, be run with up to around 24 people, working in three groups of eight, or (usually better) four groups of six.

The composition of the groups also benefits from some careful thought, with participants well-matched to the focus of attention that each group will be addressing. Sometimes the best composition is as broad as possible, mixing all levels of experience and different roles; sometimes, peer groups are preferable, for example, when discussing aspects of organisation structure and experience – so in an academic research environment, a topic such as 'The role of the post-doctoral research assistant' might be examined by a group of post-docs, who can describe their own experience; by a group of PhD students, who experience a key aspect of that role in that post-docs are quite likely to be their supervisors; and also by a group of faculty members, who themselves are, organisationally, 'above' the post-docs and so have a third, different perspective. Structuring this workshop as three peer group discussions taking place separately and simultaneously in different rooms helps safety, especially for the PhD students who might not feel comfortable expressing what they truly think in front of those more senior. And once the three discussions are complete, the whole team can assemble in plenary, with each group presenting their findings, with no specific attribution of any particular point to any particular person.

Very importantly, all participants must have an understanding of the principles of 'deliberate creativity', including the Target Diagram, Koestler's Law, learning and unlearning, and the ***InnovAction!*** process, as usually obtained by attendance at a training programme. That applies to the most senior people too – even though they are senior and busy. For in the absence of that understanding, the participant will become very frustrated, if not angry, at many aspects of the process, not least, right at the start: when everyone else is sitting 'individually and in silence, describing', that person is wondering why the room is so quiet and starts talking accordingly, usually blurting out their specific 'solution'. Then, the discipline of the share, in which each person makes one point in turn, is alien; and the question 'how might this be different?' makes no sense at all. This is annoying to the person who does not understand the process, and highly disruptive to everyone else. So best avoided. In fact, at the workshops I run, I have a rule: if you haven't attended a training session, you can't attend a workshop. Even if you are the boss.

12.4 How workshops are structured

My preference is to convene idea generation workshops residentially, off-site, over two days. There are many other possibilities of course, from a half-day session in the office to off-site events over as many as five days. I'll discuss the two-day off-site format in some detail first, and then turn to the others.

12.4.1 Why off-site, and why two days?

The two questions are related, for such an event, in my experience, is much more productive than two consecutive days on-site, and certainly more than twice as productive as compared with a one-day event.

A key reason why two off-site days work better than one (and better than two on-site days too) is that something 'magic' happens overnight – participants' brains don't stop working, and everyone has, quite literally, 'slept on it'. So the most productive session is usually the morning of the second day. Also, of course, two days allow for a much greater range of material to be explored.

And there are three benefits of being off site. The first is 'being away from the office' and its distractions, and temptations; the second is attributable to a communal evening meal, especially if the team attending the workshop do not usually work together in the same physical space; and the third is the flexibility on timing that the first day offers. If a two-day event takes place on-site, with participants going home in the evening, then, by 5 pm or thereabouts, quite naturally, people are looking at their watches, thinking 'I'll need to be off soon', and groups inevitably fall apart as individuals leave. On a residential event, however, there is no 'home' to go to at 5 pm, and the only 'fixed point' is the evening meal, probably scheduled for around 7:30 or 8:00 pm. The afternoon session can therefore extend, without duress on any individual, and without people drifting away, towards 6:00 or 7:00 pm. Indeed one of the signs of success is when I have to go to groups and say, 'it's coming up towards 7:00 pm and the meal is scheduled for 8:00. Why not move towards a close, and you can continue in the morning if you wish...'.

12.4.2 The workshop agenda

When people assemble at the start of the first morning, there is a brief introduction – say, 10 or 15 minutes – followed by an ice-breaker activity if the participants don't know each other particularly well. I then start the first idea generation exercise, in which, as already discussed, participants work in (usually) pre-determined groups of 6 to 8 people on a particular focus of attention. So, if there are 20 participants, that would normally be two groups of 7 and one group of 6, or thereabouts.

Although the activity starts in silence, once the share gets going, and especially whilst ideas are being generated, there is a lot of noise, so the groups can work in separate rooms, or, if in a single room, well-spaced from one another. In fact, my preference is for one large room, for some 'leakage' from one group to another can often enrich the 'Koestler's Law' mix.

With three groups working in parallel, it is possible for each group to work on the same focus of attention – in which case, everyone in the team has been able to contribute to the same issue, and the groups can compare-and-contrast their findings. Alternatively, each group could address a different focus of attention, giving the benefit of wider coverage. So when planning the workshop it's worth thinking that through.

My experience is that a good discussion of a well-chosen focus of attention lasts at least three hours, sometimes more – but half-a-day is a good planning assumption. I take a very low-tech approach – which seems to work well, despite the lure of technology – with groups working around flip-charts, which have the benefit that, once each sheet is full, it can be taken off the pad and attached (safely!) to a nearby wall, partition or other surface, which might have been pre-covered with brown paper. And as the process proceeds, more and more wall-space is filled, making a powerful visual image of intense, creative, activity. Also, participants are encouraged to complete 'meteor cards' to capture any idea, question or other 'interesting' thought that happens to pass through their minds, as can happen at any time. These too are put on the walls so that others can see and read them – any one might be that key Koestler's Law fragment.

My preference is not to organise formal tea or coffee breaks, but rather – if the workshop venue permits – to have refreshments available 'on tap', so participants can have a drink or a biscuit whenever they wish.

After about three hours, each group will have done a lot of work, with detailed descriptions, and – usually – many ideas too. And if the event started in a morning, by then it's lunchtime.

After lunch, the groups feed back to each other. My usual practice is to for the whole team to assemble around a particular group's flip-charts, so that the group can describe what they have done, covering both the initial descriptions and the resulting ideas. The feedback from the team is usually unrehearsed, and without the preparation of a PowerPoint slide show – it is 'natural', and 'from the heart', drawing on the discussions that the team has had, and the intrinsic knowledge of, and insight into, the focus of attention.

The whole team will be familiar with each group's focus of attention, and will be keen both to listen and to contribute to the discussion. Accordingly, the feedback from each group inevitably triggers a lively discussion with the whole team, in which more ideas emerge – and, importantly, are written down.

My experience is not to hurry this or to time limit it unduly. I do not say 'the presentation must last only [so-many] minutes', and I do not hold up cards saying '2 minutes to go': there is no reason to do this if the discussion is rich, for the whole purpose of the workshop is to achieve that very objective. And when the discussion naturally dies away, the team then moves to another group's area for another feedback session, until all groups have been heard. The event has a natural, unforced, rhythm, and a rhythm that might be rather longer than one might think – any one group's feedback can sensibly and legitimately last for an hour or 90 minutes. So if there are three groups, that's three to four hours... which is the whole afternoon.

The first day therefore has a very simple agenda: a brief introduction; idea generation in groups; group feedback and discussion; dinner.

And the second day is very similar – in the morning, another group idea generation session, usually with individuals in different groups; lunch; group feedback and discussion, and then a 'conclusions and next steps' plenary discussion before the close.

12.4.3 Longer, and shorter, durations

Sometimes, especially if participants are assembling from distant locations, it is appropriate to extend the workshop for a third day, or even longer, with more group work and covering more themes. By the end of the third day, though, my experience is that not only have a huge number of ideas been generated by that time – usually far too many for any organisation to progress, but including some real 'nuggets' – but in addition participants are by now quite tired. So if the event continues to a fourth or fifth day, that additional time is more effectively spent doing something different, for example, a social event, a discussion of another matter of interest to the team but not involving any particular creativity, or to begin the process of evaluation, classifying the ideas generated on earlier days and then formally evaluating the best ones, as described in chapters 14 and 15.

A one-day event allows for one idea generation session in the morning, followed by the group share in the afternoon, and can certainly be effective; a half-day idea generation event can be effective but allows only very limited time, if any, for a group share, so this works best for a community of up to about eight people, working in a single group, and generating ideas for a single focus of attention.

12.5 The idea generation group briefs

To maximise the productivity of the group discussions, and to reinforce the effectiveness of the *InnovAction!* process, the topics discussed in the workshops I run are each defined by a one-page brief specifying the corresponding group's focus of attention, and offering some guidelines as how to approach the topic. And rather than constraining the discussions, my experience is that a clear, perceptive, and insightful brief is a major contributor to the liveliness and quality of the group interaction, and the range and value of the resulting ideas. I therefore take considerable care in their drafting, and I ensure that the drafts are reviewed by the workshop's sponsors, revised accordingly, and then finally agreed, well before the workshop starts; I also prepare many more briefs than are likely to be used, so that there is flexibility as the workshop progresses.

The following few pages show some examples of workshop briefs written to encourage idea generation using *InnovAction!*. For the focus of attention defined by the exercise title, each brief broadly follows a similar structure:

- An initial activity in which participants, individually and in silence, describe what they know.
- The share.
- A collective activity asking 'How might this be different?'
- The preparation of a brief presentation, summarising the group's discussions and conclusions, to be given at a subsequent plenary session.

Importantly, each brief does not tell participants what to think, but rather guides what participants should think *about*. This helps structure the group discussions, for all participants have been thinking about the same issues in broadly the same way. Furthermore, since each participant has been writing their own notes, everyone had become engaged, and so everyone can contribute to the group discussion in a more considered way than might often happen at meetings in which (some) people talk without the benefit of having thought about things first.

If a different process is to be encouraged – for example, the use of random words – or if the purpose of the discussion is different (for example, to evaluate an idea that has already been generated), then the brief does of course have a different, and appropriate, structure. But whatever the wording and structure of the brief, the overall objectives are the same:

- to secure everyone's engagement
- to help provide a framework within which everyone can contribute, even those who are naturally quiet, whilst suppressing a tendency for dominance by the more noisy
- to help make the discussion focused, well-structured, and productive.

Example creativity workshop brief – Content

Improving [this product] / [this item of equipment]

Here is an opportunity to improve our main product…

Working firstly individually and in silence, deconstruct [this product]/[this item of equipment], identifying all the individual component parts, their materials, their function, their cost (cheap, moderate, expensive), how they are assembled…

When everyone has finished, share your thoughts with each other.

Then, systematically, for each item, ask 'how might this be different?' – for example, 'What if it were a different material?', 'Bigger', 'Smaller', 'A different size and shape', 'Totally absent', 'Configured differently relative to other components'… and see what happens! In doing this, remember you are looking for as many possibilities as you can think of, rather than the 'one right answer'.

How many ideas can you generate?

As a team, prepare a 15-minute presentation entitled 'Some ideas to improve [this product] / [this item of equipment]'.

Example creativity workshop brief – Content

Super-sensitive instrumentation

What would happen if the instruments we used were 1,000x more sensitive?

Working firstly individually and in silence, describe, in as much detail as you can:

- ➤ Each of the signals we detect.

- ➤ The instruments we use to measure each of these signals, identifying how the instruments are constructed, and the principles on which they work.

- ➤ The factors that determine the sensitivity of each instrument (please define 'sensitivity' – but it's something to do with the minimum signal that can reliably be distinguished from the background noise).

Supposing that any instrument might be 1,000x more sensitive, identify as many reasons as you can why this might be a 'good thing'? What would this enable that can't be done now?

What, specifically, needs to be different about our current instruments to make them 1,000x more sensitive?

When everyone is ready, share your individual thoughts. Then, as a team, explore 'how might this be different?', with the intention of identifying what needs to be done to achieve a significant increase in instrument sensitivity.

As a team, prepare a 15-minute presentation entitled 'Some ideas for super-sensitive instruments'.

Example creativity workshop brief – Content

Oops, no [whatever]

Political turmoil in Wherevera has caused the supply of [this material] to cease. We just can't get it, at any price...

Individually and in silence, make some notes on

- What, currently, we use [this material] for...

- ...and, in detail, how and why they work.

- What would we no longer be able to do if we no longer are able to use [this material]?

When everyone has finished, share your individual thoughts with each other. Is there a natural consensus on some points? Is there variety?

Then, as a team, explore these questions:

- How many ways can you think of that would enable us to achieve the same outcomes, but by different means?

- How would we test those out?

What ideas does this generate?

As a team, prepare a 15-minute presentation entitled 'Some ideas on alternatives to [this material]'.

> **Example creativity workshop brief – Process**
>
> # Operations
>
> How might we improve our operations, and make them faster, more effective, less costly?
>
> Working firstly individually and in silence, draw a detailed process map of how we currently [do this], specifying all the activities from [start] to [finish].
>
> When everyone has finished, share your thoughts with each other, and compile a collective, well-structured, and complete process diagram.
>
> Then, working as a team, for each step in the process, ask 'how might this be different?', with a view to identifying as many ideas as you can to improve the process, for example, halving the time, doubling effectiveness, halving the cost…
>
> As a team, prepare a 15-minute presentation entitled 'Some ideas to make our operations faster, smarter, cheaper.'

Example creativity workshop brief – Structures

Our structure

A key feature of today's world is that our structure comprises [three Divisions].

How might this be different?

Imagine a 'world' in which we are a single unit.

Working individually and in silence, make some notes on these questions:

What, specifically, would this 'world' look like? In particular, what would be *different* about this 'world' as compared to our current, real world? For each difference, is the difference better, or worse, than the current situation – or neutral?

When everyone has finished, share. Is there a natural consensus, or some disagreement - in which case, resolve the disagreements to compile a common view.

Given that the real world is one of [three Divisions], what might we do differently to reap all the benefits of the imaginary world, whilst avoiding all the problems?

What are your recommendations?

As a team, prepare a 15-minute presentation entitled 'Optimising our current structure'.

Example creativity workshop brief – Relationships

Working together

Working firstly individually and in silence, describe, in as much detail as you can, what has really happened when you have worked with others, as triggered by these prompts:

- ➢ **Think of some examples in which working together was really good. Why was it good? Think through some of the details such as who did what, how decisions were made, roles, what caused any conflicts to arise, how conflicts were addressed…**

- ➢ **Think of some examples in which working together was not so good. Why so? Think through some of the details such as who did what, how decisions were made, roles, what caused any conflicts to arise, how conflicts were addressed…**

When everyone is ready, share your individual experiences, and, as a team, draft some 'best practice' guidelines for working well together.

> **Example creativity workshop brief – Relationships**
>
> # Which team?
>
> We all want to be good 'team players'.
>
> The *BIG PROBLEM*, though, arises from the fact that each of us is a member of several teams simultaneously.
>
> And when the demands of two different teams are in conflict, we have to choose… and that means letting someone else down…
>
> Working firstly individually and in silence, make some notes on these topics…
>
> - What are the 'teams' of which you are currently, or have recently been, a member?
> - Identify as many different conflicts, or types of conflict, that have arisen as a result of multiple team membership.
> - How, and why, have these conflicts arisen?
> - How have these conflicts been resolved – or indeed left unresolved?
> - What have the consequences been?
> - What are the implications of all this for 'team us'?
>
> When everyone has finished, share your individual thoughts for each of these topics with each other.
>
> Then, as a group, identify as many ideas as you can whereby these conflicts could be avoided in the first place, or resolved more satisfactorily if they are otherwise unavoidable.

Example creativity workshop brief – Relationships

Are 'wholes' necessarily greater than the 'sum of the parts'?

'The whole is greater than the sum of its parts' is a familiar cliché. But it is true? Or rather, what are the conditions under which it's true – or indeed not…?

Working firstly individually, and in silence, make some notes on these topics:

- **What do you think 'the whole is greater than the sum of its parts' might mean in an academic context? How might any 'greatness' be identified?**
- **Can you think of any examples (not necessarily academic ones) – from your own direct experience or otherwise – where the whole is, or has been, greater than the sum of its parts…**
- **…and where it isn't, or wasn't?**
- **Identify as many benefits as you can for a situation in which the whole is greater than the sum of its parts…**
- **…and where it isn't.**
- **Thinking of an academic community, compile two lists: the first of the features of that community when it behaves as greater than its parts, the other when it doesn't.**
- **What fundamental principles can you identify which explain the differences between these two lists?**

When everyone has finished, share your thoughts with each other. Imagine you a team of consultants, asked to advise an academic department at a university in England on what, practically, they could do to enhance their 'wholeness'.

Prepare a 15-minute presentation for their senior management.

Example creativity workshop brief – Strategy

They don't know what they're missing!

Many other researchers who aren't in our discipline might well benefit from collaborating with us.

But if they don't know who we are, or what we do – or if they don't even recognise that our skills can help – they will never get the benefit of working with us.

Your task is to generate as many ideas, and as much content, as possible as to how we might make ourselves more 'visible' to communities and disciplines with whom we rarely interact, and even more rarely collaborate with.

So, individually and in silence, make some notes on:

➢ What those communities and disciplines might be.

➢ All the different media and methods that might, in principle, be used to make contact accordingly.

➢ The key content of the 'message' we would wish to put across – making sure to use language that would be immediately accessible to someone not in our discipline.

When everyone has finished, share your individual thoughts with each other. What ideas does this generate?

As a team, prepare a 15-minute presentation entitled 'Some ideas to enhance our visibility'.

Example creativity workshop brief - Strategy

Optimising our state function

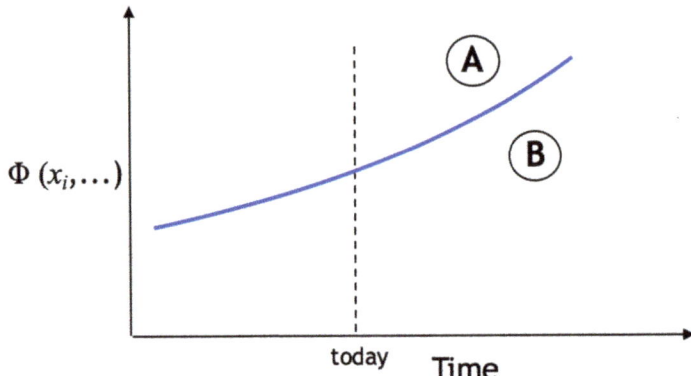

$\Phi(x_i, \ldots)$ is a function defining our overall 'state'.

So far, we have been following the trajectory shown in blue, and if we were to continue on our current trajectory, we would follow the blue path into the future. There are, however, other possible states: some in zone A, which are 'better'; others in zone B, which are 'worse'.

Assuming all future possible states are potentially accessible, working firstly individually and in silence, make some notes on…

- What are the key characteristics of states in zone A? How, specifically, are these states different from the extrapolation of the current trajectory?
- Why, precisely, are these states 'better'?
- What are the variables x_i? Which variables are the most important, namely, those for which $\partial\Phi/\partial x_i$ is relatively large and positive?
- What therefore needs to happen, now, to reach states which are both ambitious and realistic?

When everyone has finished, share your thoughts with each other. Is there a natural consensus on some points? Is there variety? As a team, prepare a 15-minute presentation entitled

'Optimising our trajectory – recommended actions'.

Example creativity workshop brief - Strategy

Our major competitor's *BIG IDEA*

Our major competitor has just hired this team to spearhead their commercial attack on us. You have an unlimited budget. Your task is to prepare the plan of what needs to be done to carry out the attack.

Working individually and in silence, make some notes on these questions:

What are our current vulnerabilities?

How might those be exploited by a truly aggressive competitor?

How would we be interacting with equipment sourced by us?

When everyone is ready, share, and discuss your collective findings. What ideas does this generate as to how we can protect ourselves against attack, and so improve our business?

What are your recommendations?

As a team, prepare a 15-minute presentation entitled 'Some ideas to strengthen our strategic position'.

12.6 Don't impose constraints on cost and resources

All ideas have implications and consequences, and in many circumstances, those involve costs and resources. The most obvious example relates to ideas for new products, for resources will be required for design, prototyping, manufacturing and marketing, and the corresponding costs will be incurred. So whilst ideas are being generated, an immediate objection is 'this will cost too much' or 'we don't have the resources to do that'.

These statements, however, are blockers. Either one stops the discussion. Stone dead.

Which, in a workshop intended to generate ideas, is unhelpful.

So, to stop this from happening, and to help the process of idea generation to flow, I always say at the outset 'during this workshop, unless explicitly stated otherwise, assume that you have access to infinite funds and resources'. Which always leads someone to respond, 'Don't be stupid – we live in a real world where of course funds and resources are limited and constrained'.

Half of this statement is true, for, yes, the real world is inevitably constrained. But the other half is not true, for the removal of the constraints at this stage of the process is not at all stupid: on the contrary, it is both wise and beneficial.

The benefit is that the removal of these constraints enables the process of idea generation to progress, and not be blocked, and perhaps a truly powerful idea will emerge. Yes, that idea might be associated with considerable costs, and, yes, to implement the idea might require resources that the organisation might not currently have.

But suppose that the idea is really good, and that the new product, say, is believed to have considerable sales potential. That generates revenue, and there is the possibility that the revenue stream might be more than enough to cover those costs, and to justify the acquisition of the required resources. So although the new idea is indeed costly, bearing those costs could well be commercially sensible. At the outset, though, it is very difficult to make that judgement, and, as will be seen, making that judgement is best done as part of the evaluation process, which necessarily takes place after the ideas have been well-formulated. So to dismiss a potential idea on the grounds that 'this will cost too much' is both unwise and potentially foolish.

And there's another important factor too. The cost of any new product is determined by a host of factors, from the costs of raw materials to the costs of marketing. But by letting the creativity process run so as to develop a good understanding of what that new product might be, that then provides insight as to how it might be designed and manufactured, and how it might be distributed and sold. Each of these – and many others too – are ideal topics for subsequent creativity workshops to generate ideas relating to, for example, 'how many ideas can we generate to reduce the cost of materials?'. This workshop would take the current specification of the new product, and then apply the ***InnovAction!*** process to deconstruct the proposed design, to scrutinise the nature and quantity of the envisaged materials, to challenge the current thoughts on how it might be manufactured, all being done with the intention of discovering as many ideas as possible that will result in a new product with appropriate functionality, but with a lower material cost. In my experience, this can always be done.

So, although it may appear strange, if not profligate, to 'wish away' cost and resource constraints at the start of an idea generation workshop (except, of course, when the focus of attention is explicitly to reduce costs or consume fewer resources), it is usually wise.

12.7 Creativity, not evaluation

It's really important to ensure that creativity workshops are about creativity, and the generation of ideas – and not about being judgemental, about evaluation. One of the key features of the Target Diagram, figure 2.1, is the distinction between 'creativity', the central 'red' zone, and 'evaluation', the 'yellow' zone, with evaluation taking place after creativity.

That can be difficult, especially for people who haven't participated in this type of creativity workshop before, for being judgemental is a very normal behaviour, and associated with many every-day speech patterns, such as 'But what about [this potential problem]?', 'I'm not sure [Pat] would like that…' and 'Yes, in principle, that could work, but in practice…'. None of these are especially aggressive or adversarial (I'm sure you can think of some comments that are!); all, however, are judgemental, and act to put the proposer of the idea on the defensive, closing the discussion down.

Whilst ideas are being generated, it is inevitably the case that the initial fragments, the first ideas, are fragile, incompletely thought-through, too expensive, flawed. So those initial glimmers can always be attacked, criticised, destroyed. But if that happens, the possibility that those rudimentary fragments might be transformed into a potentially powerful idea is thrown away, and the benefits that the idea might have brought are foregone.

So, as it says in Step 5, 'Let it be…'. Don't be judgemental. Be alert to language. Be patient. And remember that this session is, intentionally, about idea generation. That is not giving a license to every idea that emerges, nor is it about making any commitments. It's just about generating as many ideas as possible – quantity, quantity, quantity – so providing as rich a supply of ideas as the team can to pass on to the next stage of the process, wise evaluation, which will take place after the workshop, as described in chapters 14 and 15.

12.8 Quantity, quantity, quantity

Workshops can be enormously productive, generating huge quantities of material. Indeed, the more ideas, and 'fragments' of ideas, the better – for the greater the number of 'components', the greater the opportunity to combine and recombine them in different patterns to generate ideas in accordance with Koestler's Law.

Many of the 'components' will be identified in the first steps, observation and sharing, and these will usually be captured on flipcharts, whilst ideas, and the 'fragments' from which ideas will be formed, will be the result of 'Let it be…', as people freely explore 'how might this be different?'. The discussions are natural, and usually energetic – but can often be somewhat unstructured. Also, good ideas never, ever, arrive as complete, awe-inspiring, 'packages' – rather, what later is recognised as a good idea can be retrospectively traced to [this] fragment combined with [that] fragment, with [this] bit grafted on and [that] removed. Each of these Koestler's Law components arose in the discussion at different times, and probably in the 'wrong' sequence. That's important, for it tells us that, usually, the best ideas emerge towards the end of the activity, when there are all sorts of 'fragments' written on flip-charts, scribbled on scraps of paper, or in people's minds – fragments that can be selected, reshuffled and recombined (as Koestler's Law describes) only later rather than sooner. And sometimes, 'later' is after the main workshop, when someone is thinking about what happened, or reviewing the workshop write-up, or is just walking around and notices something.

It's therefore important that as many of these 'fragments' as possible are recorded, and so are available for future reference, rather than forgotten. So, during the workshop, participants should be encouraged to capture their thoughts, and the content of their discussions on flip charts, as notes or drawings. Also, given that many thoughts just flash through the mind, like a meteor, and can be so easily forgotten, as mentioned in section 10.5.4, I encourage participants to write them down, succinctly, on brightly coloured fluorescent cards – 'meteor cards' – that can then be stuck on the wall for others to read: I like using cards, but of course any method will do – many use post-it notes.

And at the end of the workshop, gather together absolutely everything – flip-charts, drawings, diagrams, people's notes, fluorescent cards, post-it notes, and 'stuff' and 'scraps' that might be on the walls or tables, for this is the 'raw material' that will be used to compile the workshop report, an invaluable record of what actually happened, and of all the ideas generated.

12.9 After the workshop

12.9.1 The workshop report

Idea generation workshops can be enormously productive, generating very many ideas, and mountains of flip-charts, notes, photographs, screen shots, 'meteor cards', and other materials. But one thing that rarely happens at the workshop itself – because there isn't time – is for this is the material to be sorted, structured and collated. That happens afterwards.

As has been noted several times, the natural flow of the workshop will rarely conform to a coherent structure. Yes, any conversation will be reasonably structured at the time, but there will be many conversations, taking place at different times, and we have all used phrases such as 'I hadn't thought of this before, but now I see that...', or 'Referring to what we were discussing an hour or so ago, it just occurs to me that...'.

But once the workshop is over, all the material can be reviewed and structured, so that the resulting write-up tells a coherent story, rather than being a transcript of who-said-what-when.

My experience is that producing a good, well-structured, comprehensive workshop report is a demanding task, requiring considerable skill in sorting similar materials together, and in telling a comprehensive narrative. And it takes time too, perhaps several days.

The purpose of the report is two-fold:

- to tell the 'story' of the workshop such that those who were present acknowledge it as a good and complete description, as well as enabling anyone who was not present to understand the key aspects of what took place
- to be a 'document of record', capturing all the content, and – in particular – all the ideas and idea 'fragments' so that they are available to the community in the future.

So it's worth taking the time, and putting in the effort, to compile a good, well-structured and comprehensive report, for that ensures that nothing is lost.

12.9.2 People keep thinking

Although the idea generation workshop is over, people do not stop thinking. My experience is that the workshops are highly energising, and participants continue to be alert, to notice things, to spot new Koestler's Law patterns, long afterwards.

So to ensure that all ideas that emerge after the workshop are not lost, it is good practice for there to be easy access – for example, by using an email address available to all participants – to an 'idea archive' that can act as a 'safe place' to which any ideas can be sent. As I will discuss later (see section 17.4), the 'idea archive' is an important feature of an organisation that has a sustainable culture of safe creativity and effective innovation, for ideas and 'Koestler's Law fragments' can have significant value – perhaps not immediately at the time they were generated, but later when another 'fragment' has been identified, or when an external event has happened.

12.9.3 The next step – wise evaluation

As noted many times, the purpose of an idea generation workshop is precisely that, to generate as many ideas as possible. Once the momentum builds, this can be hugely productive, with perhaps one or even two hundred ideas, of all scales and qualities.

But which are the truly good ideas, the ideas that merit further time, effort, and investment to develop, and ultimately to implement?

That's the key question for the next step, the second zone of the Target Diagram, wise evaluation, as will be discussed in chapters 14 and 15 in part III. But before that, and to complete the discussion of 'deliberate creativity', the next chapter explores some current 'case studies' of creativity in science and engineering, contributed by true experts in their respective fields.

References

[1] www.stanforddaily.com/2020/02/04/the-bent-the-emperors-new-clothes/
[2] https://sites.pitt.edu/~dash/type1620.html—links

IOP Publishing

Creativity for Scientists and Engineers
A practical guide
Dennis Sherwood

Chapter 13

Creativity in science and engineering

13.1 What this chapter is about

The examples given so far have been historic, from Kepler's discovery of the elliptical orbit of Mars to the determination of the structure of DNA. This chapter too presents a number of 'stories' about scientific creativity, but these are different in two respects. Firstly, they are all recent; secondly, they have all been contributed by eminent people in their own fields of science, engineering and industry, so they are right up-to-date (as at the time of writing, towards the close of 2021), and they are truly real. And I sincerely thank each of the contributors!

So, to the first story…

13.2 Detecting gravitational waves

Professor Sir James Hough OBE FRS and Professor Sheila Rowan CBE FRS, both of the Institute for Gravitational Research, in the University of Glasgow's School of Physics and Astronomy

On the morning of 14 September 2015, researchers at The Albert Einstein Institute in Hannover, Germany, were monitoring the signals from the twin newly-commissioned aLIGO (Advanced Laser Interferometer Gravitational-wave Observatory) instruments in the USA – one in Hanford, Washington State; the other, some 3,000 km distant, near Livingston, Louisiana. Shortly before noon (German time), two oscillating signals were seen: the first from Livingston, recorded at 04:50:45 (local time in Louisiana); the second, following by about 7 milliseconds, from Hanford (see figure 13.1). There was then some very careful checking, double-checking and triple-checking to ensure that those signals were not spurious, accidental or coincidental. Five months later, on 11 February 2016, it was officially announced that gravitational waves had been detected for the first time [1, 2].

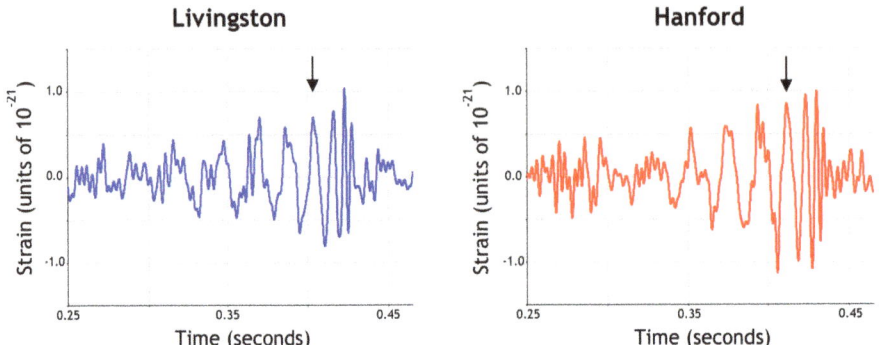

Figure 13.1. The original signals detecting gravitational waves, firstly at Livingston and then at Hanford (the 7 ms delay can be inferred from the time difference between the two equivalent peaks indicated by the arrows). The vertical axis is the strain $\Delta l/l$ measured by the interferometer as explained in the text; the time axis is relative to 04:50:45 Louisiana time, 09:50:45 UTC.
Data downloaded from www.gw-openscience.org/events/GW150914/.

The story of gravitational waves goes back more than a century, an especially significant milestone being a brief (five page) paper, published on 5 June 1905, in which the French mathematical physicist Henri Poincaré deduced, from the Lorentz transformation equations, that accelerating masses would give rise to gravitational waves, just as accelerating electrons generate electromagnetic waves [3]. Albert Einstein published his first paper on special relativity some three months later on 26 September 1905 [4, 5]; Einstein's prediction of the existence of gravitational waves was first proposed in 1916 as a consequence of general relativity [6]. But it took 99 years before the first gravitational wave was detected.

Why did it take so long?

There are many reasons, not least the fact that Einstein himself doubted that gravitational waves could exist [7]. The primary reason, however, lies in the answer to the question 'if they do exist, how big is their effect?'. Or, in this case, how small. To detect a gravitational wave, you need to measure the simultaneous changes Δl in two initially equal distances l in two mutually perpendicular directions, both in a plane transverse to the direction of the wave. The theory of gravitational waves predicts that, as the gravitational wave passes, one length l will extend by $+|\Delta l|$, whilst at the same time the other will contract by $-|\Delta l|$, such that $|\Delta l|/l \leqslant 10^{-21}$ [8]. If the detecting instrument has two perpendicular 'arms' each 1 km = 10^3 m long, then the magnitude $|\Delta l|$ that needs to be measured is $\leqslant 10^{-18}$ m. That's small. Exceedingly small. For comparison, the diameter of the nucleus of hydrogen – a single proton – is about 10^{-15} m [9].

To detect a gravitational wave therefore requires an instrument that can reliably measure oscillatory differences in distance that are three or four orders of magnitude smaller than the diameter of a proton, such that any measured oscillations at frequencies above a few cycles per second must be attributable, unambiguously, to a gravitational wave, and not be caused, for example, by a small earthquake, or a rabbit that has just been hopping across a nearby field. And of course there is the problem that any measuring stick will also experience the wave, and so might expand and contract too…

The detection of gravitational waves is therefore a story about the most wonderful creativity, alongside prodigious skills in engineering, as well as great scientific insight, in the design, construction and use of what must surely be the world's most sensitive measuring device. And the wonder is not that it took so long, but that it happened at all.

There is an intriguing scientific 'symmetry' in the fact that the instrument that discovered the first gravitational waves is in essence the same as that used to provide the key evidence for special relativity – the Michelson interferometer. In 1887, Albert Michelson and Edward Morley attempted to measure the speed of light in each of two mutually perpendicular directions – two speeds that should, according to the theory at that time, have been very slightly different [10]. Their equipment comprised a half-silvered mirror that split an incident beam of light into two mutually perpendicular beams, which were then reflected back by distant mirrors to recombine and form an interference pattern. Study of the structure of that pattern then enabled the difference in the speeds of light in each of the perpendicular beams to be determined. As we all know, no meaningful speed difference was detected, a result that remained a puzzle until finally resolved by special relativity (see section 5.13).

The aLIGO instruments at Livingston and Hanford [11] – and also GEO600 [12], near Hannover in Germany, Virgo [13], near Pisa in Italy, and KAGRA [14] in Japan – are still known as Michelson interferometers, although all are (of course!) vastly more sophisticated. But the fundamental principles are the same: an incident laser beam is split into two mutually perpendicular paths, and then reflected back-and-forth many times in each arm in a Fabry-Pérot resonance cavity between two mirrors, so enabling a much longer distance to be travelled by the light, as compared to the physical dimensions of the instrument [15]. The beams ultimately recombine to form an interference pattern, and changes in that pattern can be interpreted as changes in the distances traversed by the two split laser beams.

Within the interferometer, the fundamental 'measuring stick' is the wavelength of the laser light, which can be defined to remain constant at 1.064×10^{-6} m even in the presence of a gravitational wave. To detect a wave, the instrument must be able to measure perturbations in distance of the order of 10^{-18} m – that's a factor of 10^{-12} smaller than the wavelength. How can an instrument be designed and built to such a stringent requirement?

Professor Hough picks up the story…

'Well, the solutions took many years to emerge, and only after much hard work. And with any number of successes, failures, disappointments, elations, funding crises – and some luck too – along the way!

Most importantly, for detection, it was absolutely essential that any movements of the mirrors due to a gravitational wave must be greater than those from any other causes. So, firstly, we had to reduce any movements of the mirrors resulting from vibrations, such as seismic events, or, more prosaically, nearby traffic. To achieve this, we suspended the mirrors 'in free space' as the lowest mass in a suitably damped triple or quadruple pendulum, as illustrated schematically in figure 13.2, and more realistically in figure 13.4. That's very subtle: the damping must be good, very good. But not so 'good' as to prevent the detection of a gravitational wave. Fortunately, however, the gravitational waves we were seeking to observe have a frequency of tens and low hundreds Hz [16], and we were able to 'tune' the suspension mechanism to allow the detection of gravitational waves whilst filtering out all other mechanical disturbances.

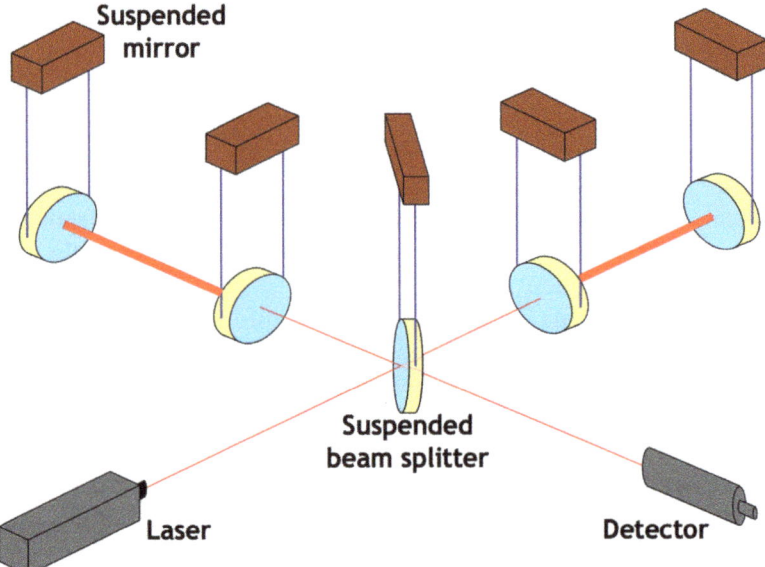

Figure 13.2. Schematic, and much simplified, representation of a Michelson interferometer as used to detect gravitational waves. The heavier red lines represent the multiple reflections of the laser beam within the Fabry-Pérot resonant cavity between the mirror pairs, so increasing the total distance travelled by each beam, within the two points fixed by the mirrors. In this simplified representation, each mirror is shown suspended as a single pendulum; in the real aLIGO instrument, the suspension mechanism is a quadruple pendulum, as illustrated in figure 13.4.
Adapted from www.universetoday.com/127286/gravitational-wave-detectors-how-they-work/.

Secondly, we had to eliminate movement attributable to fundamental physical events, such as the Brownian motion of the molecules in the mirror material, and dissipative thermal losses in the mirror suspension wires. The mirrors are therefore engineered extremely precisely from ultra-pure monolithic fused silica: the surfaces are nanometre smooth, and the low-water-content material minimises the absorption of the infra-red laser light used to illuminate the interferometer, which would otherwise cause local heating. And all the optical components are within a vacuum chamber at a pressure of around 10^{-11} atmosphere [17], approaching that of CERN's Large Hadron Collider [18].

One particularly challenging problem concerned the material used to suspend the heavy masses of the fused silica mirrors. The original LIGO instruments, which started operation in 2001, used steel wires, but the more recent aLIGO instruments use fused silica fibres: silica has the benefit of combining low mechanical loss with high mechanical strength, as well as being much less susceptible to thermal noise than steel [19]. Using silica to suspend masses is not new but goes way back – for example, in the apparatus used by Sir Charles Boys to measure the gravitational constant, 'Big G', as long ago as 1895 [20]. More recently, during the 1990s, Vladimir Braginsky at the University of Moscow was studying silica fibre suspensions [21], and Rai Wiess had earlier been experimenting at MIT with rather thicker silica rods [22]. Thin fibres are in principle better, for they tend to be more pure, and so less susceptible to inclusion flaws that cause weakening. But there was a **BIG PROBLEM** – how can a very thin fibre be reliably attached to a massive object such as an interferometer mirror?'

Professor Rowan continues (referring to Sir James Hough as Jim!):

'In 1994, I was finishing my PhD under Jim's supervision in Glasgow, and had the opportunity to give a presentation at a conference to be held at Stanford. I say 'presentation' – in fact, I had just five minutes. Why bother to go all that way for just five minutes? Yes, Stanford is a lovely place, and Palo Alto is pretty good too. But five minutes? As it turned out, though, that five minutes was an important stepping stone to success in detecting gravitational waves. I didn't realise that at the time, of course, but my part of the story starts there. After my five-minute talk, I happened to have a conversation with one of the Stanford scientists, Eric Gustafson. That led to an introduction that opened an opportunity for me to visit Stanford, which I took up in 1995. And there's been a great collaborative relationship between Stanford and Glasgow ever since.

Also at Stanford at that time was Professor Francis Everitt, whose team was using fused silica to make the most perfect spheres then known for the gyroscopes to be used in the 'Gravity Probe B' satellite, which was being designed to carry out experiments in space to test some predictions of general relativity [23]. A property of fused silica, which was exploited to great effect, is that it can form very smooth surfaces – that's how the spheres could be so perfect. But smooth surfaces are needed for interferometer mirrors, and so those mirrors are manufactured from fused silica too. And the same material – fused silica – can also be drawn into fine, strong, threads with a very low coefficient of thermal expansion.

Jim is fond of pointing out from student days in the chemistry lab that you were told never to keep sodium hydroxide in bottles with ground glass tops because the NaOH reacts with the glass to bond the top so it can't be opened. Well, Francis's team had realised that alkali hydroxides, sometimes with an additive of colloidal silica, can be used to form a strong, secure, bond between two fused silica components. And as part of their research, they had developed this into a novel process, which they had patented [24], for bonding the silica optical components within Gravity Probe B [25].

When a Stanford colleague, Professor Dan Debra, heard about this, he suggested that this new 'hydroxide catalysis' method might be the solution to the problem of bonding fused silica support fibres to the silica mirrors in the GEO600 interferometer, then being built as a collaborative project between Glasgow and the Max Planck Institute for Gravitational Physics – the Albert Einstein Institute – near Hannover [26]. We therefore asked our Gravity Probe B colleagues if we might be able to try that method out; they agreed, and, that's what we did [27].'

Jim Hough explains further:

'Yes, hydroxide catalysis was the key idea as to how to bond the fused silica fibre to the mirror, but to get that to work required the team to solve any number of problems. So, for example, our group developed a totally novel carbon dioxide laser system [28] to heat the silica from which the thread could be drawn, to polish the resulting fibres, and to weld these fibres to the suspension 'ears' shown in figure 13.3 [29].

Figure 13.3. The suspension 'ears'. Two fused silica fibres are securely welded to a block of fused silica, 60 mm wide, the flat surface of which is bonded to one side of the mirror; a similar component is bonded to the other side of the mirror, so enabling the mirror to be suspended by the four fused silica fibres, as shown in figure 13.4.
Source: Courtesy Caltech/MIT/LIGO Laboratory https://advancedligo.mit.edu/graphics/5318_20121116150521_DSCF0374_llo_itmx.jpg.

It was also critical to eliminate disturbances attributable to, for example, motions due to internal thermal currents. To do this, we exploited the very special property of fused silica fibre that the effects of thermal currents can be exactly countered by corresponding changes in the material's Young's Modulus – but only if the dimensions of the fibre, and its tension, are 'just right', which took a lot of careful calculation to determine.

Back to Sheila Rowan:

'Yes, a host of problems to solve. And solved by combining research from many branches of physics – optics, condensed matter, surface physics, thermal physics, thermodynamics, materials science, finite element analysis – the lot! And in the end, it all worked!

Everything came together in the early 2000s with GEO600, which has two arms each of 600 metres, and uses a triple pendulum with fused silica fibres to suspend the mirrors, each of which has a mass of 5.6 kg [30]. In fact, the fused silica fibres worked so well that, during the mid-2000s, as Jim just mentioned, when plans were being made in the United States to upgrade the original LIGO interferometers to advanced LIGO, the decision was taken to replace the steel suspension wires by fused silica fibres [31], even though aLIGO is much bigger than GEO600: aLIGO's arms are 4 km long [32], and each mirror has a mass of 40 kg, suspended within a quadruple pendulum by four fused silica threads of only 0.4 mm diameter, as shown schematically in figure 13.4 [33]. This upgrade, along with improved external seismic isolation, increased the sensitivity of measurement by about a factor of about 10, much of which was attributable to the fused silica suspension. And it was this enhanced sensitivity that enabled aLIGO to detect the signals of September 2015 – signals that would not have been detected by the original LIGO instruments [34].

Fused silica fibre suspension, and the use of hydroxide catalysis bonding, are just two very small details within the whole project, for it took the combined efforts of over a thousand scientists from multiple countries around the globe to get to the point of detecting gravitational waves. But the fibres were a critical piece of building an interferometer that could detect those extremely small fluctuations. And putting together the pieces of the silica technology puzzle came out of giving a five-minute talk at a conference when I was still a PhD student, and then having the good fortune to meet Francis Everitt's team. Is that luck? Perhaps. But it certainly is a wonderful example of noticing things and making the right connections. Which is, of course, the essence of Koestler's Law, combining knowledge from one domain – the bonding of silica by sodium hydroxide – with another – the use of this in a space mission – and then spotting that this might solve the problem of suspending the interferometer mirrors.

I think if you look, you'll find this kind of story over and over again in the field of gravitational wave detection – where the right ideas and inventions at the right times came together to make progress.

So now, whenever I speak to graduate students, I say, 'If ever you have an opportunity to visit another lab, or to go to a conference, or to give a talk – even if it's a long way away, even if it's only for five minutes – do it! You never know what might happen...'

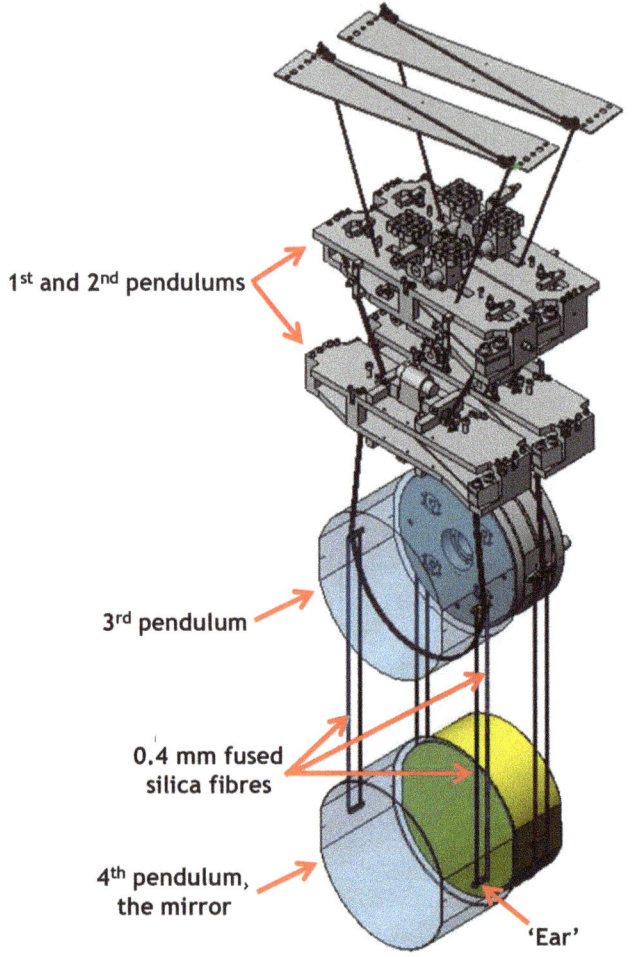

Figure 13.4. Schematic representation of the four-pendulum suspension of the aLIGO mirrors. Based on: www.ligo.caltech.edu/page/vibration-isolation.

That's just one story about the creativity associated with gravitational waves. There are many, many more…

13.3 Building Nemo

Adam Middleton, President Siemens Energy SAS France and Chair of the Management Board & Managing Director, Siemens Energy B.V., The Netherlands

'Tuesday 8 November 2016. 5.30 pm. My phone rings.

It's Mike Elmer, customer Project Director for the Nemo Link high voltage power connection project between Belgium and the UK.

I've known Mike for years, so he comes straight to the point. 'I'm worried about safety on site. Something could go drastically wrong. I'm worried someone could get hurt.'

We had two near miss incidents in two days on our construction site in Zeebrugge, Belgium, both capable of causing a major accident. Worse still, same incident, same cause. We had learned nothing.

Something had to change. Something had to be different. Or rather, many things had to be different. Not things concerning the physics of high voltage transmission. Not things to do with the technology of transformers, cables and switches. Not things to do with the concrete or the construction. But things to do with people's, attitudes, behaviours, actions. Most importantly, people had to change their minds, so that those attitudes, behaviours and actions are natural and spontaneous, rather than enforced by continuously-on-your-back supervision or the threat of punishment. As we all know, changing people's minds is really difficult to achieve. And if achieving that isn't 'creativity', I don't know what is.

And we delivered great results. Everything was completed in advance of the scheduled date. The link became fully operational on 31 January 2019. At the UK end, more than a million hours were clocked up without a lost time accident – for which the project team was awarded the *Sword of Honour* by the British Safety Council for excellence in the management of health and safety risks at work. The whole team is really proud of that. And rightly so. The team based in Belgium also recorded more than a million hours without incident.

Some words of context. Although the UK is geographically an island, the UK National Electricity Grid has been interconnected with the grids on mainland Europe since the 1960s, allowing the UK to import power when UK demand exceeds supply, and to export it when there is a surplus in the UK and a requirement on the continent. The development of wind farms in the North Sea substantially increases the UK's generation capacity, so the need for interconnectivity has grown, and will grow in the foreseeable future. Accordingly, the BritNed interconnector, linking the Isle of Grain in Kent with Rotterdam, was commissioned in 2011 [35], and Nemo Link, from Richborough in Kent to Zeebruge, as I've mentioned, went live in 2019 [36]. The most recent is IFA2, between Folkestone in Kent to Sangatte in France, which came on stream in 2021 [37]. Three others are nearing completion or under construction: the North Sea Link between England and Norway is scheduled for late 2021 [38]; the ElecLink cable runs inside the Channel Tunnel between France and England and is expected to be switched on in 2022 [39]; and Viking Link, from England to Denmark, in late 2023 [40]. Oh… and to complete the picture, there are also two interconnectors between Great Britain and Ireland [41].

Back to Nemo Link. Some numbers. The link is high voltage – that's 400 kV – and direct current, so that implies the need for all the equipment to convert between AC and DC at both ends. The cables are 140 km long, and supply 1 GW of power, which corresponds to the needs of about half-a-million homes. To build the two convertor stations, we poured 13,000 m^3 of concrete (that's about five Olympic swimming pools) and laid 970 km of fibre optic cable [42]. I could go on.

But you get the message. A big project, involving more than 1,400 engineers and project specialists across five partner organisations: in the UK, National Grid plc and the main building contractor, Murphy; in Belgium, the Belgian transmission system operator Elia and the building contractor ENGIE Fabricom (now EQUANS); with ourselves, Siemens Energy, taking overall responsibility for the construction of the two converter sites. The cable was supplied by J-Power Systems, a subsidiary of the Japanese company Sumitomo, and all the marine work was carried out by the Norwegian-owned company DeepOcean1. There were a host of sub-contractors too. As I've said before, a big project…

So after that phone call, what happened? What changed?

Let's start with what didn't change. Same project, same scope, same technology, same teams, same contractors, same customers, same leadership. Same people.

Clearly many things *did* change to make the difference: leadership, ownership, example, collaboration, communication, resilience, ambition, drive, respect, equality, creativity, responsibility, trust, teamworking. Simply, the team built a winning, learning culture and team chemistry. No individual would be seen to win. Only the team. For the overall project to succeed, there would be no competition between the two ends of the power cable in Richborough in England and Zeebrugge in Belgium: just sharing of great ideas so that the other team could further improve, and remove risk.

The project leadership met regularly, every three months, by rotation on each site. We worked together, we trained together, we shared the tough home truths together – normally informally over dinner. This built an absolute commitment not only to safety, but also to 'good housekeeping'. You've probably visited construction sites that are a tip. Lots of mud and mess on the ground; lots of muddle and untidiness in the site Portakabins. Yes, there's a lot of mud. But we instilled a culture of order, cleanliness and tidiness, which was applied everywhere, so that people took pride in their surroundings, their work, and themselves. Exactly the frame of mind that builds in safety too.

We looked after every person working on each site to the same standard, no matter whether customer, main contractor or supplier. Everyone's facilities were cleaned every day. Everybody enjoyed the same (high!) standards of food. Each site had daily briefings at the start of work every day, followed by coffee and buns. We celebrated success together, we had exchange visits between the two sites, to share best practices – or share learning when things went wrong. An example: there was a particular problem at Zeebrugge, a problem that might also happen at Richborough. But rather than trying to cover things up, and keep quiet about it (in fact, no one had been injured, but it was a near-miss none the less), the Belgian team not only reported it, but built a digital model of exactly what had happened and sent it to the English team so that they could understand the incident, and take measures to prevent it from happening there. The Belgian team were not told to do that, nor was it in their performance measures, nor was there a threat of punishment for not doing it. They just did it. Because they knew it was the right thing to do.

Our contractors and partners participated equally. We chose our new contractors on the basis of how they delivered safely. We made safety tours together with the Executive Management of all of the stakeholders – live, dangerous and totally visible, open for everyone to see.

And something else as well – something that I had not experienced in the same way before, anywhere. We involved people's families. We had open days, so that families could visit the sites and see what their fathers, mothers, husbands, wives, were doing every day. Not just that, we involved the children too: one of the truly creative ideas was the 'come home safe' campaign, which encouraged children to draw pictures of their mum or dad working safely on-site. The results were just wonderful – as you can see in figure 13.5 here:

Figure 13.5. Some examples of children's pictures from the 'come home safe' campaign. Source: Courtesy National Grid, ELIA and Siemens Energy www.siemens-energy.com/global/en/news/magazine/2020/intereuropean-hvdc-link-nemolink.html.

Those pictures had – and indeed still have – real emotional power. No parent wants to cause their child anguish. Nor – as one of those pictures reminds us – does any grandparent! Can you think of a better way of creating a culture of safety?

But it wasn't just safety. A mindset that values safety also values quality, doing things right-first-time. And high quality, right-first-time work implies there are fewer errors to correct, the costs associated with fixing problems aren't incurred, much less time is spent re-working, people are less stressed. And because there is more time and less stress, those problems that do crop up can be dealt with more calmly and effectively. Most importantly, there is more time to think. Those are huge benefits, all resulting from that fundamental commitment to safety.

The project was blessed to have some really great people, working in close collaboration. But it was no holiday camp.

Repeatable? Better to have learning, adaptation and improvement. Use the repertoire, rather than copying the recipe. And yes, the follow-up projects in

Belgium, the UK and beyond have continued to move the needle upwards in terms of performance.

On the back of the Nemo Link performance, Siemens Energy was best placed to win the subsequent project, Viking Link, and deliver the project with the same team with all the lessons learned from Nemo Link.

Is this creativity in action? I believe so: 'Creativity without delivery is like having the menu and smell of the coffee without the Segafredo arriving', to misquote Koestler...'

13.4 Synthetic synapses

Professor Anatoly Zayats, Head of Photonics & Nanotechnology Group, Department of Physics and London Centre for Nanotechnology, King's College London

'Have you ever wondered why stained glass is so sublimely beautiful, and has maintained those vibrant blues, reds, yellows and greens, even though sunlight has been streaming through the windows for perhaps many hundreds of years? When sunlight falls on painted works of art, things are very different – the paint fades, and the colour washes out. Yet the blues of the glass in a medieval cathedral are as bright today as they were when the glass was first installed. That's because of the way in which the colours are formed. For painted works of art, the pigments are chemical molecules that absorb different wavelengths of light, and the colours we see are determined by the wavelengths that are reflected. Over time, the pigment molecules can degrade, especially as a result of ultra-violet irradiation, and as that happens, the colours absorbed and reflected change, and in general we see this as fading.

But in stained glass, the colours are formed by a quite different process. Within the glass are nanoparticles containing metals such as gold, silver or copper. When light falls on a nanoparticle, this causes the free electrons on the surface to move together, vibrating in phase with one another as a 'plasmon'. These plasmons influence the scattering and the absorption of light, resulting in the colours we see. Change the size, shape or material of the nanoparticle and the colour is changed. And because the colour is formed by a physical process, the plasmon, rather than by the absorption of light by a particular molecule that can degrade, the colour is stable over time. The nanoparticles in stained glass, of course, are far too small for us to see directly. But the optical properties of plasmons are around us, helping, for example, to detect explosives, test for pregnancy, develop drugs, and detect and destroy tumour cells [43].

My team is actively engaged in plasmonics research, which involves both how light can excite plasmons by driving electron oscillations, and also the other way around – how a flow of electrons, as caused by an electrical potential difference, can excite a plasmon, so causing the emission of light. To study these effects, we use a variety of nanostructures and techniques, including arrays of gold nanorods – 'pillars' of gold about, say, 500 nm high, around 50 nm in diameter, and stacked vertically some 100 nm apart on a substrate. The result is rather like the bristles on a hair brush, but very much smaller of course! We call this a 'metamaterial' as we can change its optical properties by changing the diameter of the nanorods and the separation between them. By doing so we can make the array of gold nanorods look green, red, grey-coloured or anything in-between; no one would realise that it is in fact made of gold.

A few years ago, in 2017, we had been looking at how to generate light when electrons tunnel into such gold nanorods, so achieving a nanoscale light source. In order to realise this, the tops of the gold nanorods were covered with a monolayer of the polymer poly-L-histidine (PLH), itself coated by an alloy of gallium and indium known as 'eutectic gallium indium', EGaIn. In this three-layer structure, the intermediate layer of PLH acts as a dielectric between two conducting materials. PLH is a manufactured chemical, but is in essence a protein, for the monomer, histidine, is a naturally-occurring amino acid, and a component, alongside the other natural amino acids, of the proteins within our bodies. The PLH plays a particularly important role in that, when a voltage is applied across the three layers, electrons can tunnel through the PLH layer. A small proportion – that's about 1% – of these electrons tunnel 'inelastically' and excite plasmons in the gold nanorod, which then decay by radiating light. But the remaining 99% of electrons tunnel 'elastically', and generate so-called 'hot' electrons in the tops of the nanorods which can then induce chemical reactions in the PLH.

Within an atmosphere of pure nitrogen, in which all the PLH and gold remain chemically unchanged, the tunnelling current stays constant, as does the intensity of the emitted light produced by the plasmon. But then we discovered that the tunnelling current, and hence the intensity of the emitted light, could be changed according to the chemical conditions of the environment: if oxygen is introduced, the emission intensity increased; on the admission of hydrogen, the intensity decreased, to increase again with more oxygen, in essence reversibly and controllably according to the levels of oxygen and hydrogen present. We interpreted this in terms of reversible oxidation and reduction chemical reactions taking place on the surfaces of the gold nanorods, and also within the PLH, causing changes in the nature of the tunnelling barrier. The light intensity and the current depended on the concentrations of nitrogen, oxygen and hydrogen, and so our nanorods could act as a sensor, which could be of use to anyone who wishes to monitor these elements, perhaps as a chemical reaction progresses; also, they could serve as a 'lab-on-a-chip' to help develop and understand novel reactions, for which precise stimulation and monitoring are absolutely essential [44].

A short time later, in 2018, I attended a workshop on nanoarchitectonics, organised by the European network FORESEEN: Frontiers between Optics and Rf Extended by Study of Extreme Electromagnetism at Nanoscale – a subject quite remote from this nanorod work. One of the presentations was on magnetic materials and some recent advances in computer memory devices. During the talk, the presenter referred to another form of memory – biological memory as held in the human brain, where the mechanism of storage is not magnetic at all, but rather a property of neural networks, and the presence of stable synapses between neurons. And then I saw on the screen an image that looked both different yet familiar. The presenter was talking about neural synapses, and showed a slide depicting a pre-synaptic axon (that's the technical term for that part of a neuron that 'transmits') at the top, releasing neurotransmitters into the synapse, which then diffuse across the synaptic gap to the receptors on the post-synaptic dendrite (the technical term for that part of a neuron that 'receives') [45] below, as on the left-hand side of figure 13.6. And, at the same, in my head, I was thinking about the EGaIn-PCH-gold nanorod metamaterials we had been using in our experiments, as shown on the right-hand side here:

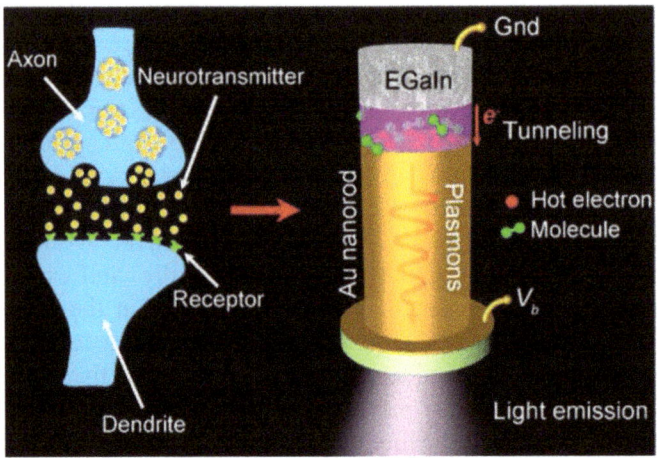

Figure 13.6. A representation of a neural synapse (left), and our three-layer EGaIn-PLH-gold nanorod metamaterial on the right.
Source: Reprinted (adapted) with permission from [46]. Copyright (2020) American Chemical Society.

Could the neurotransmitters diffusing through the synaptic gap be acting in a manner analogous to electrons tunnelling through the PLH? Is the EGaIn like a pre-synaptic axon? And the gold nanorod like a post-synaptic dendrite? The structure looks very similar in the cartoon, but will they behave in the same way?

My brain was ablaze! For this was a really startling Koestler's Law bisociation – the accidental coming together of neurobiology and plasmonics. So when I returned to the lab, I had a host of ideas to discuss with my research team.

As was described during the presentation, an important property of a synapse is the increase in the synapse's strength with repeated activity – this of course being fundamental to Donald Hebb's theory of learning. In terms of our metamaterial, the analogy would be that the greater the degree of tunnelling that had happened in the past, the stronger the tunnelling current 'now', and hence the greater the intensity of the plasmon-induced light emission. If this could be demonstrated, then our metamaterial is behaving as a synthetic synapse...

So that initiated a programme of work which came to fruition in 2020.

One way of thinking about the progressive strengthening of a biological synapse with repeated 'firing' – Hebb's theory of learning – is by analogy with an electrical circuit. If the strength of the synapse is interpreted as the current through a circuit for a given potential difference, then as the circuit 'learns', the current increases – or the resistance steadily decreases. So a 'memristor' is a device whose resistance decreases as the result of the progressive passage of current – in essence, the instantaneous value of the resistance is a 'memory' of the device's cumulative 'history'.

The BIG QUESTION was therefore 'can our EGaIn-PLH-gold nanorod metamaterials behave as memristors?' To which the answer turned out to be 'yes'! Building on our earlier work, we explicitly sought to demonstrate synapse-like properties, which we did: as the environment surrounding the PLH changes to become more oxidising (and therefore less reducing), or more reducing (and therefore less oxidising), the corresponding chemical changes cause the resistance to increase and decrease respectively. Furthermore, if the environment is pure nitrogen, this 'freezes' the resistance, so 'holding the memory' without any expenditure of energy [46].

That's all very exciting, and is the basis for a lot more inter-disciplinary research, bringing physics, photonics, nanotechnology and plasmonics together with neuroscience, neurophysiology and biology. So great scope for much more creativity!

13.5 Biomimetic adhesives
Charles Williams, Senior Technology Manager, GGB LLC

'I really enjoy being outdoors. It gives me time to be with my family, to relax, to think, to look around. One day I was walking by the shore. I was out on a pier, looking down into the water. I could see some fish, lots of green stuff, and some mussels – or maybe clams, I don't know what species they were – attached to the pier supports. A little later, I was strolling along the water's edge, and saw some more molluscs, clinging really tight to some rocks, and staying attached whether they were submerged in the water, or in the air when the tide was lower. I also noticed some boats that had been beached, with more molluscs stuck to the hulls. That got me thinking about how difficult they are to remove from the bottoms of ocean-going ships, how tightly they are attached. One other thing too – the materials to which they attach are very different. The wood of the pier support, metal areas in marinas often covered with seaweed or marine algae. The minerals of a rock. The antifouling coatings of a boat hull. The molluscs don't seem to care what the material is, so how can they stick so tightly to so many different materials?

All that was ringing all sorts of bells in my mind, for a few days earlier, I had been at a creativity workshop, asking 'how might this be different?' for one of our product ranges. We didn't have a particular problem to solve, we were just exploring. The products we were thinking about were the bearings (sometimes known as bushings) that provide lubrication for surfaces in relative motion allowing systems to operate smoothly and with the minimum of friction. And when I say 'bushings', I'm talking about bushings of all shapes and sizes, for GGB manufactures a huge range [47], from quite small hinge bearings used, for example, for car doors, to the truly massive ones that are in the bridges of numerous shipways.

Hinges, of course, have three visible components: the plate fixed to, say, a door post, the interlocking plate on the door, and the pin that holds these two plates together, allowing the door to rotate. But what you can't usually see are some other components, the bearings: friction-reducing 'sleeves' that sit between the pin and the plates. A door is usually fastened on one side, and so as it opens, there is a bending moment, forcing the plates against the pin. In the absence of the bearings, the plates would be in metal-to-metal contact with the pin, and each time the door is opened or closed, that would create wear and probably make an annoying screeching noise, as well as being subject to friction. What a bearing does is to keep the two metal surfaces apart, and to introduce contact between metal and a low-friction, very tough, polymer surface. This reduces wear, doesn't make a noise, and minimises friction.

Structurally, a bearing is a steel cylinder, of a size appropriate to the hinge, with the interior surface coated with a low-friction polymer composed largely of PTFE. PTFE, though, is chemically very unreactive, so to bond the PTFE to the steel we use sintered bronze, which adheres to the steel, and which can also hold the PTFE, as shown in figure 13.7. We've been making our bearings that way for years, and it works. We know how to do it, we do it well, and our products are very successful in the market place.

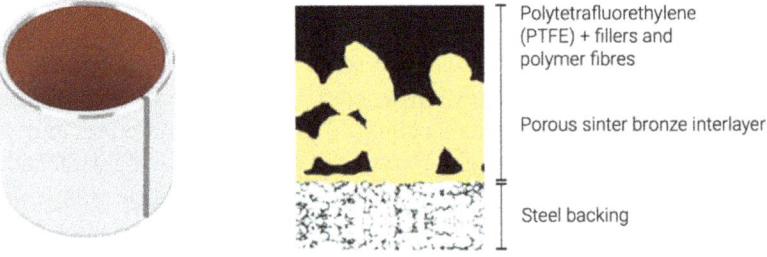

Figure 13.7. On the left, a typical hinge bearing, a steel cylinder coated on the inside with a PTFE-based low-friction material. This material adheres to the steel by means of an intermediate layer of sintered bronze, as shown on the right.
Source: www.ggbearings.com/sites/default/files/2021-12/GGB-DP4-and-DP4-B-Metal-Polymer-Self-Lubricating-Lead-Free-Bearing-Solutions-Brochure-English.pdf.

OK. That's all fine. But how might things be different? And in particular, since the (expensive and heavy) bronze is used only to bond the tribological agent, the PTFE, to the steel substrate (which determines the bearing's size and shape, and provides structural strength), how might we adhere the PTFE using something other than bronze? That question, of course, isn't new, and we'd asked it many times before. And always given the same answer. As we all know from Chemistry 101, PTFE is chemically inert. So it doesn't bond with any of the commercially available adhesives. Yes, you can use techniques such as corona or plasma treatment, or etching with some really nasty toxic chemicals, but those are expensive, potentially hazardous, and don't necessarily result in a stable, durable outcome.

So there I am on the shore, looking at molluscs clinging like limpets (or maybe there were limpets!) to all sorts of materials. How do they do that?

That got me to check out the academic literature. And then I came across some research carried out by Professor Phillip Messersmith, who, firstly at Northwestern University [48], and more recently at the University of California, Berkeley [49], has been studying how marine mussels naturally synthesise what he refers to as an 'underwater super glue' [50]. It turns out that a key component in this natural adhesive is derived from dopamine, a neurotransmitter and a member of the class of compounds known as catecholamines (see figure 13.8).

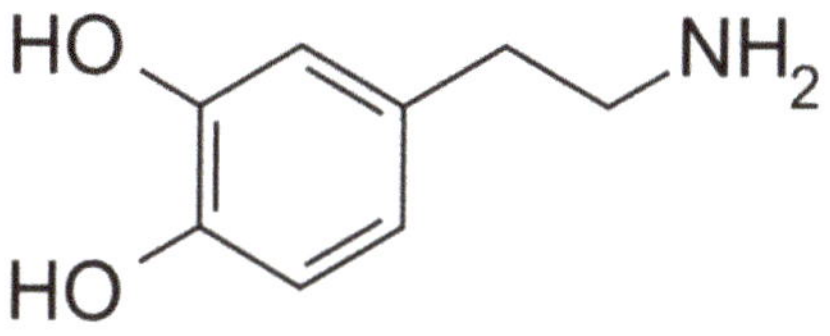

Figure 13.8. Catecholamines are molecules with two hydroxyl OH groups on adjacent 'corners' of a benzene ring, and also a side chain amino NH_2 group. The example shown is dopamine. Molecules of this general type can act as bonding agents because both the NH_2 and the pair of OHs are highly reactive.

In mussels, the active catecholamine is incorporated into a natural protein polymer to make it sticky [51], so that triggered the idea that something similar might work for the PTFE blends used by GGB.

So I conducted discovery-learning experiments to challenge the paradox of PTFE reactivity in my lab at GGB, working with catecholamines and PTFE. And after experimental comings-and-goings, it worked! I was able to synthesise a catecholamine-based adhesive, with a unique fluoropolymer backbone, that can bond securely and directly to both the structural steel in the bearing and also the PTFE coating, as illustrated in figure 13.9.

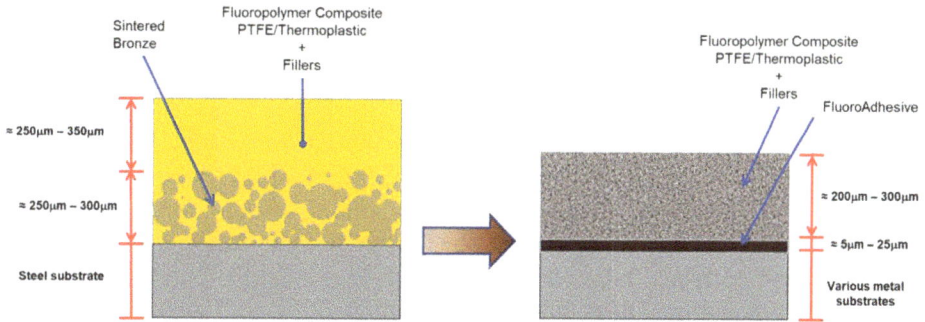

Figure 13.9. On the left, schematic of a bearing with the PTFE tribological surface bonded to the steel using sintered bronze; on the right, the structure using the new biomimetic adhesive.
Source: Charles Williams.

Not only that. It also allows other materials to be embedded within a PTFE powder to produce a blended composite material with improved overall tribological – and temperature – performance. And the synthesis isn't too difficult, and has the great benefit of taking place under very friendly conditions at room temperature. So the reactions don't need a pressure vessel, or high temperatures, and work within an aqueous system and without the need for any dangerous chemicals.

This new biomimetic adhesive gives us a viable alternative to sintered bronze, making our bearings lighter, with a cost structure that is much less volatile (the price of the copper in the bronze can really fluctuate), and, once production gears up, cheaper to manufacture too.

As a chemical engineer, I'm always striving to make things environmentally better, to solve engineering problems. And I'm always looking to nature for ideas – after all, the natural world has had millions of years to solve all sorts of engineering problems. So I'm always asking 'how might this be different?'. And I observe, observe, observe. Great!'

13.6 The magic colouring sheet

Dr Hermione Cockburn OBE FRSE, Scientific Director, Dynamic Earth, Edinburgh

'A good example of creativity in my branch of science? Well … I'll go for what I call the 'magic colouring sheet'. Let me explain. From the web, you can download a black and white line drawing of a marine animal – say, a catshark or an octopus, set in a sea-scape – which you can print at home and colour in. Seemingly a simple colouring sheet. So far, so ordinary. It's the next bit that's different. If you scan the sheet in an app on your mobile phone, then the magic happens. The catshark becomes animated, and swims around in a virtual space which is a mixture of your own home and some marine features like corals. You can click to retrieve information about the animal, the environment where it lives, and a major research project that is studying it. It's proved to be a great way of engaging young people – and older people too! – in deep-sea science, and it's a wonderful example of successful outreach. The augmented reality app – the 'proper' name is *Spectacular!* – is truly creative, and the back-story of how the colouring sheets were developed is a great tale too, all about doing things differently.

As a Science Centre, Dynamic Earth is very much committed to outreach – in our case, relating to all aspects of environmental science [52]. Conventionally, this type of 'outreach' refers to engaging with schools, local communities, and members of the public, which is all at the heart of what we do, but we are also very active in 'outreach' in a different direction – towards scientific teams in universities and research institutions, encouraging them to bring their interests, passions and enthusiasms to us, for we can provide the platform and the opportunities for them to reach the public. For example, we invite researchers to showcase their work to families and help them develop the skills to do this more effectively. Dynamic Earth therefore supports researchers to achieve their outreach objectives, and – hopefully – along the way, we add engagement capacity into the academic system, for our team has much experience, and expertise, in how to put often complex scientific messages across in meaningful, accessible ways.

In 2015, one of the academics we had been working with, Professor J. Murray Roberts, then at Heriot-Watt University and now at the University of Edinburgh, invited Dynamic Earth to be a partner in a research proposal to be submitted for a major EU Horizon 2020 grant to fund the ATLAS project, the most extensive assessment of North Atlantic deep-sea ecosystems yet undertaken [53].

We all know that every grant application asks for an 'impact statement'. And we also know that, in many cases, this might be something that is tacked-on to the end of the project. Murray's view was different, for he is committed to public engagement, and together with the ATLAS steering committee, he wanted outreach to be embedded within the project from the outset, and to be an integral part. Involving us, as outreach specialists, as a full partner alongside all the academic teams was different, and would make the proposal distinctive. ATLAS was very much aligned to the 'Galway Statement', an undertaking by the US, Canada and the states of Europe 'to increase our knowledge of the Atlantic Ocean and its dynamic systems', 'to improve ocean health and … the sustainable management of its resources' and 'to promote our citizen's understanding of the value of the Atlantic by promoting oceans literacy' [54]. That last point is quite unusual in 'charters' of this type, and so the idea of having Dynamic Earth as part of the ATLAS team was very much in harmony with the Galway ideal, even though we are not 'academic'.

The proposal – which was a lot of work – was successful, and over the four years of the programme, from May 2016 until April 2020, we sat alongside 24 other partners in a consortium involving universities and research institutes in Britain, Denmark, Spain, France, Germany, Portugal, the Netherlands, Ireland, Norway, the USA and Canada. At the start, some of the scientists were sceptical to see me around the table, wondering what Dynamic Earth might contribute, but by the end, I think all the academic teams welcomed the fresh ideas we were able to bring to discussions about how their science might best be communicated.

Within the overall grant, we had our own budget, and so had considerable freedom – and opportunity – to develop all sorts of educational materials to convey the discoveries of this fascinating study to audiences of young people, and adults, across the world.

Most children find fish, and other marine animals, from squids to whales, totally absorbing, but the marine environment is not generally well understood. For example, if you ask anyone 'where is the nearest coral reef to the UK?', the answer most likely to be given might be 'the Great Barrier Reef', or perhaps 'in the Red Sea'. In fact, the answer is 'off the Outer Hebrides, near the island of Mingulay' [55], for that's where you can find a wonderful cold-water coral reef that was only discovered in 2003. You can find cold-water corals in deeper water all around the UK, with a particular species of coral, *Lophelia pertusa*, forming the reef framework – not unlike the more well-known reefs in tropical waters.

Like their tropical cousins, cold-water reefs support a vibrant ecosystem, and to make that more accessible, one of the engagement ideas that my project officers and I came up with was to produce a floor mat, printed with a high-resolution composite image, bringing together animals and other features observed at one of the ATLAS study sites: the coral ecosystems of the Rockall Bank in the NE Atlantic – as shown (in small scale!) in figure 13.10.

Figure 13.10. The Rockall Bank cold-water reef mat.
Source: Reproduced from [56] CC-BY 4.0.

The images were captured by an ROV – a 'Remotely Operated Vehicle', an underwater robot, in water depth of about 600–800 m. The resulting picture can be downloaded and printed [57], ideally on a vinyl sheet about 3 m × 1.5 m – that's the size of a dining table. Because the mat is so large, several children can sit on it at the same time, and use magnifying glasses to 'peer into the deep', exploring the habitat, just as if they were surveying the reef for real. It's a great learning resource, especially when used in combination with quizzes (how many squat lobsters can you find?) and tick-sheets. And it gives an excellent representation of the whole cold-water coral ecosystem, with *Lophelia* providing the structural framework, upon which all the other species rely, creating a biodiversity hotspot. It's so simple to grasp, and it leads to such rich exploration, not only by young people, but by adults too.

The reef mat can be printed locally and used in a whole variety of engagement settings including schools and community events; on a rather bigger scale is another idea we had, the ROV simulator game [58], which can be experienced at Dynamic Earth. This is a video-game-style simulator that a child can use to 'fly' an underwater robot, just like a 'real' ROV pilot. We developed this with a specialist company, Marine Simulation, and one of my greatest pleasures is to watch people's faces light up in amazement as they explore deep-sea hydrothermal vents, as discovered by the ATLAS project, and try to guide the ROV to sample the gases coming out of them – harder than you might think!

We produced many more conventional learning resources too, all under the banner of 'Atlantic Adventures with ATLAS', such as information sheets, activity packs and a scripted workshop, and these were designed to be used across all the partners, in all their countries, so many elements were translated. But it's the 'magic colouring sheet' – the *Spectacular!* app [59]—that's my favourite. And a great example of Koestler's Law, bringing together two pre-existing 'components' to form a new 'whole'.

We'd already been using some commercially available augmented reality colouring cards developed by QuiverVision, a company based in New Zealand. And then my colleague Emma Paterson and I had the idea of asking QuiverVision if they might be willing to develop some bespoke cards, depicting some of the ATLAS case-study locations. They, very pleasingly, said 'yes', and so we combined QuiverVision with some particular stories from the project, and the result was the ATLAS portal on the *Spectacular!* app: augmented reality deep-sea creatures swimming around you in a colour pattern of your choice!

The app took about 18 months to develop, during which we worked closely with the ATLAS scientists and also the QuiverVision developers. We thought very hard about the contexts for each of the drawings, choosing three contrasting locations so as to broaden the scope of the learning: in the end we chose the Mingulay reef, the Hebrides Terrace Seamount, and the LoVe Observatory off Norway, populated by local species that might be rather unfamiliar, such as the blackmouth catshark, Richardson's ray, redfish, and bubblegum coral. So each portrayal is scientifically correct, and as well as being visually compelling (the necessarily static image in figure 13.11 can give only a hint...), and the app acts as a portal to relevant information about what can be seen on the screen.

The development of this app also demonstrates a benefit of our being a single, programme-wide, partner dedicated to outreach, in contrast to dividing the outreach budget across the academic partners for them each to spend locally, which is the conventional way of doing things. No individual partner would have been able to afford the cost, but collectively the programme could. And we had the time and the energy to make sure it was done. So that's a real example of teamwork, of 'wholes being greater than the sum of their parts'.

So, back to creativity. Yes, the floor mat, the simulator and the *Spectacular!* colouring sheets are all creative in their own ways. And all so much fun! For when you're inspiring people about science, fun really helps with the engagement. Enlightening too I hope. And it's great to receive emails saying things like 'my daughter has just been chasing a catshark around the kitchen, and is loving it!' The scientists, too, get real fulfilment when they can see their work being appreciated well beyond the academic journals by young people in particular.

But what was truly different was the context within which all these were developed – primarily because Dynamic Earth was a partner organisation, working fully alongside the academic researchers, throughout a major scientific

Figure 13.11. A still from a video of the *Spectacular!* app, showing a coloured version of Richardson's ray. Source: Reproduced with permission from www.youtube.com/watch?v=fHpZGUjiEOU © Hermione Cockburn.

grant, from proposal to completion. And that happened because Murray Roberts and the rest of the ATLAS principal scientists had the creative insight to involve us. I'm sure we added real value, doing things the scientists don't have the experience, or the time, to do themselves. And we, of course, couldn't do what we did without drawing on the science. It's all about teamwork. But that doesn't happen by accident. So, thank you, Murray, for inviting us to be part of the ATLAS team!'

13.7 Quantum entanglement, single-pixel cameras, and novel endoscopes

Professor Miles Padgett OBE FRS FRSE, Kelvin Professor of Natural Philosophy, School of Physics and Astronomy, University of Glasgow

'I am totally fascinated by light – and in particular how to shape it, and how to use it in novel ways. What is also fascinating is that light has been studied scientifically since ancient times – Ptolemy, for example, made measurements relating to reflection and refraction around two thousand years ago – yet optics, photonics and plasmonics are still very active research fields today. And full of creativity too – just think of Newton's interpretation of the visible spectrum, the wonderful unification of Maxwell's Equations, lasers, LEDs.

And, to me, one of the best examples of Koestler's Law – how apparently different concepts can some together to form a surprising new 'whole' – must be when orbital angular momentum 'met' quantum entanglement.

Orbital angular momentum [60] is a property of light, but a subtle one, in that its detection and measurement requires that the light beams are shaped in a very particular way, with the phasefronts propagating helically. To do this requires some specialised optics, with a laser operating in what is called 'Laguerre-Gaussian' mode. When this happens, the intensity of the beam at the centre falls to zero because of destructive interference, and so the beam itself is an annulus. In a paper published in 1992 [61], some researchers at the Institute of Physics in the Ukraine described how to form such a beam in the laboratory, but made no reference to the beam's orbital angular momentum.

That link was made in 1992 by a team at the University of Leiden led by Les Allen [62], who proved theoretically that helical phasefronts carry an orbital angular momentum of $l\hbar$, where l is an integer, and $\hbar$ is, as usual, Planck's constant divided by 2π. As illustrated in figure 13.12, a value of $l = +1$ corresponds to a right-handed helical phasefront; $l = -1$ to a left-handed one; $l = 0$ to a sequence of planes; and values of $|l| > 1$ correspond to $|l|$ intertwined helices.

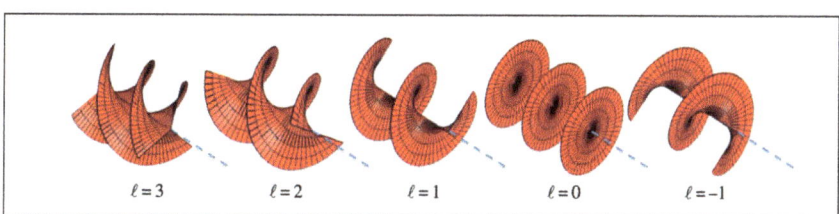

Figure 13.12. When laser beams have helical phasefronts, the beam carries an orbital angular momentum of $l\hbar$ per photon.
Source: Reproduced from [60].

The next key event was the experimental verification that the orbital angular momentum is indeed quantised as $l\hbar$, in accordance with the 1992 theoretical prediction: that was achieved by a team at the University of St Andrews (of which I was a part!), working with Les Allen (at that time at the University of Colorado), and reported in a paper published in 1996 [63].

So much, for the moment, for orbital angular momentum; let me now summarise some of the key points relating to quantum entanglement [64] – that (macroscopically most strange) phenomenon whereby the quantum states of two particles are not independent, but knowledge of the one implies knowledge of the other, even if they are physically distant from one another. Any quantum particles can become entangled, but by far the most accessible quantum particles are photons – so we're back to optics. And in 1995, a team led by Anton Zeilinger, who at that time was at the University of Innsbruck, developed a method of producing a high-intensity source of entangled photon pairs [65] that was then able to be used to test many theoretical predictions, and to explore applications such as 'quantum teleportation', in which information concerning the quantum state of one particle is transferred to another, entangled, particle, whilst at the same time erasing the state in the first particle – just as if the original particle has been teleported!

The 'big breakthrough' came in a paper [66], published in 2001 by Zeilinger's team, demonstrating entanglement of the quantum states associated with the orbital angular momentum of photons. Up to that point, entanglement had been demonstrated with quantum states that had only two possible values. But orbital angular momentum states are quantised as $l\hbar$, with no constraints as to the (integer) value of the quantum number l. That increases – enormously – the information that might potentially be coded in, or decoded from, an appropriate signal, so making it much more possible that applications such as quantum computing, and quantum cryptography (which enables messages to be absolutely secure), could become realities.

The 'Koestler's Law' components relating to orbital angular momentum and quantum entanglement were published in 1992, 1995 and 1996, and so were available to every physicist on the planet since then. In principle, anyone might have read the appropriate papers, and thought 'I wonder what might happen if I mix those ingredients together?' From a 'pure' Koestler's Law standpoint, that's true. But in practice, I wonder... why would 'anyone' do that? They just wouldn't, for they'd have no reason to; no background, no context. Only someone in the field, someone with not only relevant knowledge, but also deep insight too – someone who could anticipate the possibility of what might actually happen – would be motivated to try. And, as it happened, that person was Anton Zeilinger, who indeed had much relevant knowledge and that essential deep insight: hence the 2001 paper. Just wonderful!

Another phenomenon that was thought to exploit quantum entanglement is known as 'ghost imaging' [67], in which an image of an object can be formed even though the light that forms the image never interacted with the object itself. To form a 'ghost image', you need two light fields, one of which falls on the object, and the other on the detector. When the fields are set up in the appropriate way, quantum entanglement across the two light fields causes the image to be detected. It turns out, though, that ghost imaging can also be demonstrated without invoking quantum effects, and this dovetails into another method of image creation, using a 'single-pixel' camera [68].

A conventional digital camera, like the camera in a smart phone, collects an image by using a sensor that contains millions of individual pixels, each of which captures the light intensity of a correspondingly small proportion of the object at the same time. The final image is compiled by assembling the spatial 'fragments' from each pixel detector in the correct sequence, and, as we all know, the greater the number of pixels, the finer the resolution of the image. So if there is only one pixel, the resolution is pretty bad! But suppose you have a camera with only a single pixel, but a pixel that can take many images very quickly in succession, images which examine different parts of the object. If those fragmentary images can be stored, and subsequently recombined in the right way, then it is possible to reconstruct an image of the full object. Which – as Koestler would recognise – is not a new idea, but a variation on the idea used, for example, in John Logie Baird's mechanical televisions, in which an image is produced by scanning an object, and then projecting elements of the image successively on a screen.

For ordinary applications, it's easier to use your smart phone, but for certain applications – such as detecting non-visible radiation, or when the cost of a conventional multi-pixel sensor becomes prohibitive – single-pixel cameras have a potentially important role to play. And getting single-pixel cameras to work well, and quickly, involves some really fun physics, so it's been a topic of interest to my research group at the University of Glasgow for some years now. It's also been a showcase of creativity, especially as regards that key question 'how might [this] be different?'.

So, for example, in 2015, we did some work [69] to improve the algorithms to process ghost imaging signals using just one single-pixel sensor, so that everything worked faster. How might that be different? Suppose we used not just one single-pixel sensor, but four? We did that too, and the result was the formation of a 3D image [70], rather than the 2D images that had been produced hitherto. We then used [71] a set-up with three light sources – one red, one green, one blue – and three single-pixel sensors – one to detect each colour – and, as you can see from figure 13.13, created 3D colour pictures! All from sensors that are just one pixel!

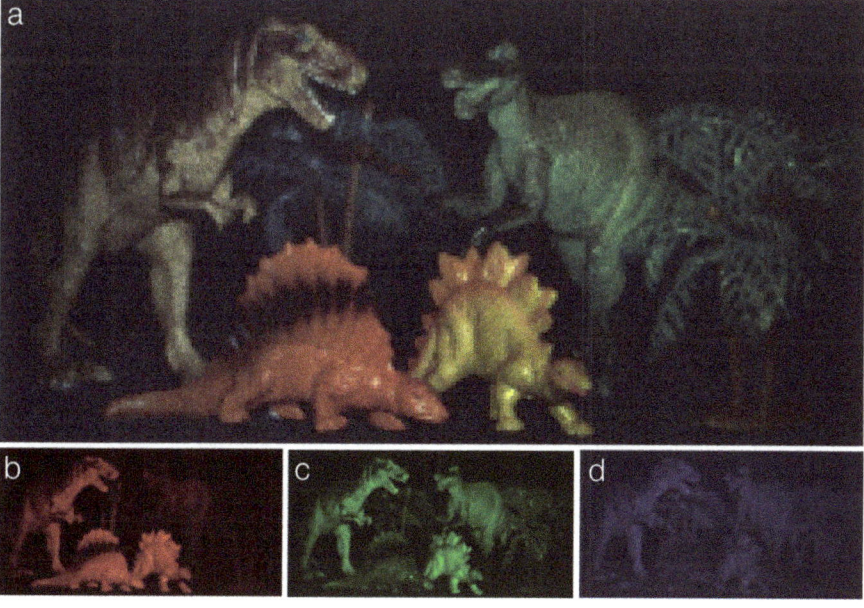

Figure 13.13. (a) is a full-colour 3D scene, reconstructed from red (b), green (c) and blue (d) single-pixel sensors.
Source: Reproduced from [71] CC-BY 4.0.

Another variation was to use two single-pixel sensors, one in the visible range and one sensitive to infra-red, which we set up within a microscope [72], and yet another used four single-pixel sensors to detect polarised signals, as is important, for example, when observing stress patterns in materials [73].

Just one more example: by using a pulsed laser light source, we were able to measure [74] the very short interval between the incident pulse, and the detection of the light back-scattered from the object by just one single-pixel sensor. Correlating the single-pixel detections with the corresponding times enabled us to reconstruct a 3D image, but in a manner different from the earlier experiment, which had reconstructed a 3D image by correlating the signals from four sensors but without measuring time. Both methods produce 3D images, but each works better in different situations. Perhaps we should combine the two and see what happens?

And talking of combinations, what happens when you combine an optical fibre with a single-pixel camera? You get a very fine endoscope. Much more slender than the ones that are conventionally used in medical diagnoses, for example. We're working on that right now [75]…

So there are a host of results, all variations on a theme, yet all of considerable scientific interest, sowing the seeds of who-knows-what possible real applications. And all great examples of how to make creativity happen!'

13.8 Keeping the UK's railways safe
Colin Haynes, UK Services Quality & Performance Director, Alstom Transport UK

The UK Railway is a complex system. Those who use the railway either for personal transportation or freight demand a high level of service. For the rail passenger, demands start with punctuality, and moves on to critical features such as seat availability and cleanliness. The effective and efficient maintenance and servicing of rolling stock is vital to enable the various companies that operate trains across the UK to meet customer demands. These activities are carried out throughout the day, but mostly at night. While most are sleeping, railway depots come alive at night.

Multiple activities are planned and re-planned to the minute. Trains must arrive and have a place to go whilst they await teams of people to service, wash, clean and replenish them. Trains need to be moved throughout the night to have their toilets emptied and their water tanks filled. A railway depot is a complex puzzle requiring the duty manager to constantly move the pieces around to achieve results. Late departures cause delay and incur penalties. Late arrivals due to disruption during the working day are a constant feature, with which the duty manager must cope.

It's inevitable that in such complex environments, incidents occur. Thankfully they are rare – due mostly to the professionalism of those who work in the depots, and to those who develop and monitor the safe systems of work. Whenever there is an incident or a near miss, an investigation seeks to understand why. All too often we find that the root cause is explained as 'human error'. And it is all too easy to stop at the error and to miss the deeper causal factors. The UK Rail industry has learned from other sectors that 'human factors' require deeper understanding.

In 2020, the world was gripped by the Coronavirus pandemic. Whilst there was disruption, and a reduction in services, the UK Railway carried on. It would be easy to believe that with the reduced pressure, life in the depots became easier. On the contrary, additional cleaning requirements with new products, and the need to protect staff with safe working arrangements, only added to the pressure.

Our management team were seeing a rise in near misses and low-level incidents and began to wonder why. I had been looking at the effect of workload and other factors on performance and I believed that the complexity of the processes on our depots might be a significant factor.

I've always believed in the imperfection of systems. Any system can have vulnerabilities, whether it's a complex computer algorithm, or a process involving people. To paraphrase Professor James Reason: 'Failure, just like success, is a team effort' [76]. Focussing on the person who took (or didn't take) the final step which caused the failure is itself a failure to properly understand what went wrong.

Over time, organisations discover vulnerabilities in their processes. Based upon investigation and diagnosis, vulnerabilities are patched, just like computer systems. So it is with maintenance processes like those in railway depots. Over time, what was often perceived as 'human error', has been patched with ever more detailed briefings, documentation, 'helpful' computer systems and so on. So why are we surprised when our well-patched systems still fail?

One reaction to mistakes in a maintenance system can be to introduce more supervision and checking. Fair enough! It was widely believed in my company that our production shift supervisors should be doing more walking about and supporting the staff in their work. So what was stopping them? Depot managers were telling me that there was little understanding of just how busy the night Duty Manager was. They told me about the 'multi-tasking' throughout the night. The constant re-planning. The phone calls explaining that a train was delayed, or going to another depot. Urgent maintenance needed due to damage. Or additional faults discovered once the train had arrived. And of course, staff shortages due to Covid-19.

It seemed to me that depot staff were over-stretched, doing too many things at once. But I was sitting in my office, reading emails and having 'Teams' calls. That wasn't good enough. I wanted to understand what it was really like. Only then would I be able to articulate my findings to the management team in a way that would convince them to direct resources to support our front-line staff.

To learn about what was really happening, my team and I shadowed night Duty Managers in three depots from London to Scotland to observe first-hand the reality of life in a busy maintenance depot.

Across all three depots the results were illuminating. The Duty Manager was constantly interrupted whilst trying to manage multiple demands, and the multiple recording systems required by different train operating companies. Layers of defence against error had become burdensome to the point of reducing the mental capacity to react.

Duty Managers were constantly switching their attention to manage this complex and changing environment. They hardly ever stopped, except maybe to grab a drink and a sandwich and carry on.

My hobby is flying (gliding mostly) and I'm fortunate that it's a sport which focuses on safety, and in particular, human factors. I'm also fortunate to be surrounded by people who are experts in that field. I was pointed to some aviation resources, notably CAP 737, the Civil Aviation Authority Flight Crew Human Factors Handbook [77], and CAP 716, Aviation Maintenance Human Factors [78].

From these publications I was reminded about the myth of multi-tasking. In fact, my Information Technology background should have told me that 'multi-tasking' is in fact high-speed 'task switching'. While a computer can store an exact copy of the status of one task whilst it switches to another, it's not so easy for the human. When we're interrupted, we can easily forget where we are in a task. It's why, in aviation, you never interrupt a person conducting a pre-flight inspection of an aircraft. Checklists help to track progress in a task, but they don't always help in maintaining that vital situational awareness that a Duty Manager needs.

Our study told us that the Duty Manager could either direct overall operations, or supervise the work being carried out. But not both. Especially not at once. Expecting – or, in practice, requiring – them to do this was asking for trouble. And the fact that, in actuality, there has been so little trouble is a testament to their hard work and diligence. But trouble would come, sooner or later. To avoid that, we needed to split the tasks.

I was aware that our study and our conclusion would mean asking for additional resources in the depots or changes to already complicated systems. Longer term we could look at the systems and processes, but how could we make the point that, in the short term, our duty managers were having to constantly switch their attention and couldn't be everywhere?

So, of course, I prepared a 'traditional' PowerPoint pack to present to the management team. My focus was on the Duty Manager and the level of multi-tasking – in truth task switching – which happened on a normal night in one depot, and on a challenging night in another. And I included some useful notes on what multi-tasking means from a psychological point of view from the aviation industry, as well as some examples about the importance of human factors in the railway industry [79].

But would PowerPoint slides and bullet points presented by me have enough impact? How could this be different? Suppose I didn't use PowerPoint? Suppose I didn't give the presentation? Who else might give it? Who might carry authority and influence that I might not have?

And then it hit me. My colleagues are mostly engineers. And many engineers are devotees of motor sport, especially Formula 1. Suppose the presentation were given not by me, but by a top driver? Someone like Kimi Räikkönen? Or Lewis Hamilton? Nobody could deny the high degree of both mental and physical skill of a Formula 1 driver. They are the epitome of mental and physical fitness. They drive complex systems – the car – in a complex system of tyre wear, fuel management and of course all the pit crews and support teams.

So I searched around, and came across a YouTube video from the 2015 Malaysian Grand Prix. It's Lap 42, and Lewis Hamilton is in second place. As Hamilton negotiates a corner, really fast, a trackside race engineer starts talking on the in-cockpit radio, telling him that 'we need to keep this pace up'. To which Hamilton replies, 'Hey man! Don't talk to me through the corners! I nearly just went off!!!' [80]. Even Lewis Hamilton, with his instincts, his reflexes, and his super-fast decisiveness can't to do two things at once.

So I incorporated that clip into my presentation. And it helped. It was a complete contrast from the rest of the presentation, but it hammered the point home. We split the tasks! We added resources with the single mandate of direct supervision and support of the staff. The Duty Manager still has a complex job to do, but instead of multi-tasking, full attention can now be paid to one-task-at-a-time, which is much less stressful – and much more effective. The additional direct supervision has been highly effective in highlighting problems with documentation and understanding. Many of the additional people are giving direct feedback via toolbox talks, and we can see the effect in our performance – and morale.

I've learned that innovation and creativity can manifest themselves in an invention or an idea. But it can also be helpful in making a point. And – in this case – a very important point indeed!

13.9 The 'Medusa Effect'

Professor Rob Jenkins, Department of Psychology, University of York

'In Greek mythology, Medusa was a monster of petrifying power. Looking directly at Medusa was enough to turn the viewer to stone. To escape this fate, Perseus avoided direct gaze altogether, approaching Medusa by looking at her image in a polished shield. The image contained all the information Perseus needed to guide him, but lacked the full potency of Medusa in the flesh.

In my work as a cognitive psychologist, I am interested in mind perception, that is, our awareness of the minds of others. I find the Medusa myth intriguing, as it has a modern analogue in Zoom and other social technologies. How do we construe the minds of people who are represented as images, as opposed to being socially present?

This particular question can be traced to my serendipitous encounter with Alan Kingstone, a Professor of Psychology at University of British Columbia in Canada. Although our paths didn't cross until 2018, I had been aware of Alan's work since the late 1990s. What I didn't realise is that Alan had spent several months at the University of York (where I now work) as a visiting researcher, a few years before I arrived. So he was quite well connected with other members of the department. In 2018, Alan was invited to give a seminar in our department while he was visiting the UK. I was quite keen to find out what he was up to now, so I made sure to attend. Beyond the specifics of his presentation, I remember being struck by the freshness of his approach to experiment design. It was all very thought provoking. I asked a couple of questions afterwards, and these turned into conversations that continued beyond the session.

At the end of the workday, a dozen of us went out for dinner. I arrived at the restaurant determined to sit a next to Alan so that we could pick up where we left off. It was that dinner conversation that launched our collaboration. As we proceeded from starter to main course to dessert, our conversation became increasingly animated. We already knew that there were obvious areas of overlap in our research, but now we were starting to uncover more interesting areas of overlap that we hadn't previously identified.

At one point, Alan mentioned something that they were stuck on in his lab. They had been thinking about pictures, and the role of depiction in our understanding of the world. They were particularly interested in everyday distinctions between reality on the one hand and representations of reality on the other. Alan had a hunch that an object, say, a vase, presented *as* a picture (for example, shown in a frame hanging on a wall) would seem less real than the same vase presented *in* a picture (for example, shown on the table as part of a dining scene). But they were struggling with how to measure such impressions. They had reached an impasse.

Around the same time I had been inspired by a *Science* paper titled 'Dimensions of mind perception' [81], which proposes a framework for measuring the mental capacities that we attribute to others. That study shows that mind perception is not unitary, but comprises two distinct components. The first is 'Experience', which refers to the ability to *feel* things, and is associated with having moral rights. The second is 'Agency,' which refers to the ability to *do* things, and is associated with having moral responsibility.

Using this two-dimensional framework, some of my undergraduate students had recently shown that vegetarians and non-vegetarians construe animal minds quite differently. For example, animals received higher ratings of 'Experience' from vegetarians than from non-vegetarians, even though both groups gave similar ratings to humans. Apparently, the vegetarians were particularly attuned to the mental lives of animals. Given that the approach was sensitive to detect such differences between viewers, I suggested that it might also reveal differences between levels of representation. Alan later described this as 'the key to unlock the whole problem'.

By this stage, I was really keen that Alan and I should follow up. We had clearly hit it off in York, and we seemed to be staring at a collaboration that could be both productive and fun. Late the same evening, I emailed Alan the *Science* paper, together with a draft report of the student project comparing mind perception by vegetarians and non-vegetarians.

When Alan was back in Canada, we had a couple of email exchanges to refine our experimental design. The basic idea was to ask volunteers to rate people in images for 'Experience' and 'Agency'. But the images had to conform to a particular structure, depicting people at different levels of representation.

An example might help. Suppose you are standing next to a poster of your face, and someone takes your photo. The new photo contains your face twice – once in the poster and once beside it. Both faces are just different regions of the same canvas. The key question is whether viewers attribute different levels of 'Experience' and 'Agency' to these two faces. Alan had roped in a student in Canada to prepare the necessary images, but we all had other things to be getting on with, and it wasn't clear that our plans would amount to anything.

Some months later, I received an email asking if we could set up a call. Alan wanted to bring me up to speed on experimental results and discuss where to go next. This was concrete progress, and it really put some momentum behind the project. After that, we set up regular calls which became weekly and remain so three years later. The main focus of those early meetings was developing the experiments that eventually became the 'Medusa Effect' paper [82]. Those experiments confirmed that abstraction dims the perceived mind. They also showed that the perceptual effects had behavioural consequences. In economic tasks involving allocation of money, participants donated the least money to a person in a photo of a photo.

We now think that these findings have implications for digital communication in a range of settings. For example, during a virtual trial, a judge may see pictures of a victim on video. Our findings suggest the judge may be less inclined to view the victim as real and vivid, which could affect how the case unfolds. Similar issues arise in remote education and healthcare.

All of this started when Alan gave a presentation in our department. Reflecting on this origin story, Alan remarked, 'One thing is for sure. This would never have happened if it had been a Zoom presentation. Not in a million years.' An interesting observation, given the basic finding of our research collaboration: depicting a person makes their mind seem less real. Which is the more credible in figure 13.14?

Figure 13.14. Which is the more credible?
Source: Courtesy Fay Osswald.

13.10 Mixing things up: ellipsometry and strong coupling
Professor Bill Barnes, Department of Physics and Astronomy, The University of Exeter

My primary research interest at the moment is a topic called 'strong coupling', in which lots of molecules interact with each other by exchanging photons. The 'trick' is that the photons involved are virtual rather than real: physicists like to describe this as 'harnessing quantum vacuum fluctuations'. We couple the (real) molecules to virtual photons by placing a thin film of the molecules inside an optical cavity formed from two mirrors that face each other. The distance between the mirrors is roughly the wavelength of light, and constructive interference of the (virtual) photons as they bounce between the mirrors selects the right frequency to interact with the molecules. If we get things right and the molecules interact strongly enough with the virtual photons inside the cavity, then new hybrid states called 'polaritons' are formed, part light, part matter; it is this phenomenon that is known as 'strong coupling'.

These new polariton modes are particularly interesting in that they can have energy levels and coherence properties that are very different from the constituents from which they are made. This opens up many possibilities, and so strong coupling is becoming increasingly fascinating to a wide range of scientists, both because this is a new area of science, and also because there are potential applications in areas as diverse as materials science and chemistry. In particular, it offers the exciting prospect of enabling new materials to be formed that are part light and part matter, and of providing a new route by which to control chemical reactions.

These hybrid states are studied using a range of techniques, primarily optical, for example by measuring extinction and reflection spectra. The field is at an embryonic stage, with many research teams trying to build a better understanding of the strong coupling phenomenon. Vital in this work is a detailed knowledge of the constituent materials that make up the samples we study, for example the optical properties of the substrates and the polymer hosts we use to form the thin film. Typically, these materials are characterised using an ellipsometer, an instrument that allows optical constants (such as the refractive index) to be determined over a wide range of wavelengths. In general, ellipsometry involves bouncing light off a sample, and makes use of both phase and amplitude information in the reflected light. And it was the use of the phase information that intrigued us, for phase had not been employed to explore strong coupling before: we wondered what would we see if we were to use an ellipsometer to study the phase signals not just from the constituent materials that form the film, but also from our completed strong coupling samples.

We were surprised and delighted by what we found [83], for what was so exciting about using the ellipsometer was that the signature we were able to measure, the phase response, offered a new way to characterise the transition between what is known as weak and strong coupling, as illustrated in figure 13.15.

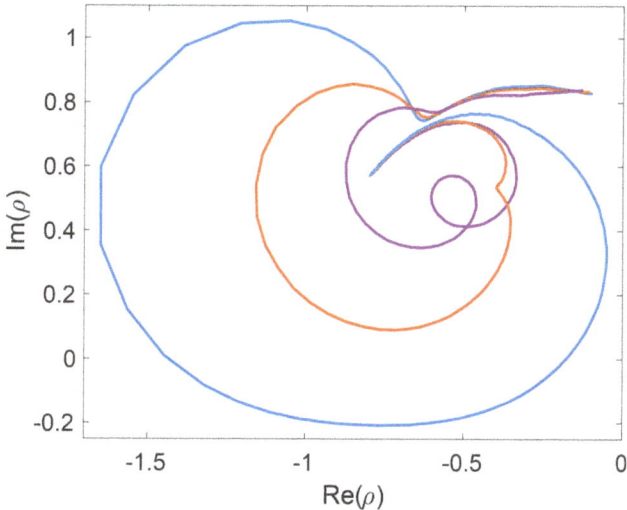

Figure 13.15. Ellipsometry involves measuring the phase and amplitude of light reflected by a thin film sample. In this work the sample consisted of a film of organic spiropyran (SPI) molecules sandwiched between two silver mirrors. Spiropyran is transparent in the visible part of the spectrum, but when illuminated by UV light it transforms into merocyanine (MC) that has a significant absorption in the visible. In the merocyanine form, the absorption hybridises with the electromagnetic cavity mode that is present due to the two silver mirrors in a process known as strong coupling. In the figure, the complex ellipsometric parameter ρ is plotted for three different levels of UV exposure. At low UV levels, the signal from SPI shows a single loop (blue curve). As the level of UV increases, SPI is progressively converted to MC, resulting firstly in a cusped curve (red), and then a curve with an interior loop (purple). The important observation is the appearance of this interior loop, which is a new signature for the onset of strong coupling. Compared to previous techniques, this new approach provides a much more 'clear-cut' signal.
Source: W L Barnes and P Thomas.

From our ellipsometric work we can now see that this transition is one of a class of topological phase transitions – so the new combination of ellipsometer and strong coupling offered some lovely new science. Ellipsometry has therefore proven to be a powerful technique that offers a complimentary way to investigate the strong coupling phenomenon.

Perhaps just as importantly, many labs around the world have an ellipsometer, so this new approach allows many people who didn't think they had the resources to explore strong coupling phenomena to get involved. It looks as though it has also acted as a stimulus to others to ask, 'what else could we use an ellipsometer for?'.

My experience of using an ellipsometer in this way has prompted me to think in a more wide-ranging way about how different pieces of equipment and different facilities might be used, and in particular how one can benefit from sharing equipment. Three things have been in my mind about this topic. First, trying to think about different ways in which one might use equipment can lead to new science, as above, and that's a real plus. Second, I have been looking for ways to reduce the carbon footprint of the science we do, and sharing resources seems obvious, but is often not practiced. Third, talking to others about what prevents people from sharing equipment more, it was apparent that there are many factors: loss of control, avoiding the 'everything left dirty' problem, reduced availability, among others. We need to encourage a culture where sharing becomes much more common, and where we see sharing as an opportunity rather than something to be avoided: sharing offers real benefits, especially by bringing people with different backgrounds and expertise together – new things easily arise in such situations. Using equipment in different ways can make the business of doing science more enjoyable, and really can lead to new exciting science. But for that to happen, maybe some people have to change their minds – the most important, and probably the most difficult, creative act anyone can do.'

13.11 Reducing noise
Mike Semens-Flanagan, Director of Global Engineering, IMI Critical Engineering

'One of the most widely-used technologies, and one most of us never notice, is the valve. Every time you turn a tap in your kitchen or bathroom, you are activating a valve to allow, or shut off, the stream of the water. Valves are everywhere, controlling flows of water, steam, oil, natural gas, saline drips in hospitals, brake fluids in a car. Valves are also a very ancient technology: the Romans used them to control water in their aqueducts. So you'd think that there would be no scope at all for creativity as regards valve design, for surely everything that could possibly be discovered about a valve must already have been discovered!

Far from it. Valve designs continue to evolve, with improved designs, using new materials, and – these days especially – incorporating ever-more sophisticated sensors to allow real-time remote monitoring. Our business dates back to 1862 [84], and, right now, around the world, there are more than a million valves installed, and in use, manufactured by us [85]. So we understand the importance of creativity and innovation in engineering!

Not many creativity stories, though, start with a herd of cows in Kansas. But this one does…

A few years ago, one of our field service engineers was talking to a farmer who happened to mention that the yield of milk from some cows that had been grazing in a particular field had recently declined. 'It's because of all that noise', the farmer had said.

When I heard about that, I was surprised, on two counts. I had no idea that the milk yield of a cow was sensitive to noise. And I was puzzled at what could be causing a noise in the middle of rural Kansas. The answer to that second question was close to hand, for the origin of the noise had been tracked down by our engineer. He had visited the farm and noticed that a long-distance gas pipeline ran along one edge of the field on the other side of a fence, and it was the pipeline that was making the noise. But the source of the noise wasn't the pipeline itself – it was a nearby control valve that was creating a very loud noise indeed, about 100 dB. To put that in context, normal speech is about 60 dB. Prolonged exposure to more than 70 dB can damage hearing, and 120 dB does immediate harm [86]. So no wonder those cows were off their milk.

Valves are noisy when the flow through the valve is turbulent, and there can be friction too – both of which can cause vibrations that trigger audio-frequency waves not only in the surrounding air, but also within the fluid itself and the pipework. And if those vibrations just happen to correspond to the resonant frequencies of the equipment, that can amplify the noise level significantly.

One way of dealing with a noisy valve is to bury it underground. That muffles the noise, but makes maintenance rather more difficult! A better way, of course, is to solve the problem by designing the vibrations out – which, although a better solution, is much more difficult, for that requires creativity to have the ideas for what possible solutions might look like, and also innovation to make the best idea real.

Across the IMI Group, 'a growth mindset that is open to innovation and learning' is one of our four core values [87], so we relish solving problems like that. And noise in valves is something we know quite a lot about, both as regards how the noise arises, and how it can be suppressed. In fact, in 1967 [88], Dick Self, the founder of one of our subsidiary companies, IMI CCI, based in California, took out a patent on a totally novel valve, which he called a 'DRAG® valve' [89]. Originally designed to control the feed of liquid hydrogen to rockets, DRAG® valves significantly reduce turbulence and friction, and hence noise. As shown in figure 13.16, this is achieved by incorporating within the valve housing a very carefully designed and engineered component that breaks the flow of the fluid into a series of more gradual, parallel streams, rather than being 'one big rush'.

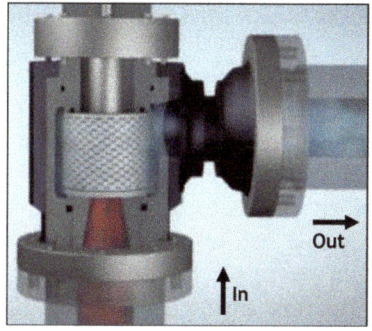 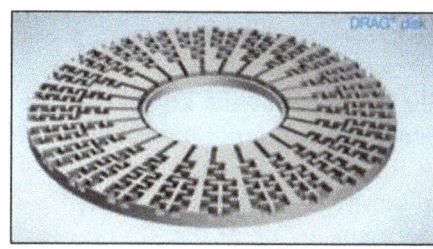

Figure 13.16. On the left is a schematic of a DRAG® control valve, the central element of which is a stack of discs as shown on the right. The fluid flows into the central void, and then through the patterned channels. This transforms the original single flow into many parallel gentler flows.
Source: www.imi-critical.com/products/drag-control-valves/.

By analogy, a conventional valve would be similar to a waterfall, with all the water plunging over a cliff; a DRAG® valve is more like a series of gradual cataracts, each of which is a 'mini-waterfall' such that collectively, all the water moves from 'here' to 'there', but in much more gentle stages [90]. As a result, there is much less turbulence and friction, and much less noise – as well as substantially reducing cavitation and the mechanical abrasion of the interior of the valve too. And, as shown in the figure 13.17, DRAG® valves can be a lot bigger than a domestic tap!

Figure 13.17. An example of a DRAG® control valve as might be installed on an industrial plant. The fluid inlet is at the bottom, and the outlet on the right; the equipment above the man's hand is the 'actuator' that determines the degree of opening of the valve.
Source: Based on brochure downloaded from www.imi-critical.com/products/drag-control-valves/.

The invention of the DRAG® valve was a true step-change in valve design – perhaps the most important single development since the invention of the valve by the Romans – for it enabled the flow of high velocity fluids to be controlled much more sensitively and precisely than had been possible before, with the additional plus of reducing noise too.

As can be seen from figures 13.16 and 13.17, the flow through a DRAG® valve forces the fluid through a right-angle, which is fine for the applications for which these valves were designed. But for some applications, for example, in long-distance pipelines, it's important that the flow is straight-through, so that when the valve is fully open the fluid continues unimpeded as if the valve were not there. To achieve this, the valves are of a different design, known as 'ball' valves, which have the important benefit that, when fully open, the aperture is exactly the same size as the pipe in which it is embedded so that the flow is undisturbed.

That set up a Koestler's Law challenge of how the key features of a DRAG® valve might be combined with those of a ball valve to result in a quiet straight-through valve that minimises disturbance to the flow when the valve is fully open. And if the new design can be manufactured in a range of different sizes so that these much quieter valves can be used not just in new installations, and also to replace existing noisy valves, all the better. That's the kind of challenge we just love, and that's the challenge we solved with our dBX SHIELD™ valve, which was granted its trade mark in October 2021.

As can be seen in figure 13.18, the dBX SHIELD™ can be positioned wherever needed along a pipeline, allowing straight-through flow with minimal disturbance, whilst reducing noise to a comfortable 60 dB [91]. Which is good not just for the cows, but for us humans too – especially those who live relatively close to above-ground pipe networks (which is especially relevant in the United States), as well as those who work in industrial environments such as power stations, factories, ships and chemical engineering plant.

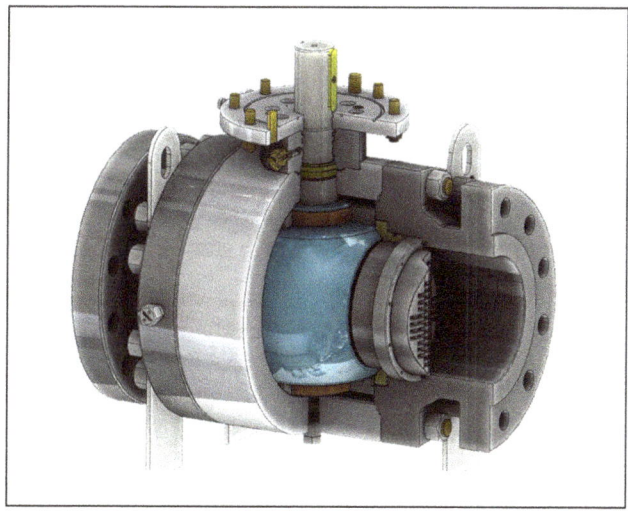

Figure 13.18. A schematic of the dBX SHIELD™ valve. The flow is straight-through, and the valve can be operated by rotating the vertical shaft. The 'slotted baffle' acts like the DRAG® discs shown in figure 13.16, breaking and controlling the fluid flow, so reducing the noise.
Source: Brochure available at www.imi-critical.com/products/dbx-shield/.

Yes, valves are very widely used, and have been around for some 2000 years. But there's still plenty of scope for creativity!'

13.12 How nanopatterns made it from a semiconductor facility to an artist's print room

Professor Nikolaj Gadegaard FRSE, James Watt School of Engineering, University of Glasgow

'How does biology interact with man-made materials – especially at the interface? Are there limits to the details that biology can 'see'? Can we use this knowledge to provide better healthcare solutions? These are big questions and ones I have been fascinated by for many years – and still am. My journey of exploration is fundamentally underpinned by a repeated process of creativity – consciously and subconsciously. I am always looking at other fields for inspiration and often try and flip the problem for a different perspective. In the following, I want to tell the story of nanostructures and cell biology.

My PhD studies in Denmark started my scientific career and my aim was to develop a method of making smarter Petri dishes for cell culture experiments. In many cases, proteins are used to coat the dishes so that the cells can stick to the surface. A commonly-used protein is collagen which has a very unique structure and shape. The protein assembles into fibres which help the cells to adhere to them and also provide mechanical strength to a tissue. When looking at the fibres in a microscope, you will see that they look like tiny worms with a regular ribbon pattern repeating itself along the fibres, as shown on the left in figure 13.19.

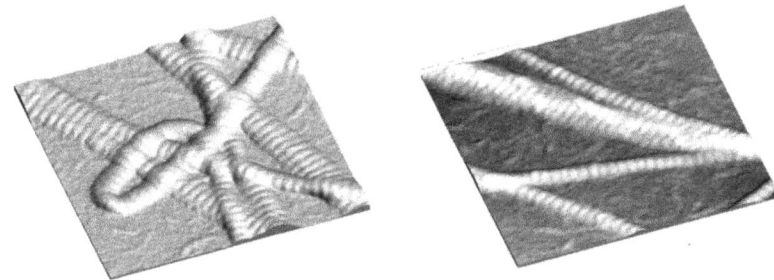

Figure 13.19. Atomic force micrographs of naturally occurring fibrillar collagen (left) and an injection moulded plastic replica (right).
Source: Professor Nikolaj Gadegaard.

The details of this pattern are very subtle. The repeating pattern is only 68 nm and the variation a mere 4 nm. It turns out that this pattern is preserved across mammalian species and so we thought that shape was important for the cells to recognise the protein. So, would it be possible to make a 'smart' Petri dish where the surface looked like collagen, hence tricking the cells to stick to it?

Petri dishes are made by injection moulding, a highly industrial process of mass-producing plastic components. How could such a 'brutal' process be used to transfer the subtle details of the protein? The information in optical storage media such as CDs and DVDs is imprinted into their surface with details of around 1 micrometer – much larger than the fine details of the protein. The discs are manufactured using polymer injection moulding against a pre-patterned master, exactly the technique which we wanted to use. The question was, would it be possible to do this for our patterns which are more than ten times smaller? And it turned out that the answer was 'yes!'.

In fact, it only took me the first couple of months of my PhD to solve this technical challenge. It looked like I was on a home-run for an easy PhD! But things took a different turn. I spent the next couple of years trying to do the biological experiment to prove my 'smart' dishes worked. At the end, it turned out that the pattern did not play a role at all in the cells' ability to adhere to the surface. So, from early success to failure, it looked like my scientific career was to have an abrupt end. Fortunately, my work on injection moulding proved unexpectedly important for future – and after all, I succeeded in getting my PhD.

During my PhD I was fascinated by the pioneering work carried out in Glasgow. Professors Chris Wilkinson and Adam Curtis were trailblazers in the field of cell response to nanopatterns – something which I didn't see the fruit of in my PhD. So when I got the opportunity to join their team, I jumped at the chance. It offered me the chance to answer the question of cells' ability to sense topographical patterns in their environment and to what the limitations might be.

To answer these questions, it was necessary to control the length scale of the surface texture – or topography – carefully and in a very controlled manner. Many researchers have developed different methods for answering this question – all with varying degrees of control. With biological cells ranging from 10–100 micrometers, which is comparable to the diameter of human hair (50–100 micrometers), I needed a tool which could make details smaller than that of a single cell, and I wanted to be able to precisely control the dimensions in a very reliable manner.

My approach was to use the power of semiconductor technologies: electron beam lithography and etching. This is an industry that is fundamentally built on reliably making small transistors (today, much less than 1 micrometer) as any variations impact on their operational performance. Moreover, the location of the individual transistors is also critical as their connections ultimately form the final chip and determine its function. This process is called micro- or nanolithography.

To test the impact of nanopatterns on cells, we designed a very simple pattern of 100 nm diameter 'dots' placed in a regular array – a bit like a crystal. With one challenge solved new ones presented themselves. The first one is that most computer chips are less than 1 cm^2 in size which is significantly smaller than most healthcare applications, such as orthopaedic implants, pacemakers and prostheses. So we looked at other disciplines for inspiration and found that the manufacturing process for 'photonic crystals', used in telecommunication, could be modified for our needs and enable much larger areas of patterns to be generated. A change to our original fabrication process meant that patterning a 1×1 cm^2 area with nano dots previously took 4 days of continuous work, this could be reduced to only 1 hour. During this patterning step, we are able to place 1 billion(!) nano dots very precisely onto our surface. In fact, we can place them with a precision smaller than the size of most proteins.

The second challenge we encountered is that computer chips are made from silicon, a material which is not generally used in biomedical applications due to its brittleness and opacity. We needed a solution to translate our patterns from silicon to more commonly-used materials in biomedical engineering such as polymers. Here my (failed!) experience with the collagen Petri dishes was put to good use. With injection moulding we are now able to make more than 2,000 samples per day fully automatic in the laboratory.

With these bottlenecks addressed, ahead of us was a new era of biological discovery no longer limited by sample quantity or size of the sample. Together with biologists, we started formulating questions about dimensions and regularity. A particularly fruitful collaboration was with Professor Matthew Dalby. In 2007 we discovered that patterns with small deliberate errors stimulated stem cells extracted from the bone marrow to specifically make bone. Through a number of control experiments, we demonstrated that the effect was entirely caused by these subtle changes in the pattern, and not by chemical or biological factors. This discovery is now finding its way into orthopaedic devices – a route that presents itself with new challenges, but that is a story for another day.

Shortly after, in 2011, we made a further discovery. When you remove stem cells from a patient and place them in an ordinary Petri dish, they quickly lose their regenerative potential and become 'regular' tissue cells known as 'fibroblasts'. However, we discovered that when cells saw a highly regular nanopattern instead of a flat Petri dish, their regenerative properties were maintained for weeks in culture. This has opened new opportunities in regenerative medicine. Both papers were published in Nature Materials [92, 93] and have been highly cited since.

I want to conclude this memoir with a very inspirational period I had with an artist in residence. With our stem cell discoveries came the question 'How could such small patterns and changes in them lead to such impactful changes in the behaviour of the cells?' And this has (and still is) a challenging question as this requires us to understand the interactions at the molecular scale. In 2005 I was awarded a European Research Council (ERC) Consolidator Award to look into these mechanisms. Using modified cells which fluoresce in specific parts of their structure and super resolution microscopy, we saw how the nanostructures were able to guide single proteins and their assembly to form adhesions between the cell and the material.

One day, I met with artist Rachel Duckhouse who was working in an adjacent laboratory on seashells. We started talking about my work, and the patterns of dots completely struck Rachel's imagination. Together with Creative Scotland and the ERC, Rachel was embedded into my research group as an artist in residence. This was a truly inspirational period of the project. Rachel asked us many 'layman' questions to understand the engineering and the biology we were looking at. Through the conversation, she kept a sketchbook of amazing illustrations, such as those shown in figure 13.20 [94].

One day sitting in my office with pens, paper and transparencies full of patterns and drawings, we started overlaying the patterns. This was to mimic the patterns of our engineered substrates and the hypothesised pattern the proteins in the cell would make. Suddenly we started to see a lot of emerging patterns we had observed in the lab. This exciting interaction between art and science informed us on the direction of our research questions. Interestingly, many of the patterns Rachel made during her residency were printed – using lithographic techniques. It seems that the research has completed the circle: starting with nanoscale lithography and ending with an artist's impression also produced by lithography.'

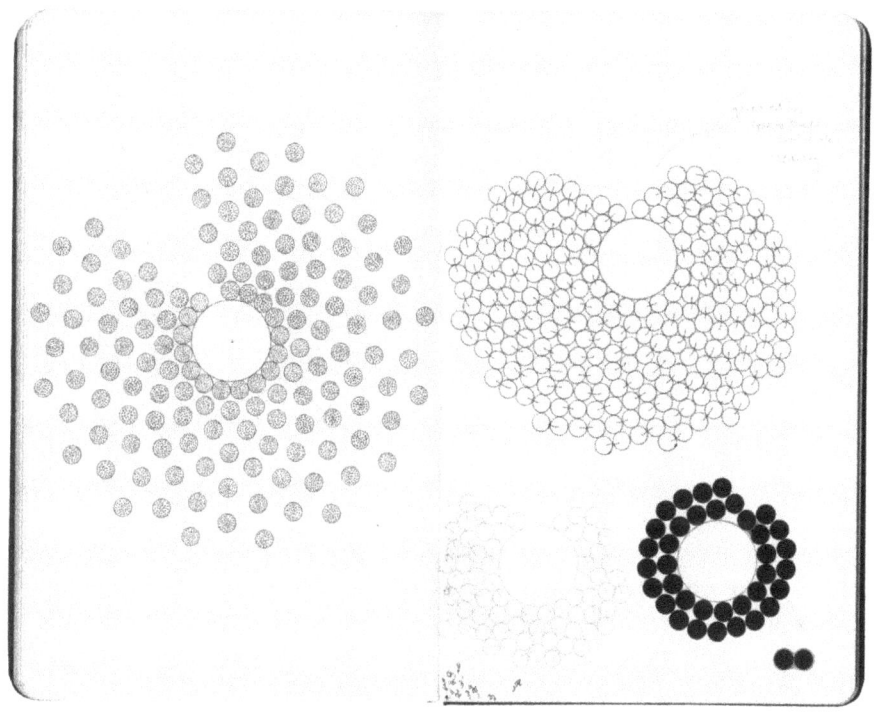

Figure 13.20. Two pages from Rachel Duckhouse's sketchbook.
Reproduced courtesy of Rachel Duckhouse www.rachelduckhouse.co.uk/patternresponse.

13.13 Newton's Rings and flat screens
Professor Harish Bhaskaran, Department of Materials, University of Oxford

'The invention of flat screens was a major breakthrough. In the past, displays for televisions and computer monitors were big, heavy, bulky cathode ray tubes; now they are slim and light. And this was of course a pre-requisite for the development of devices such as laptops, tablets and smart phones.

Pretty much all the different types of flat screens currently in use rely on some form of internal light source at the back of the screen. This light then passes through a number of layers of different materials, resulting in the image that you see. So – simplifying things somewhat – an LCD flat screen has four principal layers. Firstly, at the back, a white light source, and then a polariser that transmits only, say, horizontally polarised light. The third layer is the important one – an array of individual liquid crystal 'cells', each of which is associated with a red, green or blue filter, and organised such that each red/green/blue triplet, with their associated liquid crystal 'cells', comprises a single pixel on the screen. Finally, the top layer is a second polariser that transmits only vertically polarised light.

The polarisers are crossed, and so, in principle, no light can pass through all the layers. The screen will therefore appear black. But that assumes that the liquid crystal layer transmits all the incident horizontally polarised light unchanged.

In fact, liquid crystals have the special property that their internal structure can be altered by applying a voltage, causing the plane of polarisation of the light passing through any specific 'cell' to be rotated by up to 90°. This enables the rotation to be continuously controlled, so determining the intensity of the light that will be able to pass through the front, vertical, polariser. But before the light reaches the front polariser, it passes through a red, green or blue filter. The colour actually seen by the viewer is therefore the colour mix attributable to the intensities of the vertically polarised components of the light transmitted through each of the red, green and blue filters associated with each pixel. These intensities are themselves determined by the voltages applied to each of the three corresponding liquid crystal cells, and so controlling the voltages of all the cells in the array results in the images we see on the whole screen, and all the colours.

Importantly, those colours all originate from the internal light source at the back of the screen – light that requires energy to create, and light that is produced by appropriate components at the back of the screen, components that take up space, increase the screen's weight, and add to the overall cost.

Most of the colours we see every day – the colours of flowers, of paints, of textiles – are not backlit, but are the result of the reflection of ambient daylight or artificial light: in general, when light falls onto a material, some is absorbed, and some reflected, and the colour we see is whatever colour mix results from the reflected wavelengths. Those colours, though, are static. For a screen, we need to be able to change the colours dynamically.

And we've all seen this happening naturally – on a soap bubble which has all sorts of colours until it pops. The colours of an oil slick are similar too, but change rather more slowly.

When light is incident on the outer surface of a soap bubble, some is reflected, and some is transmitted into the soap film. As that light passes through the film, some is absorbed, and then some is reflected back at the inner surface, later emerging through the top surface. The colours we see result from both the absorbtion, and also the interference between the various transmitted and reflected wavelengths as determined by the local thickness of the film – just like the ring-shaped interference pattern first identified by Newton in 1666.

So that raises an intriguing question. Can the concept of Newton's Rings be combined with that of the flat screen to create a display that doesn't require its own backlight, but that works under natural ambient light? In principle, that's a good idea, for the resulting screen would, in general, have fewer components, be even thinner and lighter, would cost less, and consume less energy too. Interesting...

A soap bubble, though, is an accidental thin film, that just happens to show whatever colours emerge from the different thicknesses at different locations. To make a useable screen requires that the material thickness can be changed in a controllable way, and over small enough areas to form a pixelated array. That seems quite hard to do. But does the property that needs to be controlled have to be the material thickness?

No. There's another important property that's relevant here – the material's refractive index. So imagine an array of pixels of the same material of the same thickness, but with each pixel having a different refractive index, and so showing a different colour. If the refractive index of each pixel can be changed in a controllable way, the result is a flat screen which works under ambient light.

A number of years ago, one of my post-docs, Peiman Hosseini, was studying an alloy of germanium, antimony and tellurium that can form thin films. Many aspects of this material were already known, but this research was the first to combine investigation of both its optical and its electronic properties, resulting in the discovery that the refractive index of the alloy could be changed, and controlled, by applying a voltage, causing the alloy film to change colour in ambient light [95].

One problem we identified early on, though, concerned the thickness of the film that would be required to produce all the colours needed, and the corresponding power consumption to drive it: our initial calculations suggested that the screen would have to be unhelpfully thick, and demand too much power. That caused us to pause, and to turn our attention to other things, but we never forgot about the central idea.

And then something accidental happened. I remember it well. One morning, Peiman and I were working at Oxford's Begbroke Science Park, and needed to return to the town in the afternoon. We had an hour-and-a-half to wait until the next bus, and so Peiman suggested that we spend that time at a talk, about to be starting in a nearby building, given by Moritz Riede, a newly appointed Professor in the Department of Physics. The subject concerned solar cells, which is not my own field, and I hadn't planned to attend, but since I didn't have much else to do for an hour, why not? During the session, Moritz talked about ultra-thin films, and a technique called 'optical transfer matrix modelling'. And then I realised that this was the key to unlock the problem of our too-thick display films! After the talk, we spoke to Moritz to learn some more, and he told us how to access the right software for our calculations, which we obtained from the website of a Stanford Professor, Mike McGehee (now at the University of Colorado, Boulder).

After about a week, Peiman showed theoretically that we could get a display to work, and very soon after that, we had our first prototype of a revolutionary new screen, illuminated by ambient light.

That's a real 'da Vinci' problem, solved. We were stuck, for there was something important missing – a 'something' that was solved by optical transfer matrix modelling. It's another wonderful example of Koestler's Law too – how combining just the right aspects of solar cell physics with our materials science resulted in a new display. There was luck too – suppose Moritz had been giving his talk on another day, a day when we weren't at Begbroke, or when we didn't have 'an hour to spare'? Even though Peiman and I were thinking about other things that day, we had never forgotten about the display possibility. It was always in the backs of our minds. So when we heard Moritz's talk, all the 'Koestler's Law components' fell into place.

We then did much more development, but all well-focused on a *BIG IDEA* that we knew would work. We filed three patents, and launched a start-up company, Bodle Technologies Limited [96], which is now located at that same Begbroke Science Park. In fact, not only does our display differ from conventional displays by virtue of using ambient reflected light, there's another significant difference too. In addition to the switchable liquid crystal 'cells', LCD displays also require the use of red, green and blue filters to produce the required colour mix. Our screens, by contrast, produce red, green and blue light directly, and so no filters are needed.

And something else too. I mentioned that a pulse of electricity can change the refractive index of certain materials. That's just one example of an electrically-induced change in the properties of a material: another is to change a material's Young's modulus, which in turn affects the material's response to mechanical stress, and how the material resonates at different frequencies. That sparks ideas about tuneable resonators, and how the tuning might be controlled, which is highly relevant to many devices, not least mobile phones. Anyway, we looked into that several years ago, but didn't get very far because the effect was too small in the materials we were using. But more recently, I came across some research being carried out by Ritesh Agarwal, a Professor of Materials Science at the University of Pennsylvania, using nanowires made from a material that changes its Young's modulus in this way. Ritesh very kindly sent us some nanowires to study, and because each nanowire is a single crystal, we found that it acts as a very effective tuneable resonator, and with a very high Q factor too. So that's another example of a 'latent' idea which didn't have much traction when it first emerged, but proves to be much more interesting some time later, when another 'Koestler's Law component' – in this case the single-crystal nanowire – comes along. Another da Vinci problem, solved!

I think these examples highlight one of the traps of academic research. You can be trying so hard to solve a particular problem that you can become far too focused, far too narrow. That's understandable. But sometimes maybe it's better to defocus, to relax your mind and do something different for a while, to be patient, to say 'there are things I just don't know that I should find out more about'. In practice, of course, that can very difficult – and maybe the more senior you get, the harder it is to say 'I don't know'. But no one can know everything, and Koestler's Law so clearly shows that the more 'fragments' you're aware of, the more likely you'll be able to put together exciting new patterns. So we must never stop learning, we must never stop collecting those 'fragments', even if they might not appear to be relevant just now. And – as the flat screen illustrates so well – one way to do that is to attend lectures that are 'not in my field'! And without the pretext of having an hour to spare whilst waiting for a bus!'

13.14 Blue Plan-it® and Water ARC®
Dr Jess Brown, R&D Practice Director, Charlie He, Chief Technologist, and Dr Justin Sutherland, Chief Technologist, Carollo Engineers Inc.

Jess Brown: 'We're an engineering consulting firm focused only on water, operating across the US. Given that every water company employs its own well-trained and highly experienced engineers, we have to offer something special, something that adds value to the in-house team and to our clients, something that's different. So for all of us at Carollo, creativity and innovation are center-stage. As you can see from our website [97], the importance and value of creativity and innovation are stressed time and again. We're proud of our track record of developing innovative solutions for our clients, and we're also proud of the many innovations we have developed within the firm itself, two examples being our dynamic decision support tool, which Charlie will tell you about, and also Water ARC®, our state-of-the-art treatability testing facility in Boise, Idaho, headed up by Justin Sutherland...'

'Yes, let me take things from there,' continues Charlie. 'Water, of course, involves extensive on-the-ground (and in-the-ground too!) engineering – the treatment works, the pipe network, the waste-water facilities. But before you can do all that construction, there are a host of analyses that need to be done: technical diagrams and calculations to ensure that the plant is well-designed and operates optimally, financial forecasts to ensure that the costs and revenues are well-understood and under control.

All engineers, of course, are trained to draw process flow diagrams and to do the technical calculations, and accountants know how to do all the financial work. Way back, this was all carried out 'by hand', but everything was transformed by the availability of easily accessible computing, and especially by the introduction of spreadsheets. So, over the years, every engineer, every accountant, every company, has accumulated drawers full of process flow diagrams, and computer files full of increasingly complicated – and often poorly documented – spreadsheets which became ever more opaque as time passed. And not just that. Any given process flow diagram or spreadsheet will do whatever it was supposed to do, but nothing else. Which is fine as far as it goes, but fails to recognise the very important fact that all water systems are highly 'joined up' – quite literally by virtue of the pipe network, and also conceptually in that any technical decision 'here' (such as an assumption about the aquifer from which source water might be drawn) not only has a technical impact 'there' (for example, the processes required at the water treatment plant to make that source water safe to drink), but also corresponding financial consequences (for example, the costs of the materials required for that treatment). In practice, each of these aspects might be investigated by different people, using different diagrams and different spreadsheets. This adds to complexity, as well as increasing the likelihood of inadvertent error, or muddle arising from the use of contradictory assumptions or inconsistent data sets.

How might this be different?

By bringing everything together in a single, integrated, decision support system, that's how. But that's a daunting task for any one water company. And it was a daunting task for us too, but, as a consulting firm, we have the opportunity to reap an economy of scale by using one well-designed decision support system for many clients to support our various projects across the industry; in addition, we can offer the system for direct use by water companies themselves, which has been much facilitated by developments in cloud computing.

So that's Blue Plan-It®, a comprehensive decision support system bringing together all aspects of water, from mass balance calculations to optimising networks, from outage analysis to 'Monte Carlo' simulations to help navigate uncertainty [98]. A planning framework that 'connects all the dots'.

Given our wide experience, we have compiled a library of technical components – from sedimentation basins to filters, from membranes to ultra-violet disinfection – and all the associated data, enabling any specific configuration to be represented by assembling the required elements very easily by 'dragging and dropping' on a screen. And once an appropriate model has been built, the resulting 'master plan' can be routinely used as a dynamic monitoring and re-planning tool, so that, for example, data from actual operations might indicate the need to make some adjustments, or when it might be helpful to forecast future impacts. Furthermore, since everything is in 'one place', this helps organisational co-operation and communication. Which are always 'good things'. And of course the current model can continually be updated and enhanced as the real assets change, or in response to – or indeed anticipation of – events that might influence demand (say, new housing), water supply (such as changes in the aquifers), technology (when improved methods are developed) or regulation (as happens when legislators revise standards, as they do from time to time).

Blue Plan-It® has certainly proved to be of great value to our clients, and its availability gives Carollo a significant competitive edge in our own market place. But that was not, of course, the situation right at the start, when Blue Plan-It® was just an idea. Back then, it wasn't clear how complex it would become, and how much effort would be required to put it all together. Nor was there much evidence of customer demand – no customer was saying 'we want something like [this], so will you please build it', largely because no customer had ever seen anything like what Blue Plan-It® has turned out to be. I guess that's like many true innovations – things that are so new that developing them has to be something of a leap into the dark, if not a leap of faith!

The original idea was nothing like as complex and sophisticated as what we have now, for over the ten years or so that we have been working on it, more and more features have been incorporated. The fundamental concept, though, of totally integrated planning and decision support, was there from the start. And looking back, I feel very fortunate that I have been working in a culture that says 'tell me more' rather than 'no', that is not profligate but is willing to take well thought-through risks, that embraces opportunities rather than perpetuating the *status quo*, and that gave me organisational support to do something new.'

'That applies to Water ARC® too,' continues Justin Sutherland. 'In fact, I remember distinctly how the idea originated. At a creativity workshop we ran in December 2015 at our Orange County offices in California, which is where I first became aware of the power of generating ideas by asking 'How might this be different?'. At that time, we had a rather poorly equipped laboratory facility in Boise, Idaho. It had become quite run down, and was well on the way to becoming a 'stranded asset'. The 'obvious' thing to do was to close it down and plough the sales proceeds back into the business somewhere else. But at that workshop, we asked 'How might this be different?', and had the idea that rather than closing the Boise site, we might do the opposite and invest in it, revitalising it into a truly state-of-the-art laboratory, offering all aspects of analytical and field testing to our clients, and possibly other services too.

I then developed the idea and discussed it with Jess, and he agreed that we should both go to Boise to take a good look at what was there and what might need to be done to upgrade everything. And in my mind the whole time was that fundamental question 'How might this be different?'. I was then given the green light to put together a formal business plan, which included many specific and practical ideas about what we might do. And as Charlie described, our culture, which goes right up to the Board – or maybe it comes down from the Board? – is one that encourages exploration and taking well-managed risks, so the Board approved the plan, and we went ahead.

We planned things carefully, and opened Water ARC® – the ARC stands for 'Applied Research Center' – in 2018 [99]. And over the three years since then, we've extended our capability substantially, including a workshop to design and build totally novel devices and items of equipment, not only for use on our assignments, but also directly for our clients who wish to benefit from our expertise. We've also taken great care as to how the facility was designed, with efficient work flow and safety as top priorities. And we've thought hard about managing data efficiently and effectively, and are developing a mobile app so that data can be collected in the field, and fully integrated with test data generated in Boise, or up-loaded from other third-party sources.'

Jess Brown concludes: 'Carollo was founded in 1933, and we hope to be around in 2033 and 2133. Because water is such a vital, enduring resource, and our work will help keep the planet healthy over that time. But our business's continuity, and our contribution to the industry, demand that we innovate, innovate, innovate. Not for the sake of innovation, and not at the risk of doing something commercially catastrophic. But safely, steadily and reliably. And so we continue to value innovation of all kinds, from small improvements associated with specific client engagements, to continuous improvement in our own services – of which Blue Plan-it® and Water ARC® are great examples – to ground-breaking solutions for the entire industry. And fundamental to all that is that magic, powerful, insightful question, 'How might this be different?'.

References

[1] Abbott B P (LIGO Scientific Collaboration and Virgo Collaboration) *et al* 2016 Observation of gravitational waves from a binary black hole merger *Phys. Rev. Lett.* **116** 061102

[2] www.ligo.caltech.edu/system/media_files/binaries/306/original/ligo-press-kit.pdf

[3] Poincaré H 1905 On the dynamics of the electron *C. R. Acad. Sci.* **140** 1504–508 www.academie-sciences.fr/pdf/dossiers/Poincare/Poincare_pdf/Poincare_CR1905.pdf (in French) https://en.wikisource.org/wiki/Translation:On_the_Dynamics_of_the_Electron_(June) (in English)

[4] Einstein A 1905 On the electrodynamics of moving bodies *Ann. Phys.* **17** 891–921 (in German) www.fourmilab.ch/etexts/einstein/specrel/www/ (in English)

[5] https://history.aip.org/exhibits/einstein/chron-1905.htm

[6] Einstein A 1916 Approximative integration of the field equations of gravitation *Sitzungsber. Preuss. Akad. Wiss. Berlin (Math. Phys.)* **22** 688–96 (in English)
[7] https://arxiv.org/pdf/1602.04674.pdf
[8] www.einstein-online.info/en/spotlight/observation-of-gravitational-waves-from-a-binary-black-hole-merger/
[9] www.falstad.com/scale/
[10] Detecting the Aether wind: the Michelson-Morley experiment http://galileoandeinstein.physics.virginia.edu/lectures/michelson.html
[11] www.ligo.caltech.edu/page/ligo-gw-interferometer
[12] www.aei.mpg.de/38488/geo600
[13] www.virgo-gw.eu/
[14] https://gwcenter.icrr.u-tokyo.ac.jp/en/
[15] www.ligo.caltech.edu/page/ligos-ifo
[16] Rowan S and Hough J 1998 The detection of gravitational waves, report number OPEN-2000-258 https://cds.cern.ch/record/454173/files/p301.pdf
[17] www.ligo.caltech.edu/page/facts
[18] www.lhc-closer.es/taking_a_closer_look_at_lhc/0.high_vacuum
[19] www.physics.gla.ac.uk/igr/index.php?L1=research&L2=suspensions
[20] https://makingscience.royalsociety.org/s/rs/people/fst01496942
[21] www.ligo.caltech.edu/news/ligo20160426
[22] https://dcc.ligo.org/public/0065/M870001/002/M870001_Redacted.pdf
[23] Everitt C W F *et al* 2015 *Class. Quantum Grav.* **32** 224001
[24] https://patentimages.storage.googleapis.com/cf/94/4f/7a8bc464dbf08f/US6284085.pdf
[25] van Wegel A A and Killow C J 2014 Hydroxide catalysis bonding for astronomical instruments *Adv. Opt. Techn.* **3** 293–307
[26] Willke B *et al* 2002 The GEO 600 gravitational wave detector *Clas. Quantum Grav.* **19** 1377-387
[27] Elliffe E J *et al* 2005 Hydroxide-catalysis bonding for stable optical sustems for space *Clas. Quantum Grav.* **22** S257
[28] Heptonstall A *et al* 2011 CO2 laser production of fused silica fibers for use in interferometric gravitational wave detector mirror suspensions *Rev. Sci. Instrum.* **82** 011301 https://pure.strath.ac.uk/ws/portalfiles/portal/86397916/Heptonstall_et_al_al_RSI_2011_silica_fibers_for_use_in_interferometric_gravitational_wave_detector.pdf
[29] https://dcc.ligo.org/public/0005/T0900447/004/T0900447-v4_Final_design_document_ETM_ITM_ears.pdf
[30] Affeldt C *et al* 2014 Advanced techniques in GEO600 *Clas. Quantum Grav.* **31** 224002
[31] www.ligo.caltech.edu/page/about-aligo
[32] www.ligo.caltech.edu/page/ligo-gw-interferometer
[33] www.ligo.caltech.edu/page/ligo-technology
[34] www.ligo.org/science/Publication-O1Noise/flyer.pdf
[35] www.britned.com/
[36] www.nemolink.co.uk/
[37] https://ifa1interconnector.com/
[38] www.northsealink.com/
[39] www.eleclink.co.uk/
[40] https://viking-link.com/
[41] www.soni.ltd.uk/customer-and-industry/interconnection/

[42] www.siemens-energy.com/global/en/news/magazine/2020/intereuropean-hvdc-link-nemolink.html
[43] Zayats A 2013 Perspective: a glint of the future *Nature* **495** S7
[44] Wang P, Krasavin A V, Nasir M E, Dickson W and Zayats A 2018 Reactive tunnel junctions in electrically driven plasmonic nanorod metamaterials *Nat. Nanotech* **13** 159–64
[45] https://qbi.uq.edu.au/brain/brain-anatomy/what-neuron
[46] Wang P, Nasir M E, Krasavin A V, Dickson W and Zayats A V 2020 Optoelectronic synapses based on hot-electron-induced chemical processes *Nano Lett.* **20** 1536–41
[47] www.ggbearings.com/en
[48] Lee H, Dellatorre S M, Miller W M and Messersmith P B 2007 Mussel-inspired surface chemistry for multifunction surface coatings *Science* **318** 426–30
[49] https://bioinspiredmaterials.berkeley.edu/
[50] https://news.berkeley.edu/2016/06/30/fetal-surgery-stands-to-advance-from-new-glues-inspired-by-mussels/
[51] https://bioinspiredmaterials.berkeley.edu/research/mussel-inspired-adhesives/
[52] www.dynamicearth.co.uk/
[53] www.eu-atlas.org/
[54] www.marine.ie/Home/sites/default/files/MIFiles/Docs_Comms/SignedGalwayStatement24MAY2013.pdf
[55] www.nature.scot/landscapes-and-habitats/habitat-types/coast-and-seas/marine-habitats/cold-water-coral
[56] Paterson E 2019 ATLAS Reef Survey Mat - educational resources (Zenodo) https://doi.org/10.5281/zenodo.4540228.
[57] www.eu-atlas.org/education/activity-mat.html
[58] www.eu-atlas.org/education/rov-simulator-game.html
[59] www.eu-atlas.org/education/spectacular-colouring-pages.html
[60] Padgett M 2014 Light's twist *Proc. R. Soc.* A **470** 20140633
[61] Bazhenov V Y, Soskin M S and Vasnetsov M V 1992 Screw dislocations in light wavefronts *J. Mod. Opt.* **39** 985–90
[62] Allen L, Beijersbergen M W, Spreeuw R J C and Woerdman J P 1992 Orbital angular momentum of light and the transformation of Laguerre-Gaussian laser modes *Phys. Rev.* A **45** 8185–189
[63] Dholakia K, Simpson N B, Padgett M J and Allen L 1996 Second-harmonic generation and the orbital angular momentum of light *Phys. Rev.* A **54** R3742–745
[64] Zeilinger A 2017 Light for the quantum. Entangled photons and their applications: a very personal perspective *Phys. Scr.* **92** 072501
[65] Kwiat P G, Mattle K, Weinfurter H, Zeilinger A, Sergienko A V and Shih Y 1995 New high-intensity source of polarization-entangled photon pairs *Phys. Rev. Lett.* **75** 4337
[66] Mair A, Vaziri A, Weihs G and Zeilinger A 2001 Entanglement of the orbital angular momentum states of photons *Nature* **412** 313–6
[67] Padgett M J and Boyd R W 2017 An introduction to ghost imaging: quantum and classical *Phil. Trans. R. Soc.* A **375** 20160233
[68] Gibson G M, Johnson S D and Padgett M J 2020 Single-pixel imaging 12 years on: a review *Opt. Express* **28** 28190–8208
[69] Sun B, Welsh S S, Edgar M P, Shapiro J H and Padgett M J 2012 Normalized ghost imaging *Opt. Express* **20** 16892–6901

[70] Sun B, Edgar M P, Bowman R, Vittert L E, Welsh S, Bowman A and Padgett M J 2013 3D computational imaging with single-pixel detectors *Science* **340** 844–47
[71] Welsh S S, Edgar M P, Bowman R, Jonathan P, Sun B and Padgett M J 2013 Fast full-color computational imaging with single-pixel detectors *Opt. Express* **21** 23068–3074
[72] Radwell N, Mitchell K J, Gibson G M, Edgar M P, Bowman R and Padgett M J 2014 Single-pixel infrared and visible microscope *Optica* **1** 285–89
[73] Welsh S S, Edgar M P, Bowman R, Jonathan P, Sun B and Padgett M J 2015 Near video-rate linear Stokes imaging with single-pixel detectors *J. Opt.* **17** 205705
[74] Sun M-J, Edgar M P, Gibson G M, Sun B, Radwell N, Lamb R and Padgett M J 2016 Single-pixel three-dimensional imaging with time-based depth resolution *Nat. Commun.* **7** 12010
[75] Stellinga D, Phillips D B, Mekhail S P, Selyem A, Turtaev S, Cizmár T and Padgett M 2021 Time-of-flight 3D imaging through multimode optical fibres *Science* **374** 1395–399
[76] http://aerossurance.com/helicopters/james-reasons-12-principles-error-management/
[77] www.skybrary.aero/bookshelf/books/3199.pdf
[78] https://publicapps.caa.co.uk/docs/33/cap716.pdf
[79] www.era.europa.eu/sites/default/files/library/docs/leaflets/human_factors_safety_management_systems_en.pdf
[80] www.youtube.com/watch?v=PuitUnRMDSg
[81] Gray H M and Wegner D M 2007 Dimensions of perception *Science* **315** 619
[82] Will P, Merritt E, Jenkins R and Kingstone A 2021 The Medusa effect reveals levels of mind perception in pictures *PNAS* **118** e2106640118
[83] Thomas P A, Tan W J, Fernandez H A and Barnes W L 2020 A new signature for strong light-matter coupling using spectroscopic ellipsometry *Nano Lett.* **20** 6412–419
[84] www.imi-critical.com/about-us/history/
[85] www.imi-critical.com/about-us/
[86] www.cdc.gov/nceh/hearing_loss/what_noises_cause_hearing_loss.html
[87] www.imiplc.com/who-we-are/our-culture-values
[88] https://dbxshield.com/resources-category/faqs
[89] https://energyfocus.the-eic.com/power/feature/great-leap-forward-drag-technology
[90] www.youtube.com/watch?v=pVVlyFSkUOw
[91] www.imi-critical.com/products/dbx-shield/
[92] Dalby M J, Gadegaard N and Tare R *et al* 2007 The control of human mesenchymal cell differentiation using nanoscale symmetry and disorder *Nat. Mater.* **6** 997–1003
[93] Dalby M J, Gadegaard N and Oreffo R O C R 2014 Harnessing nanotopography and integrin–matrix interactions to influence stem cell fate *Nat. Mater.* **13** 558–69
[94] www.rachelduckhouse.co.uk/patternresponse
[95] Hosseini P, Wright C and Bhaskaran H 2014 An optoelectronic framework enabled by low-dimensional phase-change films *Nature* **511** 206–21
[96] www.bodletechnologies.com/
[97] https://carollo.com/
[98] https://carollo.com/innovation/blue-plan-it/
[99] www.youtube.com/watch?v=08v8V-YmN1w

Part III

How to evaluate ideas, wisely

IOP Publishing

Creativity for Scientists and Engineers
A practical guide
Dennis Sherwood

Chapter 14

Evaluation in context

14.1 Why wise evaluation is important

So you've run an idea generation workshop, and you have a stock of ideas, some of which you really like.

What happens next?

Unless the idea is relatively 'small', or unless you are an author, musician, artist or academic, and your idea is for a poem, a pop song, a painting or a modest research project, what does not happen next is the activity of making the idea real, of beginning to develop the idea into something you can touch, feel, experience.

That's because – with the exceptions noted – most ideas cannot be made real by an individual, usually the individual who had the idea in the first place, working by him- or herself. To make the idea happen, other people need to be involved, and resources beyond those of the individual have to be allocated and consumed. So, for example, a designer may have a great idea for a new product, but for that product to come to market, it needs to be manufactured, marketed and sold, and so the factory has to make capacity available, perhaps install new plant and re-train staff; the marketing team need to develop all the marketing materials; the sales people need to be briefed on all the product's benefits and features; and the accountants need to assess all the costs and determine the price. Although the idea was generated by a small team, to make that idea real requires the commitment of resources from the whole enterprise.

At the same time as moving from the domain of the individual and small group to that of the enterprise-as-a-whole (or a substantial part of it), the organisation is moving from a context of low risk (running some idea generation workshops does not involve much risk, for the worst that can happen is that some people might feel they have lost some time) to one of possibly very high risk indeed – developing the idea will consume resources of time and money, resources that could be deployed elsewhere, and resources that might ultimately be wasted if the idea turns out in practice to be less good than the originators had hoped.

The transition from idea generation to idea development is therefore a step-change in involvement, commitment and risk, and so, from an organisational perspective, 'policing' that transition is very important: resources are finite, and so should be allocated to the most promising opportunities, for if a risk goes sour, that could damage the organisation significantly. For small organisations, failure of a single innovation project could bring the business down altogether; for larger organisations, a failure might be damaging, but not fatal – indeed, one of the most important benefits of organisational size is the cushion that can be provided against the risks associated with innovation.

That risk, however, is real, and the purpose of wise evaluation is to guard the transition from idea generation to development, ensuring that the organisation does not squander resources on poor ideas, whilst maximising the benefit of good ones. But at this stage, just after the idea has been generated, the idea is just that – an idea. It's not a reality that can be seen, experienced, touched, felt. Yet a judgement has to be made on 'goodness' or 'badness', a judgement that has to be made on limited evidence.

That's not easy; but nor is it overwhelmingly difficult, as this chapter, and the next, demonstrate. And given the importance of 'wise evaluation', you might think that sophisticated organisations, organisations led by experienced managers, many with MBA degrees from prestigious universities, would be very good at doing it.

But that's not necessarily the case…

14.2 A very bad idea indeed

Hoover is known around the world for consumer goods such as refrigerators, washing machines and especially vacuum cleaners – indeed, by the 1980s, Hoover was so well-known that the brand name 'hoover' was often used as the generic word for a vacuum-cleaner, regardless of the manufacturer.

In the early 1990s, there was a global recession, and the British Hoover business was, understandably, seeking to boost sales. The idea of a 'sales promotion' – an incentive to buy a particular product by offering something else too, such as a price discount or an 'extra' – was then, and still is, in every marketing manager's toolkit, and it works, as evidenced every time you purchase, for example, '3 for 2'.

But giving away two free dust bags with every vacuum cleaner, or extra-large boxes of soap powder with every washing machine, are pretty obvious, unoriginal and rather dull promotions. The Hoover team wanted a promotion that was more substantial, something different, a real pull. They were then approached by JSI, a small travel agency, with the *BIG IDEA* of incentivising the purchase of Hoover products costing at least £100 by giving two vouchers for free return airline tickets from Britain to six appealing destinations in Europe, including Rome and Vienna. Hoover would sell more domestic appliances, and JSI would handle the bookings [1]. This was the kind of

BIG IDEA that Hoover were looking for, and so the scheme was launched in August 1992 [2].

The initial results, were, from Hoover's standpoint, successful: sales of consumer goods increased substantially, and relatively few vouchers were cashed in, perhaps because there were a number of conditions, deliberately incorporated by Hoover, to make it quite cumbersome to do so.

Later in 1992, riding on this tide of success, Hoover had a second *BIG IDEA* – to extend the two free tickets promotion to New York and Orlando, Florida, home of the global tourist attraction, Disney World.

The prospect of a trip to meet Mickey Mouse, or to explore the 'Big Apple', proved far more alluring than a trip to Rome, and so sales boomed – so much so that the stock of Hoover consumer goods ran out, causing the employment of additional manufacturing staff, so increasing costs. And many more vouchers were cashed in, especially when people who wanted a trip to the United States noticed that it was cheaper to buy a £100 Hoover product, even if they didn't want one, and receive two free flights, than to buy the airline tickets in the normal way, for which they would have paid (at that time) around £600.

As a result, Hoover was inundated: some 200,000 tickets were claimed, the equivalent of about 500 Boeing 747 jumbo jets [3]. As a comparator for that number, the 747 entered service in 1970, and by 1993, Boeing had manufactured a total of 1,000 [4]; also, in 1993, British Airways had a fleet of 56 [5]. And it was costing Hoover a fortune. In an attempt to limit the damage, Hoover tried to 'manage' the demand by making it increasingly difficult for the flight vouchers to be redeemed. This, however, provoked a very public customer backlash, with widespread adverse media coverage, and a series of court actions which continued until 1998 [6]. And to make matters worse, the second-hand market became flooded with unused Hoover appliances, so depressing demand for newly-made products.

Things did not end well. Several top executives – Hoover Europe's President, the Vice-President of Marketing, and the Director of Marketing Services – were fired [7], and the European Hoover business was sold by its American parent, Maytag, in 1995.

The Hoover free flight promotion turned out to be a very bad idea indeed. Yet some very senior managers – people with much business experience – had believed it to be good. Was the failure of the idea just 'bad luck'? Or could the problems have been anticipated in advance? In which case, had something gone very wrong with their evaluation process?

I'll leave those questions dangling for the moment; after reading this chapter and the next, you will be able to answer it for yourself...

14.3 Not all ideas are good ones...

As the Hoover example demonstrates, not all ideas are good ones, and some are very poor indeed. It's very easy to see how the idea was generated, for example, had Hoover used ***InnovAction!*** (which is unlikely!), a feature of sales promotions as they were at the time might have been 'we offer two large boxes of washing powder with each washing machine'. Asking 'how might this be different?' might have stimulated ideas such as 'we could offer three boxes', 'five boxes'... At which point, someone might have said, 'Yes, but those are all more-of-the-same... can't we think of something different, something that people would really value and so make our product really attractive?'. There is then a pause, after which someone says, 'Mmm... most people think that vacuuming the floor and doing the washing is a drudge. If we could associate our products with something joyful, maybe that would give our range a psychological edge over our competitors?' To which someone else replies, 'That's a great concept! What do people really enjoy? A good meal, a holiday?'. And then someone says, 'Holiday! You just said 'holiday'! How about doing a deal with an airline to bulk purchase some seats – at a very good discount of course – and offer free flights?'

That would be a natural discussion during an idea generation workshop, and the idea should be logged, alongside all the others.

But what should happen next is a formal process of 'wise evaluation', in which all the ideas are examined, so as to distinguish between the strong ideas and the weak ones. And it should have been during this process that the free flight idea should have been rejected, if only by considering the question 'What is the profit made on the sale of each washing machine, and how does this compare to the price we would have to pay for two airline tickets, if the corresponding voucher were to be cashed in?' – which, I would argue, is a rather obvious question to ask.

A fundamental truth of creativity is that most ideas are poor. That could be as many as 99 out of 100. Much depends, of course, on how each 'idea' is counted, and whether or not all the individual Koestler's Law fragments that, with hindsight, came together to form the 'final' idea are logged separately. But whether 99 out of every 100 ideas are poor, or just 20, doesn't matter. What does matter is that the general point that 'not all ideas are good ones – especially mine' is acknowledged, with the point about 'especially mine' acting as a reminder of how easy – and treacherous – it can be to 'fall in love' with your own ideas.

Hence the importance of the process I call 'wise evaluation' – the formation of a judgement as to whether any particular idea is 'good' or 'bad'.

In practice, wise evaluation is difficult, for the rather obvious, but nonetheless fundamentally important, reason that the evidence that can be brought to bear to support the judgement of 'goodness' or 'badness' is inevitably limited.

To make that real, suppose that you wish to purchase, say, a domestic vacuum cleaner, and you wish to determine whether it is a 'good' one or a 'poor' one. Firstly, you need to have your own criteria of what 'good' looks like – for example, the cleaner's weight, ability to access corners, power consumption, ease of use, price, whatever. You may then have the opportunity to try using the cleaner, perhaps at a shop, perhaps at a friend's house who already has one, perhaps at your own home. You will also read reviews, and the opinions of others. And you can also try a number of different models and compare them.

Overall, this provides a comprehensive body of tangible, real, evidence on which you can make your decision.

Now suppose you are the Finance Director of a company that makes domestic vacuum cleaners, and your Senior Designer has just given a presentation on their latest idea for a new, improved product – a product that does not physically exist, that cannot be tried, that cannot be compared directly to any competitor offering, that cannot be reviewed. It's just an idea. And the question you are being asked is 'Will the company commit to an expenditure of [a big sum of money] to invest in this idea?'

It is of course most likely that the Senior Designer's presentation will be compelling and persuasive, for no one would stand there and say, 'The Design team had just spent umpteen months coming up with this idea for a new product. We all think it's rubbish, but we thought we might as well see if you'd invest in it…'

How, then, can the Finance Director, and the senior management team, take a wise decision, especially when an idea might have not only its advocates, but its opponents too?

14.4 …and even good ideas can be fiercely opposed

Most people today would, I think, agree that the idea to abolish the slave trade was a good one. And the idea that, in a free democracy, women should have the right to vote. Yet, as history tells us, both these ideas faced the fiercest of opposition, and it was not until many, many years after the ideas were first proposed (the red zone of the Target Diagram) that they were implemented (the blue zone). And it required immense tenacity and courage for the advocates of those ideas to keep going against that opposition and to achieve final success.

The opposition is attributable, fundamentally, to the vested interests to whom the idea is not 'good' at all, but – in their eyes – very 'bad' indeed: the slave owners who relied for their wealth on the exploitation of slave labour; men who feared loss of political power.

Or, as expressed by Niccolò Machiavelli in Book VI of *The Prince:*

> It should be borne in mind that there is nothing more difficult to arrange, more doubtful of success, and more dangerous to carry through than to take the lead in the introduction of a new order of things. The innovator makes enemies of all those who have prospered under the old order, and receives only lukewarm support from those who may do well under the new. This coolness arises partly from fear of their opponents, who have the existing laws on their side, and partly from the incredulity of men, who do not readily believe in new things until they have had a long experience of them. In consequence, whenever those who oppose the changes can do so, they attack vigorously, whilst the defence made by the others is only lukewarm. So both the innovator and his friends are endangered together.

Those words – the first few of which were quoted in section 8.5.5 – were written in 1513, and they are as true then as they are now, more than 500 years later.

14.5 How do you, and your organisation, evaluate ideas now?

14.5.1 Evaluating ideas wisely is really important

Yes, evaluating ideas, and forming wise judgements about what is 'good' and what is 'bad', can be very difficult, both conceptually and – perhaps more importantly – politically. So it's important – really important – that this is done well.

But how can you evaluate an idea – which is necessarily intangible – wisely?

In fact, how do you do it now?

This is a question I always ask, at every organisation I work with. Here are some of the responses I receive...

14.5.2 Some frequently-used methods

- 'If it's the boss's idea, it's good; if it's anyone else's, it's bad.'
- 'Mmm... that's a good question... I'm not sure our organisation has a 'formal process'.
- When an idea is suggested, it's discussed, and we take a decision...'
- 'When any idea is put on the table, it's immediately attacked. This kills many ideas, but not all – some survive, and they are then implemented.'
- 'If I like an idea...'
- 'If I trust, and like, the person who suggested the idea...'
- 'If I feel it is politically advantageous to me to support someone else's idea, that's what I'll do, even if the idea is awful.'
- 'If I feel it is politically advantageous to me to oppose someone else's idea, that's what I'll do, even if the idea is pretty good.'
- 'Our organisation is very rigorous. Every idea needs to be accompanied by a comprehensive list of factors in favour of the idea ('pros'), and factors against ('cons'). These are then carefully considered, leading to a balanced judgement as to whether the idea is to be accepted or rejected.'
- 'Our organisation is very rigorous. Every idea needs to be accompanied by a comprehensive cost-benefit appraisal, a key feature of which is a very detailed financial analysis. Only those ideas that meet some very stringent financial criteria are approved.'

You may recognise these, and there are indeed others. But these are very common, and so a few thoughts on each...

14.5.3 The boss knows best

'The boss's ideas are by definition good' has the merit of being a very clear, and simple, rule which everyone understands. If the boss has a continuing sequence of good ideas, the organisation might indeed prosper. But there are other organisational consequences too: not least the likelihood that other people who have – or rather had – good ideas resigned long ago in frustration.

14.5.4 (Apparently) no process

The second response, 'we don't have a formal process' is more common than you might think, but is usually a simplification of a deeper truth about who has power and who exerts influence...

14.5.5 'Trial by ordeal'

...as made more explicit by the third response, the 'medieval trial by ordeal'.

Those are dramatic words, but this is probably the most frequent evaluation process. Think about some recent instances when an idea was first suggested. What happened next? What were the first words that were uttered? If those words were, for example:

- 'Mmm... I'm not sure about that...'
- 'Yes, but...'
- 'That won't work because...'
- 'Mmm... that's interesting..., but I don't think the boss will like it.'
- 'That's a great idea, but have you any idea how much it will cost?'
- 'Have you thought of [this problem]?'...
- ...or it's more aggressive variant 'If you had thought of [this problem], you wouldn't have suggested that!'

then you are experiencing 'trial by ordeal', even if the phraseology is reasonably polite.

Immediately, the person who suggested the idea is forced onto the defensive, a position made progressively more difficult the greater the status of the attacker.

14.5.6 Likes and dislikes

The next responses, concerning 'likes' and 'dislikes', are also very common, if not ubiquitous, in that it is most natural that we all react emotionally to an idea as soon as we hear it. Immediate responses of 'I like it' or 'I don't like it' – even if those responses stay only in one's mind – have great power. A fundamental truth, however, is that whether or not anyone likes an idea is of little significance in judging wisely whether that idea is good or poor... and of even less significance is whether or not 'I like the person who suggested the idea', which, in many organisations, is more of a political statement than one relating to the idea, as made explicit by 'if I feel it is politically advantageous...'.

14.5.7 'Pros' and 'cons'

The last two responses both begin 'our organisation is very rigorous', and are associated with two widely-used approaches to the evaluation of ideas, an analysis of 'pros' and 'cons', and the use of a financial criterion such as the rule that 'any project that can recover the costs of the required investment within [so many] years is accepted, otherwise it is rejected'. Some organisations use one or the other, others both together.

The analysis of 'pros' and 'cons' usually involves the compilation of two, often lengthy, lists, the first being of the arguments in favour of the idea, or of the idea's advantages, the second of the arguments against the idea, or the risks associated with the idea. Once the lists have been completed, an individual, or perhaps a panel, will review the lists and come to a view as to whether or not the arguments in favour outweigh the arguments against, resulting in a decision to approve, or reject, the idea.

In principle, this is can be rigorous, and therefore sound. In practice, however, my experience is that the method is prone to a subtle, but nonetheless real, bias. This can be appreciated from the two key phrases 'arguments for' and 'arguments against', for a moment's thought will verify that there is no such thing as an 'argument for' or an 'argument against' in an abstract, dispassionate sense; rather, there are 'arguments used by a person who is in favour of the idea' and 'arguments used by a person who is against the idea'. Accordingly, whoever is compiling the list of 'arguments for' and 'arguments against' is quite likely to enter into each column the reasons why he or she believes the idea to be good or bad. Someone else, reading that list and seeking to form an opinion, might not appreciate that, and regard those lists as being 'absolutes', and so be influenced by the compiler's bias.

An example of the misuse of 'arguments for' and 'arguments against' took place in March 2020 in connection with the decision, taken as a result of the disruption caused by the outbreak of Coronavirus-19, to cancel the school exams, called GCSE (for, usually, 16 year-olds) and A level (for 18 year-olds), that would normally have taken place across England that summer. The regulator of school exams in England, a body known as Ofqual (the somewhat Orwell-esque acronym for the 'Office of Qualifications and Examinations Regulation'), was then tasked with proposing a method to determine the grades that students would be awarded in the absence of exams – grades that are important as regards, for example, a student's eligibility for a university course, or that might be used to support an application for a job.

In a document [8] dated 16 March 2020, Ofqual describe 11 possible options, three of which are evaluated according to their perceived 'advantages' and 'risks', and the other eight according to 'arguments in favour' and 'arguments against'.

One of these options was for teachers to issue, on behalf of each of their students, a certificate detailing the teacher's assessment of the student's performance. These are two of the items identified as 'arguments against':

- Schools are likely to expect a refund of exam fees.
- This would call into question the future of GCSE exams.

It is indeed likely that the recipient of exam fees might not relish refunding them, and from that perspective, this is an understandable 'argument against'. From the standpoint of a school that has paid fees for exams that did not actually take place, however, the refund of the fee is quite likely to be an expectation consequent on the decision to cancel the exams.

Likewise, the possibility that the issue of a teacher certificate in 2020 could be a precedent for issuing certificates thereafter certainly might 'call into question the future of GCSE exams'. This would of course be a threat to those with a vested interest in their continuation, so, from that standpoint, this is an 'argument against'. But over recent years, many teachers and educationalists have been arguing for the abolition of GCSE exams, if only on the grounds that a key reason for their introduction in 1988 – as a way to determine a student's achievement prior to their leaving the English educational system at age 16 – is no longer relevant since all young people now remain within that system until age 18 [9]. One reason why the GCSE abolitionists have not prevailed is the absence of a proven alternative. But if the cancellation of the exams due to the virus provides an opportunity to verify that an alternative can work, then this strengthens the argument to dispense with GCSEs very considerably [10]. So from the standpoint of the abolitionists, the prospect of 'calling into question the future of GCSE exams', and so proving the alternative of teacher certificates, is a very strong 'argument for'.

'Arguments in favour' and 'arguments against' are therefore very easily biased, and in my opinion, this approach is a very poor way to evaluate ideas wisely – and it is indeed a great pity that this approach was used by the top managers in such an important public body as Ofqual, the regulator of school exams in England, in such an important context.

14.5.8 Evaluation by numbers

It might be thought that the use of 'rigorous' financial criteria, such as the 'pay-back' time, the 'internal rate of return' or the 'net present value', as used largely by accountants in cost-benefit analysis, might be better [11]. Possibly; but certainly not always, for these measures can always be 'manipulated'.

One feature of all such methods of investment appraisal is that, within any organisation, the 'rules' are well-known – for example, that the pay-back period must not exceed, say, 5 years, or that the internal rate of return must be at least 6%.

One consequence of this is that any manager submitting a cost-benefit analysis knows that if the pay-back is shown as, say, 6 years, the proposal will be thrown out. So no manager would dream of submitting such a document. As a result, there is a strong incentive on the manager to 'manipulate' the data so that the pay-back figure, as shown in the submission, is, say, 4.7 years.

Which is usually quite easy to do with a spreadsheet, for all financial forecasts, however complicated, end up with two key sets of numbers: the first is the projection of the total quantity of cash coming into the organisation (for example, from product sales) in each time period (say, every month, quarter or year), and the second, the total quantity of cash leaving the organisation (for example, to pay for all the costs incurred) over the corresponding period. Inevitably, both these sets of numbers are forecasts or projections, and so matters of judgement, which may or may not be well-substantiated. It is therefore usually very easy to 'tweak' some of the numbers so that they remain plausible yet end up with the 'right' answer for the pay-back, internal rate of return, or net present value. And given that there is usually more evidence to support the forecasts of costs than revenues, it's the revenue forecast that is likely to be more easily 'tweaked'.

Also, cost-benefit analyses inevitably have 'boundaries', beyond which any costs and benefits are ignored. And in many cases, these are costs, rather than benefits, for example, the costs associated with any atmospheric or environmental pollution that may result from the manufacture of the 'great idea'. This implies that the results of the cost-benefit analysis are falsely more favourable than they should be.

So be suspicious of all cost-benefit analyses, and be sure to challenge the 'boundaries'. And beware financial forecasts that 'prove' what a success the new product will be, for spreadsheets that show a 'good' pay-back or whatever are more likely to be evidence of the ability to engineer spreadsheets to give the 'right answer' and of the organisational political skill to get away with it,rather than of the intrinsic merit of the idea.

That's quite a long list of how ideas should *not* be evaluated; the next chapter takes the much more positive stance of presenting my suggestions for how ideas, of all scales, from the modest to the truly *BIG IDEA*, can be evaluated wisely.

References

[1] https://thehustle.co/the-worst-sales-promotion-in-history/
[2] http://news.bbc.co.uk/1/hi/business/3704669.stm
[3] http://allbestidea.blogspot.com/2018/10/hoover-free-flights-promotion.html
[4] https://aircraft.fandom.com/wiki/Boeing_747
[5] www.planespotters.net/airline/British-Airways
[6] www.linkedin.com/pulse/hoover-flights-fiasco-free-travel-offer-worst-promotion-chris-rogers/
[7] www.upi.com/Archives/1993/03/31/Hoover-top-executives-fired-for-free-flights-offer/6398733554000/
[8] Ofqual, Contingency planning for Covid-19 – Options and risks 16 March 2020 https://committees.parliament.uk/publications/2437/documents/24161/default/
[9] www.bbc.co.uk/news/education-47149808
[10] www.theguardian.com/education/2020/sep/20/private-and-state-schools-bid-to-kill-off-gcses
[11] www.chathams.co/1311-2/

IOP Publishing

Creativity for Scientists and Engineers
A practical guide
Dennis Sherwood

Chapter 15

How to evaluate ideas wisely

15.1 Features of a wise evaluation process

15.1.1 What does a well-designed evaluation process look like?

Having criticised other ways of evaluating ideas, I now describe what I believe a 'good' process should look like. In my opinion, a well-designed process that will result in the wise evaluation of ideas should exhibit all of these features:

- Consistency
- Fairness
- Balance
- Completeness
- Speed
- Pragmatism
- Transparency
- Openness

15.1.2 Consistency

Young children quickly learn which parent to approach – if not nag – to say 'yes' to whatever they might wish for right now. If that's 'mum', what happens next is the child goes to 'dad', and says, 'mum says…', at which point 'dad' is stuck.

People in organisations play the same game. They go to whichever senior person is likely to say 'yes', and then exploit that to influence others. That is understandable human behaviour, for we all seek to build a coalition of support. But this becomes organisationally dangerous when different parts of the enterprise use different processes, or apply different criteria, when judging between 'good' ideas and 'weak' ones. One feature of a 'good' evaluation process is that is exactly the same, and is used in exactly the same way, everywhere.

15.1.3 Fairness

This requires that any idea, no matter who originated it, or wherever it was originated, is treated in exactly the same way. This counters 'authority bias', in which any idea suggested by the boss is 'good', and any by the receptionist or a shop-floor worker is just ignored.

15.1.4 Balance

The originator of any idea is, quite naturally, an enthusiast – indeed the originator usually has to be to get other people to take any interest. So of course the originator looks 'on the bright side', and will emphasise why the idea is so good.

At the same time, it's quite likely that there will be others who feel that they might be disadvantaged by the idea, or who are playing political games against the idea's originator. Those people will seek to identify as many reasons why the idea is really bad, and will challenge accordingly.

Both these standpoints are biased, emphasising some aspects of the idea whilst deliberately ignoring, or at least underplaying, others. So any evaluation process that does not counter this tendency is likely to result in wrong outcomes – ideas that are weak will be judged good if the advocates are more powerful; ideas that are good will be judged poor if the opponents win.

I believe that a 'good' evaluation process must ensure balance – that all aspects of the idea are identified and considered, without bias in favour or against.

15.1.5 Completeness

This is the organisational equivalent of the legal obligation to 'tell the truth, the whole truth, and nothing but the truth', with the emphasis on 'whole' – for we all know how easy it is to present a case which doesn't contain anything untrue or wrong, but just happens to omit a 'detail' that might support a different view.

The requirement for completeness attempts to counter this: a 'good' evaluation process ensures that as many aspects of the idea as possible are taken into consideration, and that nothing is missing – either inadvertently, or deliberately.

15.1.6 Speed

'Head Office are still looking at it'.

When you submitted the idea months ago, and see that reply to your email asking 'how are things going?', yes, your heart sinks. Or when you suggest an idea to your boss, who says, 'That's a great idea, but it needs funding, so we'll have to wait until the next budget cycle. Let's think about it again in nine months' time.'

Yes, organisational processes do take time, and have their own timetables. But some ideas are time-critical, and over-long delays in deciding whether to progress an idea or not can be a huge disincentive to those who might be thinking about suggesting an idea, only to end up not bothering – a theme I will develop in the discussion of the impact of organisational culture in chapters 15, 16 and 17.

A 'good' evaluation process is speedy, but not rushed.

15.1.7 Pragmatism

Some ideas are 'big', some 'small'. It is therefore clumsy to use a very thorough process, suitable for 'big' ideas that require significant resources and pose considerable risk, to evaluate 'small' ideas that are much less resource-demanding and risky.

A 'good' evaluation process is pragmatic, in that it is based on sound principles, but can be applied in different ways, depending on the 'size' of the idea.

15.1.8 Transparency

'Oh, I didn't realise you needed that!' is a very frustrating statement for me to make when I have been informed that, as a result of the evaluation process, my idea has been rejected, primarily because I had not supplied the right information, or because I had not understood, or known, the criteria of judgement.

A 'good' evaluation process therefore makes both visible and explicit the nature and details of the process, and the criteria that are used – the process should be transparent to all.

15.1.9 Openness

It is understandable that anyone whose idea is rejected by the evaluation process would feel disappointed. But if the originator is just told 'We're sorry, no' then that disappointment can result in the thought 'Well, there's no point in my suggesting another idea, is there?' This will then act to slow down the flow of ideas as people feel increasingly disenfranchised – which is usually a bad thing to happen.

Openness is about providing full information to the originator – and others – as to why an idea was rejected or indeed accepted, so that the originator can understand fully what might be done differently in the future to increase the likelihood of acceptance, or what has been done well so that this can be replicated. So rather than an originator thinking 'I'm never going to do that again', the thought is 'Yes, that makes sense – I'll make sure that's covered next time'.

15.1.10 How does your organisation's evaluation process rate?

These eight characteristics can serve as a benchmark against which you can assess the evaluation process currently used in your own organisation. So, for example, the organisation that uses the rule 'the boss's ideas are good, all others are not' will score highly as regards consistency, speed and quite likely pragmatism, but not so well for the others. The ideal process scores highly on each, so here is one that does...

15.2 An ideal process for wise evaluation

15.2.1 Overview

The accompanying box presents an overview of my six-step process for wise evaluation; subsequent sub-sections discuss each step in detail. As will be seen, this can be a lengthy and laborious process, but a vital one none the less, especially for ***BIG IDEAs*** which demand large resources and incur high risk. A simpler process – 'Evaluation Lite' – based on the same principles but more appropriate for 'smaller' ideas, will be described and discussed subsequently.

> **My process for wise evaluation**
>
> 1. **Imagine that the idea has successfully been implemented.** What does that 'future world' look like? And what is *different* as compared to today?
>
> 2. **What are the consequences?** Which of those differences make the 'world a better place' as compared to today, and better for whom? And which consequences are not-so-good? And for whom? How are different individuals, communities and constituencies likely to feel?
>
> 3. **The journey: what issues had to be addressed, and resolved, to implement the idea successfully?** This includes all issues likely to be encountered, for example, technical difficulties, the availability (or otherwise) of resources, and political and organisational issues such as rivalry, jealousy, both within and beyond the organisation.
>
> 4. **How might all the problems best be solved?** The not-so-good consequences, and the issues-to-overcome, are all problems that might be 'blockers'. But not necessarily… What ideas can be generated to solve and overcome these problems? How might the idea be enriched so that these difficulties are significantly reduced, if not eliminated?
>
> 5. **The numbers:** Now that the idea, and its implications, are well understood, how can the benefits, and all the costs, best be quantified?
>
> 6. **The decision:** Is the idea accepted, rejected, or is it worth spending some time and money to reduce some uncertainties, so providing more, and better, insight to inform the yes/no decision?

15.2.2 Step 1: Imagine the future

When an idea is first suggested, the originators are enthusiastic and excited, whilst others might be more measured or sceptical. Everyone is forming a judgement, based on their own prejudices, beliefs, self-interests and political alignments, and everyone seeks to persuade everyone else to their point-of-view. These prejudices and beliefs are based on each individual's personal 'picture' of what they imagine the idea to be, their own concept of what the 'world might look like' once the idea has been implemented, and their own personal judgements of 'good' and 'bad'.

And it is a perhaps unfortunate, but fundamentally valid, truth that different people will often imagine the idea in totally different ways, and react and behave accordingly. Nowhere has this been more true in the UK over the last several years than in relation to the idea that the UK should leave the European Union, the event – or rather protracted process – known as 'Brexit'. To the day this is being written – in November 2021, and nearly one year after the UK left the European Union – there are still huge variations in what different individuals, lobby groups and the media believe the idea actually was, and what the world really should be looking like.

So the purpose of Step 1: 'Imagine the future' is to ensure everyone has the same concept in mind. To do this, those who have a view on what the idea is – or might be – are invited to describe it in as much detail and context as possible, feature by feature. This is just like the second step of the *InnovAction!* process for generating ideas, 'Individually and in silence, write down everything you know about the agreed focus of attention' of (see pages 9-3 to 9-6), but imagining the future, not describing the present. And, as with *InnovAction!*, the (quite possibly different) descriptions of different individuals are shared to compile an aggregate list.

Yes, this can be laborious, but it is absolutely essential, especially for complex ideas, and ideas that are about organisations, structures and power. During the discussion, expect to hear people say things like 'I don't see it that way', 'I didn't expect that', 'Are you sure? Isn't it like this...?'. Those statements are important, for they indicate people's different perceptions, and the sooner these differences are resolved, the better. Sometimes, though, differences can remain, for there may be valid reasons why the future might look like [this] or like [that], depending on other circumstances. In which case, that's fine, and note that too.

Very importantly, while this vision-of-the-future is being compiled, DO NOT give value judgements of the form 'Oh dear, that's bad!', or 'Wow! I didn't think it would be as good as that'. Keep to 'flat' emotion-free descriptions: so 'The factory in [wherever] will be closed' not 'It will be a pity that the factory will close'; 'It's possible that solving [this problem] will be a step along the way to discovering [cold fusion]', not 'That's great! I'll win a Nobel Prize!'

15.2.3 Step 2: What are the consequences?

In the absence of the idea, the *status quo* would be perpetuated – so, for example, if the idea for a new product were not be pursued, the existing product would continue; if the idea for running a conference in a different way were not implemented, the next conference would be run in the same way as last year.

Step 1: 'Imagine the future', has created a 'picture' of what the 'world might look like' if the idea were successfully implemented, and so the purpose of Step 2: 'What are the consequences?' is to compare that vision of the future with the perpetuation of the *status quo* – or perhaps its extrapolation if that *status quo* itself might change.

In exploring the consequences, the fundamental question being asked is 'will [this feature] of the implemented idea be better, or worse, than the corresponding feature of the 'world' if the idea were not implemented?' The identification of 'better' or 'worse' immediately requires a value judgement, and also the identification of the individual or community who would take that view. So, to take the example of a new product, the likelihood that the product will sell well is good for the manufacturer, and the likelihood that the product will benefit the customer is good for the customer. But from a competitor's perspective, the introduction of a rival, potentially more competitive, product is a challenge, threatening their market share, so causing them to retaliate. What might the competitor do? And how might that itself be countered?

It's therefore important to compile two lists, as long and as comprehensive as you can, of, firstly, the benefits associated with the implementation of the idea, and secondly, of those features of the future 'world' that are not so good. Ideally, the first list is longer than the second, but there are always some things that are 'not so good', and these should not be ignored. But the expected benefits should be far more valuable than the costs and consequences of the problems – for if they are not, the idea should be scrapped without any further analysis.

15.2.4 Step 3: The journey

Step 2: 'What are the consequences?' imagines the 'world' after the idea has been successfully implemented, and, in essence, addresses the question 'is the resulting 'world' better or worse than an extrapolation of the *status quo*?' This is in contrast to the perspective of Step 3: 'The journey', which examines the likely path from 'here' to 'there', for although 'there' might indeed be a 'better place', the 'journey' itself might be very tough. But even if it is, the benefits of the end-game might be worth the effort.

The objective of this step is therefore to identify, as perceptively as possible, all the issues that need to be addressed and resolved so that the idea can be implemented successfully – the work that has to be done, the problems that need to be tackled, the resources that need to be deployed, the political battles to be fought, the risks to be managed, the likely feelings of those affected.

In general, this results in the identification of some benefits (for example, the acquisition of new skills) and also a (quite possibly long) list of things-to-be-done and problems-to-solve, identifying all the issues that need to be addressed and resolved so that the idea can be implemented successfully. The resulting list can appear to be quite daunting, and some might feel that it should not be drawn up because of the possibility that a long list of the problems of implementation might appear to be so overwhelming that the idea will be rejected. This fear can lead some people deliberately to ignore or down-play these tasks and potential difficulties, but in my view, this head-in-the-sand stance is not only dishonest, but only leads to bigger trouble later – if a problem needs to be solved, refusing to acknowledge that the problem exists does not cause it to go away. On the contrary, the problem will still surface, but since you will be less well-prepared to address it, the severity of the problem's impact is likely to be considerably greater, and the problem will be even harder to solve.

15.2.5 Step 4: How might all the problems be solved?

Step 2: 'What are the consequences?' will quite possibly identify some aspects of future 'world', after the idea has successfully been implemented, that are not as good as today's 'world' (or the extrapolation of the *status quo*), and Step 3: 'The journey' will certainly result in a probably long list of problems-to-solve. At this point, it's very easy either to say 'it's all too difficult' and scrap the idea, or to pretend that the problems aren't there, only to go ahead regardless and pay the price later.

The right action at this point, however, is neither of these. Rather, it is to be creative! In particular, to be creative as regards asking 'how many different ideas can we generate as to how to solve [this problem]?' For in fact, almost every problem can be solved (except those that 'break the laws of physics'!!!), given sufficient resources of people and money. And if the benefits associated with the idea's implementation are sufficiently large, then those sums of money required to solve the problems might well be worth it. But if the solutions to those problems are not even searched for, the idea will certainly be rejected – and who knows how many ideas that were in fact very good were abandoned simply because no effort had been taken in being creative.

15.2.6 Step 5: The numbers

By now, the idea, and its implications, are well-understood, specifically:

- What the world will 'look like', when the idea has been successfully implemented, identifying both the resulting benefits and also any remaining problems (that's the outcome of Steps 1 and 2).
- The key characteristics of the 'journey from here to there', and the nature of the (often many) problems that need to be successfully resolved to bring the idea to reality (Step 3).
- How the various problems identified in Steps 2 and 3 might be solved (Step 4).

The objective of Step 5: 'The numbers' is to quantify all these, so answering the questions:

- How might the benefits be measured?
- What resources will be required to deliver the idea successfully?
- What is the cost of implementing this idea? Or, better, what is the range of likely costs?
- How long will the implementation take?

The numbers are important, for, in general, the value of the benefits need to outweigh the costs.

The need for the benefits to be greater than the costs is of course known to both the idea's advocates, and its opponents – so the advocates will seek to emphasise, and bid up, the value of the benefits whilst minimising the costs, and the opponents will do the opposite, downplaying the benefits and discovering yet more costs. In many organisations, the role of the numbers is as much political as it is analytical.

Compiling the numbers is often difficult, for there is usually little 'hard' data, and there is therefore much reliance on forecasts, some of which will be more uncertain than others. The actual number-crunching is most likely to be done within a spreadsheet, which makes it much easier to explore the impact of, say, a number of different sales forecasts (which is a good exercise to carry out), but sometimes the precision implied by the numbers presented on a spreadsheet can suggest that those numbers are much more accurate than they really are (which is not so good).

There is often a strong temptation, if not organisational pressure, to start working on the numbers almost as soon as the idea is deemed 'interesting'. My advice, however, is that it is preferable to do this later, rather than sooner – and certainly not before Step 4: 'How might all the problems be solved?' has been completed, for the solutions identified in that step will all be associated with costs that need to be taken into account.

Some other matters that merit consideration are these:

- Some benefits – such as future revenues – are amenable to numeric forecasts, others – such as the benefit to a student of an improved educational curriculum – much less so. Such 'intangible' benefits are often ignored in financial spreadsheets, and so the benefits will then be under-estimated. In contrast, the costs, for example, of the development of the new educational curriculum can be estimated as, say, [so many] months of time devoted by experts who are paid [this much] per month. If the benefits are not quantifiable or underestimated, it is then very easy for the costs be seen to be a higher number, so causing an intrinsically good idea to be deemed bad.

- A further problem is the 'boundary' within which benefits, or costs, are identified. So, for example, if the 'cost-benefit' analysis of a new powerplant fails to recognise the cost of environmental damage, this will make the building of that power plant appear to be more favourable than it truly is – a problem exaggerated by the possibility that the environmental costs might not be borne by the factory developer.

- Since everybody knows that the 'right answer' is for the benefits to exceed the costs, this can lead the idea's advocates to attempt to 'engineer' the spreadsheet to make that happen. So, for example, the benefits can be 'adjusted' upwards to ensure they exceed the costs, and as long as the associated sales forecast (or whatever) is not obviously implausible, this can often be done. Alternatively, the costs can be reduced by 'forgetting' to include some cost categories – which is often harder for the idea's advocates to get away with, but can happen.

The overall message is clear. Be honest. Be wary. And most importantly, be wise. The numbers are important buy they are never the whole story. Which leads to...

15.2.7 Step 6: The decision

Ultimately, an idea is either deemed 'good', and accepted for development, with a view to implementation (as represented by the transition from the yellow 'evaluation' zone of the 'Target Diagram' into the green 'development' zone) or it is deemed 'weak', and no further attention is paid to it. There is, as I shall discuss shortly, a half-way house of 'we're not sure yet, we need more information', but for the moment, I'll explore what is in essence a binary yes/no decision...

...the most important feature of which is that this decision is necessarily taken by a single human being, or (in my opinion, better), by a small team. At least at the moment, and for a few years to come. In the future, perhaps, a machine learning algorithm might do this, and be shown to take wiser decisions than humans, but at present, and for the foreseeable future, it's a human process. That said, given that human beings can take very poor evaluation decisions indeed, perhaps the bar for using artificial intelligence isn't that high...

Humans take decisions of this nature using their personal judgement, and every individual does this in his or her own way, a way that the individual might not be able to describe or even fully understand. Some people are reflective, take their time, and seek as much data as they can; some are much more instinctive, and take decisions quickly, often with much self-confidence; for some others, being forced to take a decision can be quite uncomfortable. There are many books examining how people do this, one of my favourites being *Thinking, Fast and Slow* [1] by Daniel Kahneman, who was awarded the 2002 Nobel Prize in Economics [2]. Very briefly, Kahneman explores the (to me highly convincing) evidence that our brains have two decision-making 'pathways', one that operates very quickly, and is associated with 'urgent' decisions, such as those required to avoid imminent danger, the other that operates more slowly, and associated with weighing up a variety of factors and influences.

The wise evaluation of an idea is rarely a 'danger' moment, and so the 'slower' pathway is to be preferred, and the 'factors and influences' that need to be 'weighed up' are detailed in the dossier of the evidence compiled as a result of Steps 1–5 in the process just described, and summarised in the box in section 15.2.1.

Importantly, the evaluation decision is not taken by a 'system'; rather, it is the personal decision of an individual, or – rather better – a small team.

My ideal is that, for any idea, the results of Steps 1–5 are documented, and presented to a small panel of, say, three – or perhaps five or even seven – people who are trusted by the organisation to exercise their judgement to take the evaluation decision. Using a panel (hopefully!) protects against any bias that even an experienced person might show, and the panel used for any particular idea is drawn from a wider community of people who are trusted by the organisation to act in that capacity.

Who, then, are these 'expert trusted evaluators'?

That's an important question, and one to which there are no easy answers. There are (to my knowledge) no qualifications or diplomas in 'wise evaluation'; rather, the key characteristics are experience (which is not necessarily the same as seniority), and, most importantly, organisational trust.

Perhaps some clues as to whether or not an individual is likely to be good at evaluating ideas wisely might be obtained from the widely-used 'personality type' classifications, such as the Myers-Briggs Type Indicator, which claims to 'provide a constructive, flexible and liberating framework for understanding individual differences and strengths' [3]. This 'framework' identifies four pairs of characteristics: Extroversion (E) and Introversion (I); Sensing (S) and Intuition (N); Thinking (T) and Feeling (F); and Judging (J) and Perceiving (P). Within this framework, any individual has a 'preferred style', this being one of the sixteen possible combinations formed by choosing one characteristic of each pair. So, for example, the INTP combination is associated with *'people who think strategically, who are able to build conceptual models to understand complex problems, who tend to adopt a detached and concise way of analysing the world, and who often uncover new or innovative approaches'* – a description that is promising; even more so, perhaps, is the 'Monitor Evaluator', who *'Provides a logical eye, making impartial judgements where required and weighs up the team's options in a dispassionate way'*, this being one of the nine team roles identified by Meredith Belbin [4].

Although there are training programmes available for the mechanics of evaluation, at the time of writing, I know of no such programmes that develop the personal skills to evaluate ideas wisely, so perhaps an entrepreneur reading this will fill what appears to be a gap in the market. Some organisations, however, do offer internal training programmes, using case studies based on ideas that the organisation has evaluated in the past, in which participants are invited to study the evidence that was available at the time, and then to give an evaluation decision, describing how they chose that decision. This then leads to a discussion amongst the participants, and with those of rather more experience.

Ideally, an organisation has a number of individuals who are trusted to analyse the evidence fairly and impartially, resulting in a wise decision, three (or for 'larger' ideas, perhaps five or seven) of whom are invited to serve on a panel to evaluate the ideas currently in the pipeline. If the flow of ideas is irregular, then a panel can be convened whenever necessary; alternatively, if the is a continuous flow, a panel can convene at regular intervals, for example, every quarter or every six months.

15.3 The half-way house

As already noted, any idea is either accepted into development, or not, implying that the evaluation decision is a binary yes/no.

In practice, however, there is a third possibility – that the idea looks 'interesting' but the evidence presented so far does not enable a yes/no decision to be made with sufficient certainty. Rather, it would be helpful to enrich the evidence, so that the idea can be re-assessed in the light of more information.

To do this, however, will require some resources of time and money, and so the result of the evaluation panel is to commit those resources, specifically to carry out some further research or analysis, which, once gathered, will enable the panel to take a more informed decision.

In essence, this is about limiting risk. If the evaluation decision is 'no', then any benefits of the idea are foregone; if the decision is 'yes', then this could launch the organisation into a possibly long and expensive programme which might fail. But by saying 'maybe', the organisation can commit a much more modest sum to discover more about the idea, so reducing the uncertainty.

The uncertainties that need to be understood more deeply will be evident from the evidence already gathered. So, for example, the benefits of the new idea – if a new product, for example – will almost certainly include 'increased sales revenue'. As a generality, that is true. But for this specific idea, is that a large increase or a small one? Just how successful might the product be?

The idea's advocates will, of course, be optimistic, and the evidence pack will almost certainly include a sales forecast. The evaluation panel, however, might take the view that the uncertainty associated with that forecast is too great, and so a 'half-way house' decision is to authorise funds to carry out, for example, more market research, or perhaps the building and trial of a prototype. That will provide more information and reduce uncertainty, providing helpful additional insights into a subsequent evaluation.

For really 'big' ideas, the 'maybe' evaluation decision can be made several times in succession. At each decision point, there is an option to 'pull the plug' and to abandon the idea, or to commit further resources to reduce a particular uncertainty. This is a form of the 'stage gate' process [5], as illustrated in figure 15.1.

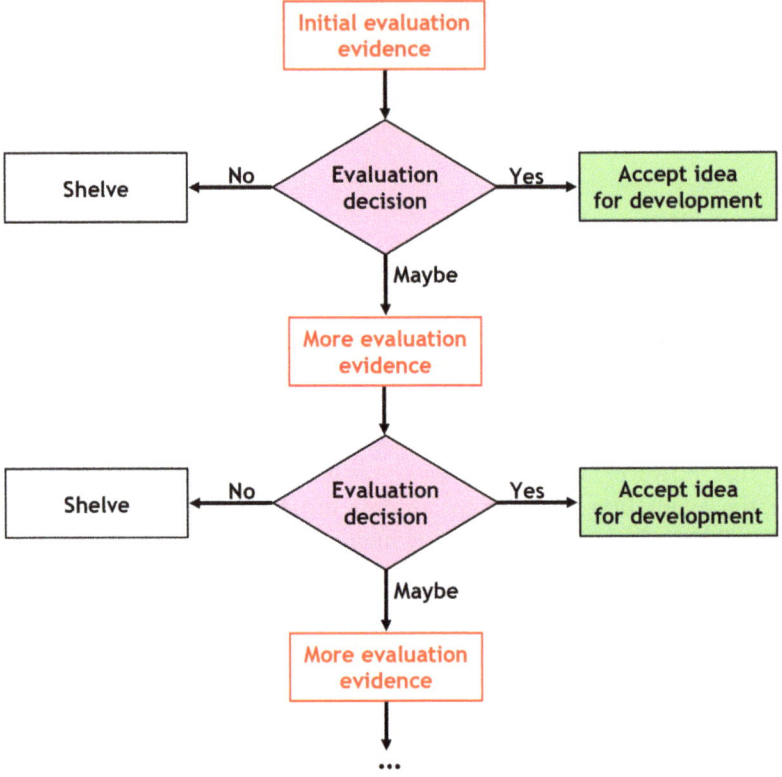

Figure 15.1. The 'stage gate' process.

15.4 Wise evaluation, Edward de Bono's 'hats', and the importance of language

I've already mentioned four of Edward de Bono's many books – *The Mechanism of Mind* (page 8-8), and, in section 11.2.3, *The Use of Lateral Thinking*, *PO: Beyond Yes and No* and *Serious Creativity*. But perhaps his most influential has been *Six Thinking Hats*, first published in 1985, in which he highlights the benefits of looking at an idea from a number of different perspectives, so countering the biases that result from taking only a single view.

To represent these perspectives, de Bono uses the metaphor of a coloured hat, so that 'wearing the appropriately coloured hat' encourages the wearer to take that perspective. As the title of the book states, de Bono identified six different hats, which he describes using these words [6]:

- White: facts and figures.
- Red: emotions and feelings.
- Black: what is wrong with the idea.
- Yellow: speculative-positive.
- Green: creative and lateral thinking.
- Blue: control of thinking.

I mention this here, for there is an overlap between de Bono's hats and the process I have described for wise evaluation, especially as regards Step 2, 'Consequences' and Step 3, 'The journey', both of which require the identification of benefits (de Bono's yellow hat), issues-to-manage (black) and feelings (red). Furthermore, Step 4, 'Solutions', is 'wearing the green hat', whilst Step 5, 'Numbers' is 'white'.

de Bono's association of a perspective with a 'hat', 'worn' to symbolise a role, is psychologically astute. At a workshop, for example, the team can all be asked to 'wear the yellow hat' and to identify the benefits associated with an idea, so obliging everyone so say something constructive and good. Amazingly, even the most habitually miserable person complies, or at least stays silent!

Likewise, inviting someone to wear the black hat can bring even the most 'head-in-the-clouds' person back down to earth, and the red hat makes it safe to think, and talk, about people's feelings – which is something that is often avoided.

And in the ordinary, every-day world, the 'black hat' is commonly worn, even when people don't realise they're doing so. All those negative remarks – 'that will cost too much', 'that's going to be a real problem...', 'that won't work because...' – are black-hat statements that are sometimes made inadvertently, sometimes with intent. Problems, however, are real, and as discussed in sections 15.2.3 and 15.2.4, they should not be swept under the carpet (as an idea's advocates might be tempted to do), nor should they be used as weapons to kill the idea (as the opponents might be tempted to do). Rather, problems should be identified and addressed professionally and thoughtfully, as associated with Step 2, 'What are the consequences?', and Step 3, 'The journey'.

Step 3 in particular is associated with the 'black hat', for its purpose – to quote the words used in section 15.2.4 – is 'to identify, as perceptively as possible, all the issues that need to be addressed and resolved so that the idea can be implemented successfully'. That language is important, and deliberately different from the words used by de Bono in his description of the 'black hat' – 'what is wrong with the idea' – which are words I don't like.

Why so?

Logically, my words, 'identify all the issues that need to be addressed and resolved so that the idea can be implemented successfully', mean much the same as de Bono's 'what is wrong with the idea', for both result in a list of problems-to-solve.

Emotionally, however, the words are very different. Talking of 'issues to address' represents a constructive, positive, realistic stance. We all know there are problems; of course there are. And by referring to them as 'issues to address', I am taking the position that, yes, the corresponding problems can be resolved and managed, even if that might be difficult. 'What is wrong with the idea' will identify the same problems, but has already categorised them as blockers, as insuperable.

To me, that is a dangerous mind set. Using language such as 'that idea is wrong/won't work/is bad/is flawed...', 'it will cost too much' and the like, indicates that the speaker has pre-empted the evaluation decision as a 'no'. Most problems – except those that 'break the laws of physics' – can be solved, given enough time, money and resources. So the word 'wrong' already dismisses that possibility, which, in the context of evaluating an idea, is, in my view, inappropriate. Far better to identify all the 'issues', and search for solutions, than to dismiss the whole thing as hopeless or impossible right from the start. And no, this is not 'mere semantics' or just 'playing with words'. Language is important, and a subtle cultural indicator, as I will discuss further in section 16.2.

That said, the 'hats' metaphor, when used properly, can be very useful organisationally and politically.

Imagine you are at a meeting, and Alex tables an idea. Sam then says 'that's really good'. Sam might indeed believe that. But Chris is suspicious, and thinks, 'That's strange. Sam usually opposes everything Alex ever suggests. But not today. What's going on? Is there some deal happening behind the scenes?'

That might be evidence of Chris's paranoia. But perhaps not; in the real world, people do make alliances, and strike deals behind the scenes… if you back me on this, I'll support you on that….

But if Sam says, 'Wearing my 'yellow hat', I think one of the benefits of that idea is [whatever]', then this is organisationally and politically neutral, and therefore 'safer'.

Likewise, many 'black hat' observations – 'That won't work because…', 'Have you thought of … ?', 'I don't think Pat will like that…' – are in general negative, especially when expressed using that sort of language. As a consequence, the individual who suggests the idea might feel under attack. Yes, of course organisational life can be 'robust'; but if the idea has been suggested by someone new, or junior, then that kind of attack, however well-intentioned, can demotivate and demoralise. Especially since there are much 'safer' alternatives, such as 'If I wear my 'black' hat, I wonder if one issue that needs to be thought about is [whatever]?'. Once again, reference to the hat metaphor depersonalises the comment, which is now explicitly about the idea, and is not an attack on the idea's originator.

Some organisations make the metaphor real – for example, a meeting room might have yellow, black and red chairs, and if you are sitting in a chair of a particular colour, then you play the corresponding role in the discussion. Alternatively, some organisations have sets of hats, in all the colours, which people wear according to the role. Yes, that might appear to be overly theatrical, but that's just a veneer to remind the team about what's going on. The underlying principles about taking different perspectives, about being balanced, about ensuring that everything is considered, about avoiding bias, about being truly professional, are much deeper.

15.5 'Evaluation Lite'

The process of wise evaluation described so far is comprehensive, and, when done thoroughly, can require a considerable time, and the corresponding resources – for example, several, if not many, man-months or more for a strategically significant idea, such as a potential merger, penetrating a new market, or a major public sector initiative. The process, however, is scaleable, and can sensibly be carried out in a few hours or perhaps a few days for 'smaller' ideas.

In practice, at its smallest scale, it is possible to focus solely on Step 2: 'What are the consequences?' and Step 3: 'The journey', and just explore the benefits (yellow hat and red hats combined) and the issues-to-address (black and red hats combined), using personal judgement. This is especially useful for screening the ideas generated at a creativity workshop, which might number 100 or more, so that no time and resources are in essence wasted thinking too hard about ideas that are inherently weak or of no consequence.

The process is very simple, and is best carried out by a small team (say, 6) so as to avoid too much individual bias:

- The idea being studied is tabled, and a few minutes spent describing the idea, so that all participants share more-or-less the same 'mental picture' of 'what the world would look like' once the idea has been successfully implemented – this, in essence, is a rapid version of Step 1: 'Imagine the future'.

- Then, working individually, each individual considers their own answers to these two questions:
    - Once the idea has been successfully implemented, are the net benefits 'high', 'medium' or 'low' (yellow hat, and 'feels good' red)?
    - Is the 'journey' 'difficult' (with many, hard, issues to address and resolve), 'moderate' or 'easy' (black hat, and 'feels bad' red)?

- ...so allowing the idea to be plotted on the 2 x 2 grid shown in figure 15.2.

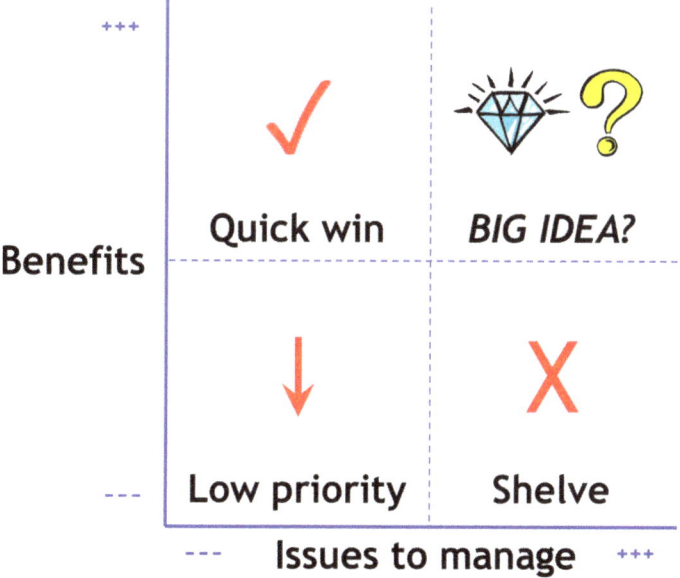

Figure 15.2. 'Evaluation Lite'.

- The views of each participant are then shared, with a discussion of what each participant had in mind in making their judgements, leading to a consensus view on where the idea should be plotted.

As can be seen from figure 15.2, the four quadrants classify any idea as:

- **'Quick win'** – ideas offering high benefit for relatively little effort.
- **'Low priority'** – ideas requiring relatively little effort, but offer low benefit.
- **'Shelve'** – ideas of low benefit and high difficulty.
- **'BIG IDEA'** – ideas that are believe to offer high benefit, but are difficult too.

'Quick wins' are ideas that the team believe to offer sensible benefits at low effort – for example, an idea that can be implemented within current budgetary constraints, and that can sensibly be added to someone's 'to do' list.

'Shelved' ideas are ones judged to offer low benefit, and are difficult to deliver. There is no point in devoting any resources to them now, but note that these ideas are designated 'shelve' rather than 'discard' – which implies that the idea should not be lost, but rather should be stored so that it is potentially available in the future. There is a very important reason why even apparently poor ideas should be archived and not thrown away. An idea is 'poor' for a variety of reasons – for example, a particular aspect of it might be too expensive. So, for example, it is quite feasible that two manufacturers of cars might, some time ago, have generated the idea of having a rear-facing camera so that a picture could be displayed to the driver on a screen, so making parking, for example, easier and safer. At that time, the cost of cameras and screens was prohibitive. But a few years later, the cost of these components fell sharply, so making the idea much more feasible commercially. One manufacturer, however, had thrown the idea away, and so it had been forgotten; the other hadn't. That second manufacturer was therefore able to react much more quickly to the changed conditions, so gaining a competitive advantage. The moral of that story is 'don't throw ideas away' – as well as implying that there also needs to be an archive in which such 'fragments' can be stored, as well as retrieved, as will be discussed further in section 17.4.

Any idea that is of relatively low benefit, but quite easy to do – so fitting into the lower-lift quadrant – might be intrinsically worth-while, if the (low) benefit exceeds the (even lower) difficulty of implementation. But only if there is capacity to pay attention to the idea: if there are other ideas that rank higher in benefit, whilst also being relatively easy to deliver, then it makes sense to implement these first. The ideas in the lower left quadrant are therefore 'Low Priority'. Accordingly, for any degree of difficulty (as represented by the position along the horizontal axis), the sequence with which ideas should be acted upon is from the most beneficial, working 'down' the vertical axis.

Which leaves the upper-right quadrant – those that are hard to implement, but are believed to hold the promise of high benefit too. If the difficulties outweigh the benefits, then the idea should be shelved. But if the benefits are really good, even though implementing the idea is costly, time-consuming and very difficult, the idea could still be worth it in the end. These are the truly **BIG IDEAs**. But at this stage, we don't know – we just have a belief that the idea might be really good, but are fearful about its difficulty. So it's these ideas that merit the investment of time and energy to carry out a more detailed evaluation, following the full six-step process described in section 15.2.

15.6 And so to development and implementation

As we have seen, the result of the process of 'wise evaluation' is either to shelve the idea or to accept in into development and implementation (subject, of course, to any intermediate iterations through the half-way house of the stage gate process).

And once the idea is passed across that boundary from the 'yellow' to the 'green' zones of the Target Diagram (see figure 2.1), the key skills required are those of project management, which are beyond the scope of this book. Save for two thoughts...

Firstly, even when an idea is in development, or in the course of being implemented, there are still any number of problems-to-solve and issues-to-manage. So creativity and wise evaluation are still needed, embedded within the overall processes of development and implementation, as represented by the 'fractal Target Diagram' shown in figure 2.2.

Secondly, there is an important, and very valuable, link between the evaluation process described in section 15.2, and the activities actually carried out in development and implementation.

During Step 2: 'What are the consequences?', Step 3: 'The journey', and Step 4: 'Solutions', the evaluation process will have identified any number of problems associated with the idea, and the ways in which these can all be resolved. Each of these represents a task that needs to be done, and perhaps also a risk that might need to be managed. Collectively, these determine both the development and implementation work-plan, and also the risk register.

So a truly wise evaluation maps out the future path to success.

References

[1] Kahneman D 2011 *Thinking, Fast and Slow* (New York: Farrar, Straus and Giroux)
[2] www.nobelprize.org/prizes/economic-sciences/2002/kahneman/facts/
[3] https://eu.themyersbriggs.com/tools/mbti/mbti-personality-types/intp
[4] www.belbin.com/about/belbin-team-roles/
[5] https://airfocus.com/glossary/what-is-the-stage-gate-model/
[6] de Bono E 1987 *Six Thinking Hats* (London: Penguin)

Part IV

Building an innovative culture

IOP Publishing

Creativity for Scientists and Engineers
A practical guide
Dennis Sherwood

Chapter 16

What is 'culture'?

16.1 The Covid-19 vaccine miracle

In February 2021, the Covid-19 pandemic had been raging for about a year, and the UK was in lockdown: schools closed, shops shut, hospitals crammed full. But a vaccine had become available just a few weeks previously, and soon millions of people had received their first jab. Yes, the vaccine – or rather, vaccines, for seven soon came to be in use around the world, and more than 50 others were at that time in clinical development [1] – are true life-savers, in every sense of the term.

Yet a year beforehand, there were no vaccines at all. And in the past, vaccines have taken five or more years to develop [2]. How was it possible to bring not just one, but seven, effective Covid-19 vaccines into operational use so quickly, with many more soon to come?

Was it because the world's scientists, virologists and immunologists had suddenly become much more creative? Or perhaps all those 'old' uncreative people had, overnight, been replaced by bright 'new' creatives?

No.

The scientists, virologists and immunologists, the lab technicians and factory workers, the administrators and regulators, and everyone else involved in this hugely complex task were, for the most part, exactly the same as those who had been in post a year, two years, previously. And they were no more creative then than they had been hitherto.

The world-changing achievement of developing the vaccines was not attributable to some magic enhancement of the intrinsic creativity of any of the individuals involved, nor on their collectively creativity.

What was dramatically different is the context in which they were working, the circumstances under which their individual and collective creativity was allowed to flourish. To the benefit of everyone on the planet.

Circumstances such as urgency. Every government, every university, every corporation, every individual, all agreed that the development of a vaccine was an absolute priority. So there were no arguments as to what was the most important activity; no one played politics wanting to push 'their' project at the expense of 'mine'. That gave organisational clarity, focus and endorsement.

Circumstances such as funding. Yes, developing a vaccine is expensive, but not compared with the cost of an international pandemic potentially on the scale of the influenza outbreak of 1918–20, let alone the plagues of earlier centuries.

Circumstances such as risk. Hitherto, the process of vaccine development had been a well-established sequence of processes, each very well-controlled and highly regulated, so as to ensure safety. Which all makes sense. But which also implied that the overall elapsed time was the sum of the end-to-end times of each individual step – together amounting to years. A key feature of the development of the Covid-19 vaccines, by contrast, has been to overlap many of these processes so that a subsequent process is started not when the preceding one has been totally completed, but as soon as possible after the preceding process has begun, provided that it is safe to do so. Yes, there is some risk in this, but the prize of development sooner than later has enormous value – and the 'conventional' serial development process is not totally risk-free either.

Circumstances such as a willingness to say 'yes', not 'no'. I wasn't around the table when, in the past, the scientists were talking about how a vaccine might be developed in the shortest possible time. So I'm making this up… but… imagine the team was talking about vaccine development, say, a year or two before Covid-19, and a young biologist, new to the team, says, 'Instead of doing [this] after [that] has been totally completed, why don't we start [this] once [that] is known to be working properly? That would allow much of the two processes to run in parallel simultaneously, and would save a huge amount of time!'

Silence.

Some around the table make mutual eye contact, eyebrows slightly raised.

The most senior person leans forward; clasps their hands together; and in a quiet voice says, 'Thank you for that suggestion. Unfortunately, the regulator won't allow that. I know you're new here, but you will quickly learn the exceptionally stringent regulations within which we work – and with which we of course comply.'

But soon after Covid-19 had struck, around that same table…

'Do you remember that idea about running processes in parallel rather than one after the other that was suggested a while back?'

'Yes… I do… but we did nothing with it then…'

'Mmm… but it's a really good idea… what do we have to do to test it out?'

'Well, we could do [this], and [this]… and we'll need [this], and probably have to stop doing [that]…'

'And we'll need to get the regulators on board too, so we should speak to them right away.'

'Right. Let's do it...'.

Those mini-stories are not about idea generation, about creativity.

They are about 'culture' – the umbrella word I use to describe all those aspects of organisational life that influence what we do, and how we do it; what we're allowed to do, and what we're not; what we're incentivised for, and what we're penalised for; what we're measured on, and what no one else cares about; what resources we can use, and what resources we are denied. Which covers pretty well everything...

16.2 Language

...and the most fundamental aspect of culture is the language we use, every day. In fact, language is so fundamental to culture, we don't pay any attention to it: it is so 'natural'. Indeed it is; but it is well worth paying attention to, and noticing, for example, how the language we use often reflects the relative importance or status of the person being addressed, how it reveals the speaker's attitudes and beliefs, how it signals what it organisationally 'right' – or 'wrong'. And tuning in to the language, and reflecting and using it in the organisationally 'correct' way, is an important aspect of how a newcomer becomes 'one of us' – or not. And it has a significant impact on creativity and innovation.

Take, for example, meetings. Next time you attend a meeting, notice the frequency of the words 'no' (and equivalent negative phrases such as 'Really?', 'I'm not sure I agree', 'Have you thought of?'...') and 'yes' (and equivalent phrases such as 'I agree', 'What do you need to make that work?', 'Tell me more about that'...). Is 'no' more frequent than 'yes'?

And notice too the relative status of those who use those words: is 'no' more frequently used by someone senior to someone junior, whilst 'yes' is more frequently used by someone junior or someone more senior?

These are tell-tales signs of the culture – the extent to which it is more, rather than less, authoritarian and hierarchical; the willingness to explore, rather than the dominance of control.

Many organisational contexts, of course, are necessarily rule-bound, where it is absolutely right that the process follows 'the book' quickly and efficiently, where compliance is essential, where creativity is just out-of-place. The numbers on a spreadsheet need to add up; goods must be manufactured to specification and at high quality; vehicles must be driven safely. So a culture that encourages creativity, where innovation flourishes, is not a chaotic free-for-all; rather, a creative culture is full of rules – as all ordered societies must be – but those rules are reviewed from time to time to ensure continued fitness-for-purpose; those rules allow for everyone to ask questions; and those rules recognise that there are times and contexts when compliance is absolutely required and others where a more exploratory approach is more suitable.

One such context is when ideas are being sought, for example, in a group discussion. And in which attention to language is especially important.

So, for example, the innocent-sounding 'What ideas have we got?' is more likely to generate silence, if not fear, than a flurry of good ideas. Far, far better, to ask questions like 'What do we know about [this]?' and 'What assumptions are we making about [this]?', for as long as everyone knows what [this] is, everyone can contribute, without fear, and – with reference to the discussion about ideas relating to chess, as described in sections 10.1, 10.2 and 10.3 – someone might say something equivalent to 'the objective of chess is to capture the opponent's king', or 'a square is either empty or occupied by a single piece'. This can really open things up, leading to creativity-stimulating questions like 'How might [this] be different?', 'How many ways can we identify of doing [this] differently?' and 'What might happen is we did [this] differently?'. Ideas will then flow.

And when ideas are being generated, remember that **all** ideas are totally ill thought-through when first articulated. So questions like 'How would that work, exactly?' can elicit only 'I haven't the faintest clue – it's only an idea'. But few of us are brave enough to say that. So, since I know that as soon as I say something, I will be challenged (which is how it feels to me, even though I know you are not being aggressive), and since I also know I don't have a response to that challenge, it's a lot safer for me to say nothing at all. So I'll stay silent. With the consequence of denying you, and everyone else, a glimpse of what might be a vital Koestler's Law fragment on which you can build, so transforming a lamentably raw, and weak, idea into something that's beginning to look good.

Even worse is 'that won't work because...', and its many variants. Yes, this is often 'explained' as 'the cut and thrust of academic debate' and 'a way of getting people to think'. But is the 'academic debate' response just a very poor excuse for thoughtless behaviour, if not arrogance ('that won't work because...' is often code for 'you're wrong', and 'I'm cleverer than you')? And if it's supposed to be a way of getting people to think, it's pretty clumsy, for the recipient of the intellectual punch just delivered is more likely to feel diminished, if not startled, and even hurt. The best way to get people to think is by asking intelligent questions, not by making demeaning declarations. So 'what are the issues that need to be fixed to get [this idea] to work?' and 'how many ways might there be of dealing with [this problem]?' – to give just two examples – work much better.

This is not, indeed, 'rocket science'. But its effect can be. If you want to build a culture of sustainable scientific creativity and innovation, if you want your 'whole' to be greater than the sum of the 'parts', pay (great) attention to language.

16.3 Observation, curiosity and permission revisited

Throughout this book – and especially in the Prologue, section 1.8, and throughout chapter 8 – I have emphasised the importance of observation, curiosity and permission as underpinning all creativity: careful observation of the-way-things-are-now constitutes knowledge, and is the platform, rooted in reality, for curiosity, as exemplified by that oh-so-powerful question 'how might this be different?' And permission provides the context for being curious, safely.

Observation, curiosity and permission, however, are deeply cultural. Observation is about noticing things, both locally and more widely. And very often, an 'outsider' might notice something that the 'insiders' haven't spotted. But if 'the way we do things around here' is all about 'keeping your nose out of other people's business', any 'outside' view is culturally rejected, and creativity withers. And if curiosity, asking questions, seeking to understand 'why', are culturally regarded either as a sign of personal weakness (people who 'know' don't ask questions, so if you do ask a question, you don't know, therefore...) or as 'stirring things up', then the *status quo* perpetuates. And in a culture of low permission, the predominant motivation is that of compliance, of sticking to the rules, and the pervasive mood is of fear, of not taking any risks at all. So the organisation's cultural attitudes to observation, curiosity and permission are all-important, as indeed was emphasised in many of the stories in chapter 13.

16.4 The wider picture – 'enablers' and 'motivators'

Language, and the organisational attitude to observation, curiosity and permission, are but two (albeit singularly important) aspects of an organisation's culture – there are many others. And that, indeed, is a major part of the 'problem' with talking about culture, and doing anything to change it – 'culture' is so big, amorphous, intangible. To make matters even more difficult, no single manager 'owns' it and so can be given the responsibility to 'fix' it. And even if the Managing Director says 'we need to change the culture around here', everyone around the table will nod and agree, but whether anything actually happens is another story...

So to make something real happen, that amorphous blob 'culture' needs to be broken down into manageable pieces – pieces that in this book I (rather loosely) refer to as 'enablers' and 'motivators' – 'enablers' being aspects of the organisation that influence the resources available for creativity and innovation or the context within which they take place, and 'motivators' aspects of the organisation that motivate people to behave in one way or another.

My check-lists of what I consider to be the most important enablers and motivators are shown in figures 16.1 and 16.2 on the following pages, and discussed in the following two chapters.

But before we look at these in detail, let's take a more 'big picture' view.

Each of the features shown – performance measures, budgets and all the rest – represent an important aspect of organisational life, and each has an often significant impact on an individual's behaviour. So, for example, if at my annual appraisal, the whole conversation is about how I did, or did not, run a project on time within budget, did or did not meet my sales quota, or did or did not have enough papers published in peer-reviewed journals, with no reference whatsoever to the ideas I contributed at various discussions, or how I encouraged my team to have ideas too, then I quickly learn that anything to do with creativity is neither

Figure 16.1. Enablers.

recognised nor rewarded. Being creative might be personally satisfying, but if the organisation couldn't care less, then why bother?

This cameo story highlights just a very small detail of the complex working of the organisational machine – that in this particular organisation, 'ideas contributed' or 'encouraging creativity in the team' and the like are not explicitly itemised in the check-list used at appraisals, and so are quite likely not to feature in the appraisal discussion. As a consequence, the person being appraised is left with the impression that creativity and innovation are not important, that the organisation certainly does not wish to encourage them, and that the organisation quite likely does not care about them at all.

Why, then, might any reference to creativity be absent from the appraisal checklist? Perhaps because 'the organisation' really doesn't care; but perhaps the reason is more prosaic – simply because the HR department (or whoever is responsible for writing the check-lists designed to help people do appraisals) hasn't updated their check-lists since 'creativity' became an organisational priority. It could well be that the check-list was originally compiled when 'creativity' was not on the organisational agenda, so its absence at that time has an explanation. But

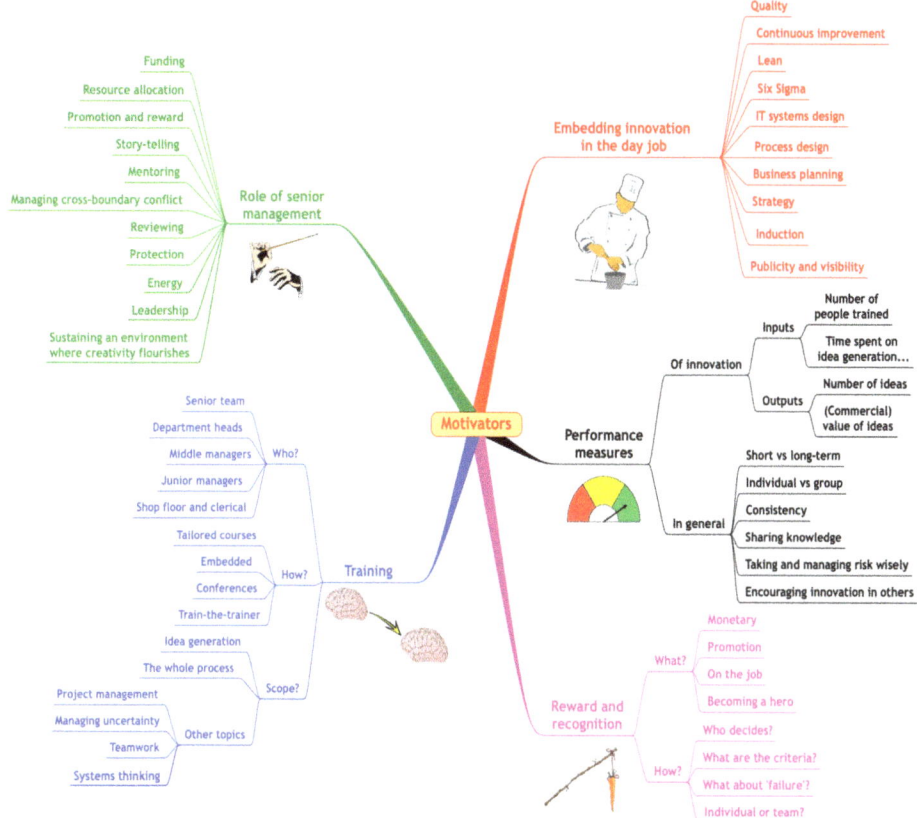

Figure 16.2. Motivators.

it is also understandable – at least to an extent – that the organisation's overall objectives can evolve, but that the appraisal check-list hasn't caught up.

The impact of that 'disconnect', however, can be profound, as the cameo story demonstrates. Importantly, this does not indicate an organisation's deliberate attempt to frustrate creativity, and to crush all creative initiative; rather it's more likely to be an accidental oversight. But an oversight that matters.

For each of the items identified in the diagrams of enablers and motivators counts; each has a contribution to make. No one item is more important than another; rather, all must be in 'harmony', in that if any one is (usually inadvertently, rather than deliberately) 'pointing in the wrong direction' – as the appraisal system was in the story – then this can jeopardise the potential success of creativity and innovation across the whole organisation.

Ideally, in a culture that is truly supportive of creativity and innovation, in which creativity and innovation flourish, all the enablers and motivators, simultaneously, encourage this, each playing its own role in helping make innovation happen. For all these to be continuously 'aligned' requires considerable organisational effort to maintain, but this is worth doing for any organisation that wishes to support an innovative culture.

In reality, very few organisations have gone the other way, deliberately 'misaligning' them with the intention of intentionally frustrating creativity. In reality, most organisations have not paid much attention to creativity, and when the various processes and systems for performance measures, budgeting, appraisal and the rest were designed and implemented, at different times and by different people, some of those processes might, perhaps inadvertently, have been supportive of creativity, some not, and some in-between.

Fundamentally, for any organisation that now wishes to build a culture of safe creativity and wise innovation, the issue is not to 'start from scratch', but rather to review all the enablers and motivators, and identify:

- those that currently support creativity and innovation;
- those that are 'neutral';
- those that currently deter creativity and innovation.

This 'triage' allows energy and attention to be focused where is it most needed right now: on the processes in the third category. So, in practical terms, the first set are left alone; the 'neutral' ones are ignored for the time being; and all the organisation's available energy and attention is devoted to a series of well-defined projects to do whatever is needed for each of the items in the third category to bring them into 'alignment'. And once that has been done, the 'neutral' ones can be improved, and then the 'supportive' ones even more so.

Overall, this makes the process of 'culture change' much more manageable and tangible: firstly, there's a project to improve the budgeting system; then the design of a training programme; after that… Each project has its own specific objectives, and each project contributes to the overall 'big picture' of encouraging and supporting creativity and innovation. Yes, this sequence of projects can take some time to complete, but as each project is finished, each will deliver benefit within its own context. And the cumulative end-game is truly worth the effort.

Much can be written about each of the enablers and motivators individually, and so, seeking to balance being informative with sensible brevity, the next two chapters discuss just a few headline points about each.

One final point: the distinction between the main headings 'enablers' and 'motivators' is not hard-edged – 'training', for example, both equips those trained with appropriate knowledge, as well as building confidence in the exercise of skills, so enabling those trained to participate more effectively in creativity and innovation. In this context, training is an 'enabler'. Many people, however, find being trained, and training itself, highly motivating, and so 'training' is a 'motivator' too. So don't get too concerned about the 'enabler'/'motivator' labels – what's important is not those names, but the way in which each of the processes described influences individual and collective behaviour, each making its own valuable contribution, so that collectively they all act together, in harmony, to make the organisation truly creative.

References

[1] www.who.int/emergencies/diseases/novel-coronavirus-2019/covid-19-vaccines
[2] www.weforum.org/agenda/2020/06/vaccine-development-barriers-coronavirus/

IOP Publishing

Creativity for Scientists and Engineers
A practical guide
Dennis Sherwood

Chapter 17

Enablers

Figure 17.1. Enablers.

17.1 Budgets

The idea generation workshop had gone really well. In a most lively day, the team had generated a host of ideas for improving the design of their key product: a small pump used in domestic water systems. As usual, some of the ideas weren't particularly good, and so were shelved, and just a few were 'quick wins', which the team could implement right away. But there were also two *BIG IDEAs*, which, to take further, required some laboratory work to test out some technical possibilities. That would take about three months, and would need some new materials, as well as an item of new test equipment. So Chris thought all that through, worked out the costs, compiled a work-plan, and sent the proposed work package to the boss, Pat.

'That's a great idea,' said Pat. 'We really need to progress it, so I'll definitely put it in for the next budgeting round.'

'Thank you,' replied Sam. 'But the next budgets won't be approved until about nine months' time, and it would be a real pity to have to wait that long until we get started.'

'Yes,' said Pat. 'I understand. But things are a bit tight, and there's just no way I can authorise that level of expenditure right now...'

You might recognise that story. The idea is great, but there's no money in the budget. And there's no money in the budget because when the budget was compiled the previous year, that idea had not yet been thought of, so there was no allocation...

That's an example of how the budgeting process can kill innovation stone dead. How does Chris feel? And what does Chris say to the team in an attempt to explain why the *BIG IDEAs* from the creativity workshop have gone nowhere. You can forgive the team for thinking 'why did we bother?' And for Chris for resigning and taking the idea to a competitor.

What's important about this story is that it is most likely that this was 'accidental', in that the accountants who designed the budgeting process all those years ago did not deliberately set out with the intention of extinguishing every creative spark that might emerge. And we can all understand why many budgeting processes seek to deter the squirrelling away of 'fat', especially in budgeting processes that are primarily 'bottom-up' (whereby more junior managers bid for budgets that they believe they need) as opposed to 'top-down' (in which more junior managers are obliged to make do with whatever they might be given). But processes that seek to squeeze out all the 'fat' can sometimes fail to distinguish between 'fat', and a legitimate need to be able to spend money on things that were not explicitly thought of at the time the budget was agreed. If Pat had those resources available, they could be allocated at once to help develop the promising idea, to the team's, and more importantly, the organisation's, benefit. But because the budgeting process had been designed without having the need to fund creative ideas in mind, that opportunity is denied.

And here's another variant.

As before, Chris goes to Pat and has a conversation about funding the initial development of the new idea. To which Pat says, 'That's a great idea! But it also has impact on Ali's department, and so I can't fund all the development out of my budget. Have you spoken to Ali?'

At which point Chris goes to Ali, who says, 'That's a great idea! But I've already used most of my budget on [whatever], so I don't have much left. And it also spills over to Pat...'

That's an example of an idea that is an organisational 'orphan' – an idea that straddles organisational boundaries, an idea that no one budget holder 'owns'. And so the only way for that project to proceed is for all the 'part-owners' to agree. Which can often be organisationally very difficult to achieve.

And once again, this is not any fault of Pat or Ali; no one is being 'difficult', and everyone genuinely wants the idea to succeed. But the structure of the organisation, and the way budget holders exercise their responsibilities, has, once again, frustrated creativity and innovation.

So, let me play that one again... imagine this conversation between Chris and Pat...

'That's a great idea! But it also has impact on Ali's department, and so I can't fund all the development out of my budget. Nor, I'm sure, can Ali. So have you spoken yet to Charlie? Many of the best ideas cross boundaries, and Charlie holds a budget specifically to fund great ideas that don't naturally 'fit'...'

Chris then goes to Charlie, who once again says, 'That's a great idea! Yes. I hold a budget for ideas just like that. Could you convene the team so I can meet them, and agree how we might best get on with it?'

You get the point.

In most organisations, the budgeting process is the main mechanism for controlling expenditure, and the local allocation of resources. Which we all respect and understand. And in many cases, the resources likely to be required can be anticipated, forecast and put into a direct context: next year, we will need [so many] staff to [do whatever they do] so the salary budget will be [this]; next year, we plan to manufacture [so many] of [these], which will require [this quantity] of raw materials, so the purchasing budget will be [whatever]. That's all fine, and can be estimated and agreed.

But the effect of creativity and innovation isn't like that. The results of a creativity workshop cannot be predicted with certainty, and sometimes ideas 'just happen' serendipitously. Furthermore, all idea development is fraught with risk and uncertainty for, by definition, it is a journey into as yet unknown territory. But what can be predicted is that each workshop is likely to produce a number of 'quick win' ideas – which are defined as being able to be implemented within normal budgets – as well as a few *BIG IDEAs*. No one can predict what those might be, or what resources might be required to progress them. So the best that can be done, at budget time, is to identify [some amount] that a local manager can deploy, as appropriate, to any *BIG IDEAs* that might emerge. And no, this is not 'fat' or 'slack'; and yes, the organisation should build into the process safeguards to ensure that the corresponding resources can only be used for suitably authorised *BIG IDEAs*, and not just used as a general slush fund. And even better if there are budgets for ideas that are organisational 'orphans' too.

And as noted in the check-list in figure 17.1, budgets are required not only to fund the development of ideas, but also for all the other processes associated with creativity and innovation, such as training, communication and knowledge transfer.

17.2 Funding

In the context of this book, 'budgeting' refers to a process that happens across the organisation, with budget holders at all levels; by contrast, 'funding' is about the overall allocation of the enterprise-wide cake, and in particular about two things: firstly, the top-level decision of how much to spend, overall, on all aspects of creativity and innovation, and secondly, decisions about funding the development of *BIG IDEAs*, where the sums involved are beyond the usual departmental budgets, and where resources need to be deployed from several parts of the organisation.

As regards funding innovation in general, this would encompass the on-going overall support of training, across the enterprise, as well as the resources required for idea generation and initial follow-up. So how much should this be? And how is that amount justified?

The second question is particularly important, for it is a constraint on the answer to the first. So, for example, suppose that someone suggests that, this year, the organisation should hold just one idea generation workshop, at an estimated cost of [this number] of days of people's time (which would otherwise be devoted to the day-job), and [this amount] of cash spent on, for example, the workshop venue, travel, accommodation and the rest. For that event alone, someone might ask 'why are we spending that amount of time and money, when there will be a penalty attributable to lost time, and the money could be better spent on [whatever]?'

That challenge is understandable, but also very difficult to answer tangibly. The event is an idea generation workshop, held in the hope that some ideas will be generated; ideas that will not have happened otherwise; ideas that we hope will have value to the organisation. But how do you *know* that some good ideas will be generated? How do you *know* that those ideas would not have happened 'naturally', whilst people were working on the normal day-job? And how do you *know* that any ideas will have value?

In fact, in answer to those three questions, the only truly honest responses that anyone can make are 'I don't', 'I don't', and 'I don't'. In which case, the challenger can respond 'so why are we wasting all that time and money?'

Indeed.

And the answer 'because I believe it's the right thing to do, a good thing to do, with the upside that we might generate some really powerful ideas which could be of great value – let alone the benefits to the team of bringing them together' just has no traction with the person who believes the opposite, that it's all just a big waste.

Therein lies an important organisational issue. Belief.

Yes, it is possible to compile a spreadsheet in which the possible commercial value of an idea is set against the costs of the idea generation workshop that discovered it, and all the follow-on costs too, so 'proving' that it was worthwhile to hold the workshop in the first place.

But does this spreadsheet have any real meaning? Or is it simply an arithmetical demonstration, with the 'authority' of a spreadsheet behind it, showing that it is always possible to construct a revenue forecast that can cover any costs you might think of, plus a bit more too?

Investing in creativity and imagination is, in my opinion, primarily a question of belief, rather than justification. If you believe that having good ideas is intrinsically valuable, and that building a vibrant culture of creativity and innovation is a 'good thing', then you will be willing to invest in making them happen. If you don't, you won't. It's as simple as that.

And if the senior management of an organisation share that belief, funds will be allocated accordingly; if not, they won't. Which is another angle on why the organisational culture is either the driver of creativity and innovation, or the killer. Culture starts at the top.

So in my opinion, the question 'how is the funding of creativity and innovation justified?' is answered either as 'because we believe that creativity and innovation are organisational 'goods' and so should be funded accordingly' or 'they shouldn't, so let's talk about something else' – in which case, that's that!.

If creativity and innovation are considered to be organisational 'goods', then the amounts of time and money to be allocated can then be determined by a sensible 'bottom-up' process in which the elements are identified and costed – [so many] idea generation workshops, each of which, on average, requires [this number of man-days] and [this level] of expenditure; [this much] to be associated with the process of wisely evaluating the resulting ideas; [this much] for local development; an allocation of [this] for awards and prizes... That will come to whatever it comes to, and the actual administration of those funds can then take place in a professionally competent manner. It's not a free-for-all, but a well-disciplined management process.

Also shown in the check-list in figure 17.1 are some questions associated with funding the development of those *BIG IDEAs* that emerge from the evaluation process, and that require a level of approval beyond the budgets held locally. This relates, for example, to the authorities held at, say, departmental, site, or divisional level, which will vary from organisation to organisation.

What's important here is the overall timeliness of the authorisation process. In general, the closer the authority to the location at which the idea is generated, the faster the process, but the lower the authority level. So the ideal is to achieve a balance between the 'size' of the idea and the speed of a decision. So, for example, it is not sensible for small ideas to be authorised, or not, by the main board;

likewise, truly *BIG IDEAs* merit higher level consideration, even if the process is rather slower.

17.3 Managing development and implementation

Idea generation is an activity that can readily fit within the day-job – many ideas can emerge whilst the day-job is taking place, and events such as idea generation workshops, which might last just a few hours, or one or two days, can usually be accommodated without any disruption. And the same applies to relatively straightforward evaluation activities too, such as 'Evaluation Lite'.

But the more thorough evaluation of *BIG IDEAs*, and many development and implementation activities, cannot take place within the normal demands of the day-job, but require dedicated resources to be committed over whatever period of time is appropriate, which may range from just a few days, to years – as required, for example, to deliver the *BIG IDEA* of landing a man on the moon.

Larger-scale evaluation, and the subsequent development and implementation of the most promising ideas, therefore require the formation of project teams and the successful management of the corresponding projects; accordingly, the 'Managing development and implementation' check-list in figure 17.1 notes the main points to bear in mind in this context. The items identified – such as the identification of the project team, milestones and the rest – are generic to all projects, and much more information on each (and many others too) will be found in (the many) books on project management and so will not be discussed here.

Save one particularly important one: 'How is the team selected?'

As an example, consider a manufacturing company that has a successful product on the market. The engineering design team is currently working on some improvements, with the intention of launching the Mark II product in, say, six months' time. Wishing to look ahead beyond Mark II, the team convenes an idea generation workshop which turns out to be most productive – some great ideas emerge for Mark III, ideas which are major improvements on the current product and the soon-to-be-launched Mark II too.

The discovery of the ideas for Mark III immediately raises two problems. Firstly, who will work on the project to do the prototyping, test the materials, and all the other technical things that need to be done? The problem arises because the key people are all working on the development of Mark II, and if they were to be pulled off to work on Mark III, then Mark II will be delayed...

One possible approach is to keep the 'key people' on the Mark II development, and assign some others to Mark III. But which 'others'? The department does not have a surplus pool of people, with nothing much else to do, just waiting for the next project. And those people who might be available might be available because they are perhaps less sought after, less able... with the result that those who might be allocated to Mark III are the 'second division'... which doesn't exactly get the development of Mark III off to a flying start...

And there's another problem too. Everyone agrees that Mark III is a significantly better product than Mark II – to the extent that everyone is thinking 'why didn't we discover [this], [that] and [the other] last year when we were designing Mark II?' The fact is that these discoveries were not made, and the current plans are to launch Mark II in six months' time. However, having discovered the materially better Mark III, would it be better to stop the development of Mark II, and go straight for Mark III, even though the earliest Mark III might be launched is at least a year away?

The idea generation workshop that discovered Mark III was a great success in inventing a new product, but a consequence is the creation of two organisational problems: how to staff the Mark III development, and whether or not Mark II should be scrapped.

We mention this because these problems are by no means uncommon. The purpose of all idea generation workshops is to discover good ideas. And once these are discovered, the development of the best ones inevitably requires resources of time, people and money. Resources that are inevitably limited, especially as regards people – very few organisations have people with nothing else to do but to wait for the results of an idea generation workshop. That suggests that the timing of such workshops should be carefully considered: if it can be anticipated that development resources will be coming free in [so many] months' time, then it could well make good sense to convene a workshop to 'stock the idea pipeline' so that some powerful ideas are immediately available.

The 'should we scrap Mark II?' question is in essence strategic. Although Mark III might indeed be better than Mark II, Mark II itself may well be better than both Mark I and also the competition. Mark II might therefore be a commercial success in its own right, to be supplanted at some time in the future by Mark III. In that case, a commercially sensible decision would be to continue with Mark II as planned, and, over time, to develop Mark III with a view to launching it in the future when Mark II is becoming somewhat 'tired', or whenever might be right to counter any competitor moves.

There is no generic 'right answer' to this type of question, for everything depends on the details of the specific context. But there is a generic approach that enables a wise answer to be determined under any specific circumstances: to carry out a wise evaluation of the idea 'We should scrap the development of Mark II now, and go straight for Mark III instead'.

17.4 The idea archive

As just noted, idea generation workshops will inevitably result in more ideas than the organisation can handle, and indeed that is their purpose. Not all ideas will be of the same quality – and as we saw in section 15.5, the purpose of 'Evaluation Lite' is to identify, quickly and easily, those ideas that can sensibly be shelved, the 'quick wins' that can be actioned right away, and the (usually small number of) *BIG IDEAs* that require considerable resource to evaluate wisely.

Importantly, no ideas should be 'thrown away', for ideas that are judged to be poor at the time they were generated might have some value in the future – they might turn out to be Koestler's Law 'fragments' of better ideas that will emerge later, or something might happen elsewhere, such as the reduction in the cost of a component, that makes a previously non-viable idea much more attractive now.

Furthermore, ideas will emerge in contexts other than at formal idea generation workshops: for example, a sales manager might have had a conversation at a customer's site, which, on driving home, suddenly triggers an idea. What happens next? What might the sales manager do with the idea? Perhaps the manager thinks 'I'll mention that to the Engineering Director next time we bump into each other'. That may indeed happen. But there is also a possibility that the sales manager will have long forgotten the idea by the time that chance meeting happens.

No one will ever know how many ideas are lost or forgotten in any organisation. And no one will ever know how many opportunities were foregone as a consequence.

That's what the 'idea archive' is all about: a formal process of capturing, recording and tracking ideas generated across the organisation, at any time, at any place; a process that the sales manager can easily tap into, so ensuring that the idea is not lost, but sent to the right person for consideration; a process which takes the outputs of all idea generation workshops, even those judged, at the time, to be poor; and the provision of a repository where ideas, and Koestler's Law fragments, may be stored, retrieved and reviewed.

But this won't happen 'naturally', by itself: the process by which ideas are captured, stored, processed, tracked and reviewed has to be well thought-through, designed, implemented, managed and used. Which is no small task, for the process must be easily accessible throughout the organisation, and not burdensome to use. It is in fact an organisational 'memory', and is potentially of great organisational value.

But is it worth the time and cost to build an archive process and maintain it? Some organisations will say 'no'; others, 'yes'. Either way, any organisation that wishes to build a culture of creativity and innovation should consider how it will prevent most of the ideas generated from being forgotten, lost or otherwise discarded. And once again, the process of wise evaluation can be very helpful in answering the question 'Is building a process to manage an archive of ideas itself a good idea?'

17.5 Physical environment

One of the consequences of the Covid-19 pandemic was to enforce physical separation and fragmentation, especially for those whose usual working environment was an office, or some other space where colleagues came together. Working from home has the benefit of saving the time otherwise spent commuting (whilst driving an increase in the consumption of biscuits!), and one of the unexpected consequences of holding meetings on-line, it seems, has been to encourage punctuality – my experience is that people are seldom late for a Zoom or Teams meeting, as compared to similar gatherings for which the venue was Conference Room 6 on the fourth floor!

With the virus no longer a public threat, organisational life is evolving to a different state, but perhaps one closer to the 'old ways' before the virus than the obligatory distancing with which we were all obliged to cope for so many months. And this is relevant to creativity and innovation, for underpinning all creativity is Koestler's insight that all ideas are patterns formed from pre-existing elements. Those elements are in each of our minds, sometimes bundled together in the larger patterns of learning, knowledge and experience, sometimes as rather looser fragments when we're wrestling with a problem-to-solve. And it's often when 'my' bundles and fragments can mingle with 'yours' that a new pattern is formed, that wonderful new pattern of a really good idea.

Formal events, such as idea generation workshops, are convened deliberately to help make this happen, especially when techniques such as the ***InnovAction!*** process are used, and these events can be hugely productive.

But ideas can also be generated in many other contexts too. Sometimes by oneself, certainly; but much more likely during a meeting, or when you happen to bump into someone accidentally, and just have a conversation. These interactions can take place on-line, but happen far more 'naturally' in a collective community environment, and so the design and nature of that environment can, and does, have an influence on creativity and innovation.

Some organisations are indeed, and rightly, famed for their creativity, for example Google and the consultancy Ideo. And both are also famed for their 'innovation-stimulating' office environments: for example, Ideo's 'clutter' – large quantities of apparently random bits of 'stuff' to serve as 'inspiration' [1] – and Google's 'quirky' meeting spaces and furniture [2].

Yes, those environments are surely 'fun places to hang out'. But in my opinion, an environment that encourages creativity and innovation does not have to be eligible for feature articles in *avant garde* design magazines; rather, the environment recognises the guiding principle of Koestler's Law, and its relevance in this particular context as regards these two questions:

- How can the environment maximise the likelihood of chance interactions, which might therefore stimulate discussions that might lead to ideas?
- In those locations where chance interactions might happen, what can be done to encourage conversation?

Some ideas triggered by these questions are, for example:

- Conventionally, the manager, or senior person, of a department has a desk close to the people in the departmental team, perhaps within an open-plan area, perhaps in a separate office. Suppose, then, that the manager of Department A is located in Department B's area. Yes, that does imply more to-ing and fro-ing. But it also means that people from both departments – and especially the managers – will bump into each other much more than they did previously… Which might have 'interesting' effects on creativity – especially if the two departments need to collaborate closely, or have some form of common interest.

- Two locations, within any building, where people naturally come together, often randomly, are lift lobbies (and the lifts themselves), and staircases. But these are often the two most 'unfriendly' places in the whole building, where people either avoid eye contact or just quickly nod if they know one another. In principle, though, these should be the locations where creativity is most likely to happen serendipitously. So what might be done to these locations to encourage people to talk to one another, and to explore ideas?

- Another is where people assemble for food and drink – such as a canteen or a restaurant, the coffee machine, or the tea room. These are inherently more convivial and communal than the interior of a lift. But the question still stands: what might be done to these locations to encourage people to talk to one another (even more!), and to explore ideas?

The Sainsbury Laboratory, Cambridge

The UK's top accolade for architecture, the Stirling Prize, is awarded annually to the UK's best new building, as judged by an expert panel appointed by the Royal Institute of British Architects. And in 2012, the winner was the Sainsbury Laboratory of plant science at the University of Cambridge.

The most obvious feature of a building is the aesthetic of its exterior. For a research laboratory, however, it's the nature of the interior that's important, especially if scientific discovery, creativity and innovation are to flourish.

Here is an extract from a newspaper article* describing some key aspects of the interior design that the panel took into particular consideration:

'Within, the building's main spaces are linked by a continuous route, designed to promote chance encounters, dotted with informal areas to sit and chat. Wooden cubbyholes line the first floor windows, while staircases are broad, allowing two people to walk and talk side by side. The labs are generous and airy, flooded with natural light from curved funnels, with glazed walls which let other people see what's going on.

It is entirely unlike most research buildings, a far cry from the usual warrens of endless corridors and closed doors behind which new discoveries are squirreled away. It is hoped that this open layout will change the way of working, fostering a more collaborative approach – and although only half-occupied so far, it seems to be working well.'

Is this not a description of the physical embodiment of observation, curiosity and permission?

Source: [4].

17.6 Behaviours

A confession. The items in the 'Behaviours' branch of figure 17.1 are in essence a list of topics that I couldn't readily place elsewhere. But they are each very important in building a creative culture in the first place, and in maintaining it once it gets going.

The importance of *language* was discussed in sections 15.4 and 16.2, and what an individual is, or is not, allowed, to say, and whether, or not, others listen are often related to the *role of hierarchy and status* within the organisation: unless the 'big boss' has a continuing stream of truly great ideas (which can happen, but rarely), top-down hierarchical organisations are much less likely to be creative and innovative. But that is by no means saying that they are unsuccessful in achieving their objectives, whether commercial or associated with the delivery of a public service. Hierarchical organisations are usually associated with top-down command-and-control cultures, and these are usually highly effective at delivering the same thing day after day, to consistent quality standards. In this environment, compliance with the rule book is everything, all the time – and this benefits consumers too, for we all want a breakfast cereal to be the same as the one we bought before, we want the light to go on as soon as we flick the switch, we want the X-ray machine at the local hospital to be available, and working properly, when we need it. This degree of reliability and consistency doesn't happen 'by magic', and command-and-control cultures are designed to ensure that this does indeed happen.

Creativity, however, is about doing something different, and that word 'different' often does not fit comfortably alongside 'compliance'. It might therefore be thought that a creative culture is the antithesis of a command-and-control culture – say, some sort of anarchic hippy commune, in which there are no rules, and everyone can do whatever they like.

But no. An environment with no rules at all is a mess, for there is no co-ordination across individuals, no teamwork, and great conflict. And – perhaps somewhat surprisingly – truly creative cultures are very well-disciplined, have many rules, and can behave in a high command-and-control manner when necessity requires it, for example, when a project has to be delivered on time within budget, as many projects must.

A hallmark of truly creative cultures is therefore flexibility – neither too rule-bound as are many command-and-control cultures, nor anarchic, but able to adapt locally, or at different times, as circumstances require. Creative cultures also recognise that all ordered societies need a set of organisational rules, which are honoured and trusted by everyone. But in a strict command-and-control culture, the rules are the rules are the rules, and cannot be questioned; in a creative culture, the rules are recognised as being necessary and helpful, but not 'sacred' – as the context changes, then maybe it's time to revisit the rules and change them if that would be a sensible thing to do.

Furthermore, creative cultures encourage a positive *attitude to challenge*, for they understand the importance of experimentation. Creativity, by definition, is about doing something that had not been done before, which is intrinsically a challenge, and necessarily risky. And yes, creative cultures have a positive *appetite for risk* too, but this is neither reckless nor foolhardy. Rather, creative cultures study risks very carefully, to understand them, to mitigate them, and to deal with them if they crystallise. Which is very different from avoiding risk altogether, and always playing safe. Indeed, the 'black hat' question in the wise evaluation process – 'what issues need to be addressed and resolved so that the idea can be implemented successfully?' – and the 'red hat' question, 'how might people feel?', are both designed to identify the risks in advance, so that they can be managed. And experimentation is an important part of that: as exemplified by Thomas Edison's trial of 6,000 materials in his search for how to make the filament for his electric light bulb [3].

And as Edison's work testifies, not everything will work, and sometimes things will go wrong. And so the organisational *attitude to 'failure'* is important. In a culture in which 'failure' is career-ending, no one who wants to have a career will take any risks at all. Creative cultures understand that, which is why 'failure' is in inverted commas – for they recognise that 'failure' can mean two, very different, things. One type of 'failure' is negligence, when someone is, perhaps, lazy, or fails to comply with a rule that everyone agrees is sensible; another type of 'failure' is when an outcome does not match an expectation or hope. Yes, Edison might have hoped that animal hair would form a suitable light bulb filament, but whether or not it does requires an experiment, which might or might not work. And if it doesn't, that 'failure' is not negligence or laziness; rather it serves to eliminate an otherwise plausible possibility. So in creative cultures, there is less *fear* of doing something wrong and more *empowerment* to discover what works.

Flexibility is also the hallmark of an innovative organisation's structure too. Recognising the importance of combining ideas as fundamental to all creativity, such organisations encourage *cross-boundary sharing* of people, resources, and most importantly knowledge. Furthermore, teams can be formed and disbanded as required, with line bosses not blocking the release of their staff, and those on project teams not fearing that their line job will no longer be available once the project has been completed.

References

[1] https://eyeondesign.aiga.org/human-centered-design-thinking-in-action-at-ideos-boston-office/
[2] www.businessinsider.com/google-zurich-headquarters-tour-2018-1?r=US&IR=T
[3] www.fi.edu/history-resources/edisons-lightbulb
[4] www.theguardian.com/artanddesign/2012/oct/15/sainsbury-laboratory-deserved-stirling-prize

IOP Publishing

Creativity for Scientists and Engineers
A practical guide
Dennis Sherwood

Chapter 18

Motivators

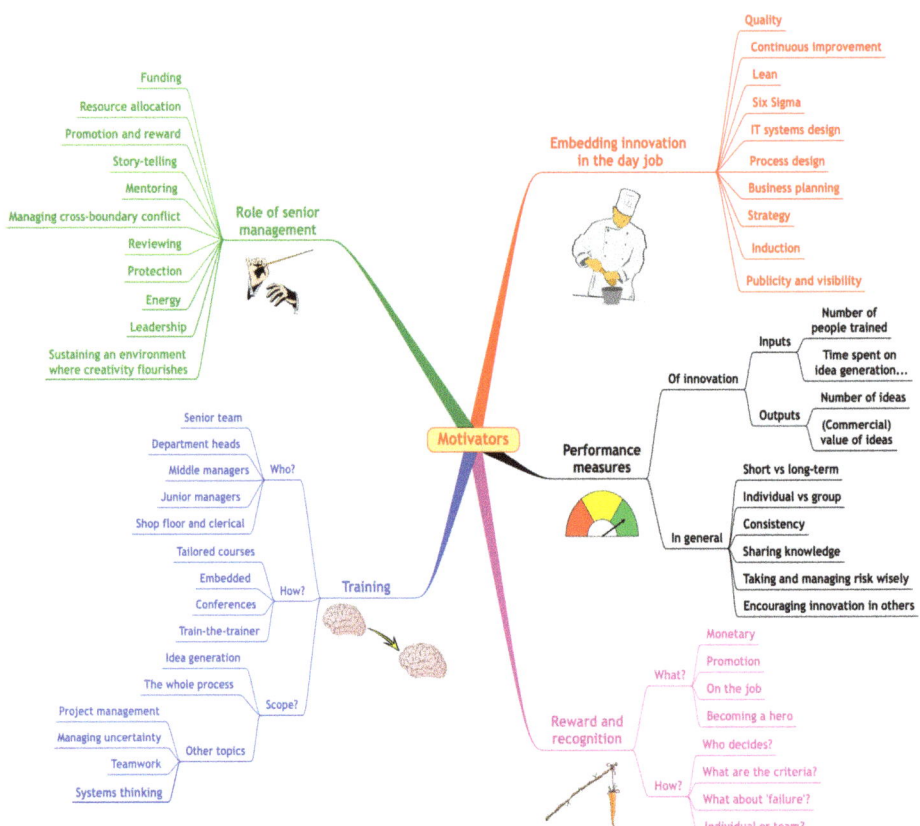

Figure 18.1. Motivators.

18.1 Reward and recognition

Every individual responds to how they are, or are not, rewarded; how they are, or are not, recognised. Accordingly, the processes for determining organisational rewards, for example, remuneration, and also recognition – such as being promoted – significantly influence behaviour.

So, for example, if I notice that someone whom I'd always thought of as a peer, and someone who always delivered on time and within budget but has never contributed a single idea in any context, gets a promotion, whilst I remain stuck at my level despite contributing ideas a lot, then I soon learn that having ideas is not valued. No surprise, then, if I stop doing it – and advise everyone in my team to stop doing it too.

Alternatively, if creativity is explicitly recognised, and rewarded, guess what – it will happen.

An organisation that wishes to encourage creativity and innovation therefore ensures that all related activities – having ideas, participating in idea generation workshops, contributing to evaluation panels... – are both recognised (most people feel good when the boss says 'well done, that was really helpful' – and means it!) and rewarded.

As noted in figure 18.1, there are some subtleties that merit some thought. For example, attributing an idea to a single person might not be fair to others, for as we have seen throughout this book, one of the key concepts is that ideas are rarely generated by the 'lone genius', but result from team interaction, perhaps at an idea generation workshop, perhaps over time as individuals talk to one other. Often, there is a 'lead' person, who perhaps puts the last piece of the Koestler's Law puzzle in place, but in a culture that encourages creativity and innovation, and in which everyone understands how Koestler's Law works, asking that 'lead' person about how the idea emerged will usually reveal who else contributed.

Another potentially contentious issue is failure. Are only ideas that prove to be successful rewarded? These are, by definition, ideas that have been generated, then positively evaluated, and then successfully implemented – ideas from which the organisation has benefited, and those who contributed surely merit appropriate recognition. But that might become clear only long after the idea was initially generated. And what about those who have generated hosts of ideas, but ideas that have never made it through evaluation?

Certainly, no organisation wishes to reward failure, but 'failure' in the context of creativity is a subtle concept – an idea that might be weak at the outset might turn out to be a critical component of an absolute blockbuster some time later. So there is a strong argument that creativity is best rewarded in two, different, ways: firstly, to reward those involved in ideas that are implemented successfully, and secondly to reward those involved in the processes of idea generation and wise evaluation, thereby giving organisational support to those who contributed to those processes, stressing their importance, even if the ideas themselves are shelved.

18.2 Performance measures

Formal mechanisms for reward and recognition – such as appraisal, remuneration review, and promotions – are sporadic; by contrast, the performance measures within which we all work apply day-by-day, week-by-week. The processes associated with performance measures, and recognition and reward, are – at least in principle – related, in that success at meeting, if not surpassing, my performance measures should result in the appropriate recognition and reward.

Performance measures, and the corresponding criteria for recognition and reward, therefore need to be mutually consistent. So, for example, if a key criterion for a manager's promotion is 'has demonstrated the ability to encourage creativity in members of the team being managed', then it's important that the performance measures of the members of the team being managed allow, for example, time for them to participate in idea generation workshops, and to work on projects compiling the evidence for wise evaluation. If the performance measures of the team members are solely about delivering 'the day-job', then this sets up a problematic, and probably unintended, conflict.

Once creativity and innovation become organisational objectives, it is likely that performance measures will be introduced accordingly – so the items in figure 18.1 identify some relevant aspects.

The distinction between 'inputs' and 'outputs' refers to measures associated with the processes of creativity and innovation (such as the number of people who have been trained in, for example, the *InnovAction!* process and the principles of wise evaluation, and the time devoted to idea generation workshops or spent on evaluation), in contrast to the results of those activities (such as the number of ideas generated, the number of ideas accepted for development and implementation, their value...). In general, it is my experience that an organisation wishing to become innovative is better advised to introduce measures of 'inputs' first, leaving the measures of 'outputs' for perhaps a year or two – that helps people to focus on idea generation, which is where everything starts, without being too worried about 'but we must generate really *good* ideas', which is often unhelpfully constraining, especially when people are not yet fully confident in the fundamental processes.

The *in general* items in the 'performance measures' branch of figure 18.1 are just a short check-list of things that can easily be overlooked, but are helpful to bear in mind. So, for example, in some organisations, 'knowledge is power' is a key aspect of the culture, for it is a source of internal competitive advantage: because 'I' know something, or someone, 'you' don't, 'I', not 'you', will get that promotion. Koestler's Law, however, stresses the importance of forming new patterns from existing components, which is all about sharing knowledge, experience and ideas. Being willing to share knowledge, and to forego 'knowledge is power', are therefore important features of the transition from a non-innovative culture to an innovative one, and so a performance measure explicitly encouraging knowledge sharing could well make a valuable contribution.

Short vs long-term is a reminder that it can take a long time for some ideas – especially really *BIG IDEAs* – to come to full fruition, and so a perhaps well-intentioned shorter-term performance measure such as 'revenue attributable to ideas that have been generated this year' or 'number of papers published this year' could drive behaviours that meet those performance measures but which deny bigger opportunities – for example, rejecting any ideas that will take more than one year to develop, or encouraging researchers to carry out more mundane work that will result in a relatively dull journal paper now, so distracting attention away from bigger, more important, more difficult problems – but problems that might not result in publishable results in the shorter term.

And *consistency* refers to the importance of ensuring that the performance measures throughout the organisation need to be mutually consistent and co-operative, rather than (usually inadvertently) in conflict – as in the example just mentioned of the more senior manager who is measured on 'encouragement of creativity in others' but whose team's performance measures make it very difficult for them to attend the required training courses.

18.3 Training

...and talking of training...

As I trust this book has demonstrated, creativity is a skill which can be learnt, and taught. Koestler's Law, for example, is not intuitively obvious, and – at least in the UK at the time of writing – learning the tools and techniques of creativity, such as the ***InnovAction!*** process, and the principles of wise evaluation, do not feature in the curricula of schools, colleges or universities. So most people enter their occupations with at best, no knowledge of how to generate ideas or to evaluate them wisely; at worst, a belief that having ideas is something that only a 'creative' person can do – which 'I' am not.

So formal training is an absolute necessity. And the answer to the question *who?* is everybody, from the most senior to the most junior. No one should be disenfranchised by being denied the opportunity to build confidence in their ability to have ideas – which is all about the importance of involving even the most junior people. Likewise, no one is too senior; despite a feature of senior people that I often encounter, arrogance. Yes, even the top people can usually benefit from enriching their skill set too – and the disciplines associated with wise evaluation can sometimes be an eye-, if not brain-, opener. And there are two other important reasons why top management need to be trained. The first is that it is usually a good idea for senior people to have a good understanding of what their juniors are being taught, so that they can at best use, at least understand, the language. But much more importantly is the role of senior management in general (of which more shortly), and of leading any initiative to change the organisational culture: the richer a senior manager's understanding of what creativity is all about, the more active, and effective, that senior manager will be in leading, and driving, the culture change programme.

Culture change can only be driven from the top. And if the top managers don't want it to happen, can't be bothered to ensure that it does happen, or are otherwise lukewarm or indifferent, then it surely won't happen. And for the top to be that engaged, they must understand. If a top manager does not know of Koestler's Law, and so does not understand its deep significance, then that person will never realise why sharing knowledge is so important – especially if he or she rose through the organisation on the back of 'my knowledge is my power'. So that person is most unlikely to support initiatives to improve knowledge sharing, or even to allow more junior people time to attend idea generation workshops.

Some other details... *embedded* is about the opportunity to incorporate relevant training, perhaps quite subtly, into other activities. So, for example, if a team is convening a meeting to discuss, for example, a particular problem – say, how to improve cross-boundary collaboration – it is both easy and natural to suggest something of the lines of 'would it be helpful if we took a few moments to think about our individual experience of how collaboration works – or indeed doesn't work – right now? We can then share our individual experiences, which are quite likely to be different, and that then gives us a very realistic framework to explore what we might do differently, to everyone's benefit.' That, of course, is a description of ***InnovAction!***, as applied to the focus of attention 'cross boundary collaboration', but expressed in that way doesn't require that people know all about Koestler's Law, yet enabling the team to have a well-focused and productive conversation.

Likewise, during a meeting, when someone suggests an idea, and the usual 'that won't work because...' discussion is about to take place, someone who understands the principles of wise evaluation can, quite naturally, intervene and say, for example, 'before we go down that route, would it be helpful if we firstly gain a richer understanding of what the idea really 'looks like', and of its potential benefits?' That achieves three important objectives. Firstly, it gets the person who suggested the idea out of a defensive position, and invites that person to explain more about the idea, which is usually good for that person and for everyone else too. Secondly, it shifts the initial focus to the benefits of the idea (the 'yellow hat' of section 15.4) before the conversation turns to the 'issues to manage' (the 'black hat' of pages section 15.4). And thirdly, it nudges those present into a general awareness of the principles of wise evaluation...

Finally, *other topics* notes that there are a number of themes associated with creativity and innovation that are also well-suited to formal training. Project management is of course an important skill in many circumstances, of which managing a project developing and implementing an idea resulting from an idea generation is just one example; creativity and innovation are inevitably associated with risk, and always involve teams; and if the focus of attention for creativity is the behaviour of a complex system, then 'systems thinking' can be of great value in describing the current behaviour of the system, so providing the platform for asking 'how might this system be structured differently, and so result in a different, and better, outcome?'

18.4 The role of senior management

I've already mentioned several aspects of the role of senior management in building and maintaining a culture of creativity and innovation. For this to happen, senior management can't just be 'neutral', leaving things to others, for this lack of interest and engagement will be noticed throughout the organisation, and will undermine the effectiveness of those trying to make things happen. Senior managers must be, and be seen to be, 'champions', showing leadership, acting as a source of energy for others, and showing the tenacity required to keep things going over what might be a relatively long time.

Protection means 'ensuring that the culture change project does not become eroded, side-tracked, reduced in scope, or allowed to peter out', especially as conditions change (as they inevitably do), and other, often urgent, priorities arise (as they inevitably do). And since changing the culture applies across the whole organisation, there will inevitably be some cross-boundary issues that cannot be resolved locally.

> ### The story of Fred, Bert and Marc
>
> One day, Fred had what he thought was a good idea. So he went to his boss, Bert, and told him about it. 'That's a really interesting idea,' said Bert. 'Why don't you write it up so we can take it to Marc for approval?'
>
> Which is what Fred did.
>
> But Marc's reaction was not what Fred had hoped for. 'What a stupid idea! We all know that won't work! No!!!'
>
> The story might had stopped there, but it didn't, for Bert said 'don't worry about that, I have some limited funds that you can use to try the idea out…'
>
> For the Fred in this story is Fred Sanger, the only person so far to win two Nobel Prizes in chemistry*. But that might not have happened, had not Bert intervened.
>
> After receiving his PhD in chemistry, Fred Sanger joined a research team at the University of Cambridge led by Professor Albert Chibnall. Fred's *BIG IDEA* was to identify the amino acid sequence of the protein insulin, and he submitted his proposal for funding to the Medical Research Council, the MRC. Fred's proposal was duly reviewed by a panel of the 'great and good' of the time - and received a resounding rejection on the grounds that all proteins, according to the then strongly-held belief, were just random sequences of amino acids, so there could be no 'sequence' to determine*.
>
> That could have been the end of the story. But it wasn't, for, at considerable personal risk, Albert Chibnall intervened, and provided funds to get Fred's research going*. And the rest, as they say, is history. Fred's discovery of the amino acid sequence for insulin won him his first Nobel Prize in 1958, and paved the way to his subsequent research to determine the sequence of the bases in DNA. This won a second Nobel Prize in 1980, so laying the foundations of all the DNA 'fingerprinting' that takes place around the world today – for example, at the appropriately named Wellcome Sanger Institute in Cambridge*.
>
> But if 'Marc' had had 'his' way, if Bert had not intervened, Fred would have done other things – perhaps winning a Nobel Prize in another field. But perhaps not. And although it was Fred who won the Nobel Prizes, the hero of this story is Bert.
>
> I'm sure we'd all like to be like Fred: winning a Nobel Prize (or two!) must feel good. But, deep down, we all know that that's pretty unlikely. But *we can all be like Bert* – to say 'yes' rather than 'no', and to build the conditions in which creativity can flourish.

For sources, see [1–4]

Yes, senior managers do indeed play a crucial role in building the conditions in which creativity can flourish.

18.5 Embedding innovation in the day-job

If creativity and innovation are regarded as something 'done at Head Office by the clever people', or an activity that happens every now and again at off-site workshops, then that's where they will remain if the organisation is lucky. More likely, creativity and innovation will wither and die.

But if creativity and innovation can become a natural part of the day-job, the usual 'way we do things around here', then they will flourish, indefinitely. And so embedding innovation in the day-job must be the overall objective of the culture change programme.

The detail of the *day-job* branch in figure 18.1 therefore notes some aspects of the day-job that are 'natural' opportunities for embedding creativity and innovation. So, everything in this book is totally consistent with, for example, *quality* and *continuous improvement programmes*, *lean* production methods, and the methodology of *Six Sigma* – the fundamental question 'how might this be different?' underpins all of these. So if programmes of this nature are taking place, they can all be enriched, and broadened, by applying the principles described here.

More generally, good IT systems do things differently and better, so creativity should be an integral component of every IT development, and all IT staff should be trained accordingly. Likewise, the objectives of process improvement programmes, new product development, and business planning too all provide contexts for that oh-so-important question 'how might this be different?', as applied to what-happens-now. And the biggest 'how might this be different?' question applies to the organisation's strategy, this being so important as to feature as its own segment on the Target Diagram, as discussed section 3.4.

A particularly effective way of incorporating creativity and innovation into the day-job is to incorporate some creativity training, and perhaps some case studies, into the organisation's induction programme, so positioning creativity and innovation as being organisationally important from the outset. Indeed, portraying the organisation as innovative, internally and especially externally through, for example, press articles and social media, can build the organisation's reputation, and present the organisation as being a good place to work, helping to attract recruits who seek to work in an innovative environment… which in turn enhances that environment when they join… and even more so as their careers progress… making the organisation even more attractive to future recruits… so fuelling a most powerful virtuous circle…

18.6 So, what next?

That completes my overview of the 'enablers' and 'motivators', the various systems, processes, written and unwritten 'rules' that collectively determine an organisation's culture. Each of the formal ones – such as the budgeting system and the processes for reward – were probably designed and implemented by different people, working at different times, and to different objectives. Each is internally self-consistent, but it is quite possible that, probably inadvertently, they pull in different ways. And some aspects of the culture, such as the attitude to risk or the way in which language is used, were not 'designed and implemented' at all, but just emerged over time.

Collectively, these pervade the whole organisation, and there is no single person – with the sole exception, perhaps, of the single person at the very top – who 'owns' the whole lot. That's why culture change is so difficult – 'culture' is amorphous, indistinct, hard to grasp. To make matters worse, there is no single branch in figures 17.1 and 18.1 labelled 'creativity', so there is no single 'lever to pull'.

Yes, culture change is difficult, especially when attempted in an unstructured manner. But it is not impossible, especially when attempted in manageable chunks, as suggested by the component elements of the enablers and motivators, as described in figures 17.1 and 18.1. For to build a culture that encourages creativity and innovation, where creativity and innovation flourish, to the benefit of the enterprise as a whole and the individuals within it, the key is to ensure that all of the enablers and motivators actively support creativity and innovation, rather than inadvertently frustrate them.

As I have already mentioned, it is very rare for any organisation to be designed, deliberately and intentionally, to crush creativity, to imperil innovation. Much more likely is that, when the various enabler and motivator systems and processes were independently designed, creativity and innovation were not high on the organisational agenda, so encouraging them was not an over-arching objective. As a result, some of those systems and processes will be 'pointing in the wrong direction', and it is those that need to be 'realigned' first.

So, to repeat the points made in section 16.4, the task is to review all the systems and processes associated with each of the enablers and motivators, and identify:

- those that currently support creativity and innovation;
- those that are 'neutral';
- those that currently deter creativity and innovation.

That can then set up a series of projects, each of which has its own objectives, each of which will deliver its own outcomes, but outcomes that collectively achieve the overall goal: to build a culture of creativity and innovation.

References

[1] www.nobelprize.org/prizes/chemistry/1958/sanger/facts/ and www.nobelprize.org/prizes/chemistry/1980/summary/
[2] www.theguardian.com/science/2013/nov/20/frederick-sanger
[3] www.whatisbiotechnology.org/index.php/exhibitions/sanger/insulin
[4] www.sanger.ac.uk/

Epilogue

That's the end! Thank you for reading this far!

In the introduction, I said: 'I trust you will enjoy reading this book, and will benefit from putting its messages into practice'. Now you're at the end, I do hope that the first half of that sentiment has already come true, and you have found the book interesting, engaging and informative. As regards the second half, you may already have tried some of the processes, and I trust they have worked, and worked well. But if not, I urge you do so, for the book will have impact only when you have tried the ***InnovAction!*** process and seen that by asking 'how might this be different?' you can generate ideas more easily and more deliberately than before; when you have approached the evaluation of ideas according to the process described in chapter 15, and have convinced yourself that, yes, it is wise; when you have noticed some aspects of your organisation's cuture that you can influence locally (such as the use of language), and have therefore helped build a more high-performing team.

For this book is about taking action, doing things differently, making 'the world a better place'. But not taking action simply as a consequence of a procedure manual's instructions to do [this], then [this], then [this]; rather, taking action based on sound principles, such as Koestler's Law, unlearning and emergence; principles that explain *why* the action makes sense, and which allow you the freedom to modify the action as circumstances might require whilst keeping the fundamental principles intact. And as I trust the many stories told in this book – from the invention of the paint tube to the 'magical colouring sheet', from Florence Nightingale's 'Rose Diagram' to the 'Medusa' effect, from Kepler to gravitational waves – provide convincing evidence that those principles are valid, and of the broadest applicability.

So over to you…

Arthur Koestler's definition of creativity
(Section 4.1)...

The creative act is not an act of creation in the sense of the Old Testament.

It does not create something out of nothing; it uncovers, selects, re-shuffles, combines, synthesises already existing facts, ideas, faculties, skills.

The more familiar the parts, the more striking the new whole.

...and mine
(Section 7.6)

Creativity is the process of forming new patterns from pre-existing component parts.

The more the resulting pattern shows emergent properties, such as perceived beauty, the likelihood of enhanced understanding, the prospect of utility, or the possibility of creating value, the better the corresponding idea is likely to be.

The Target Diagram
(Sections 2.1, 3.1 and 3.8)

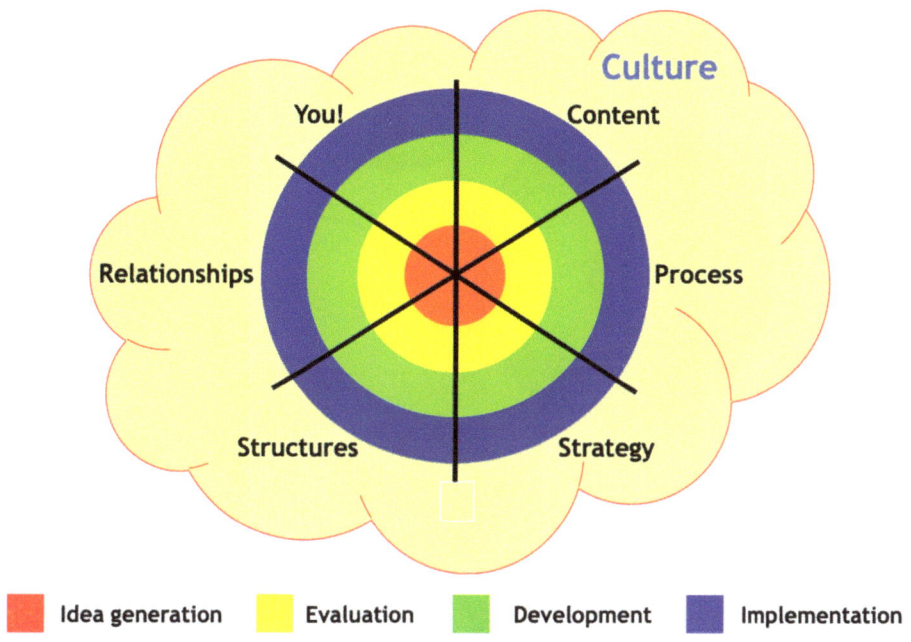

The 'fractal' Target Diagram
(Section 2.2)

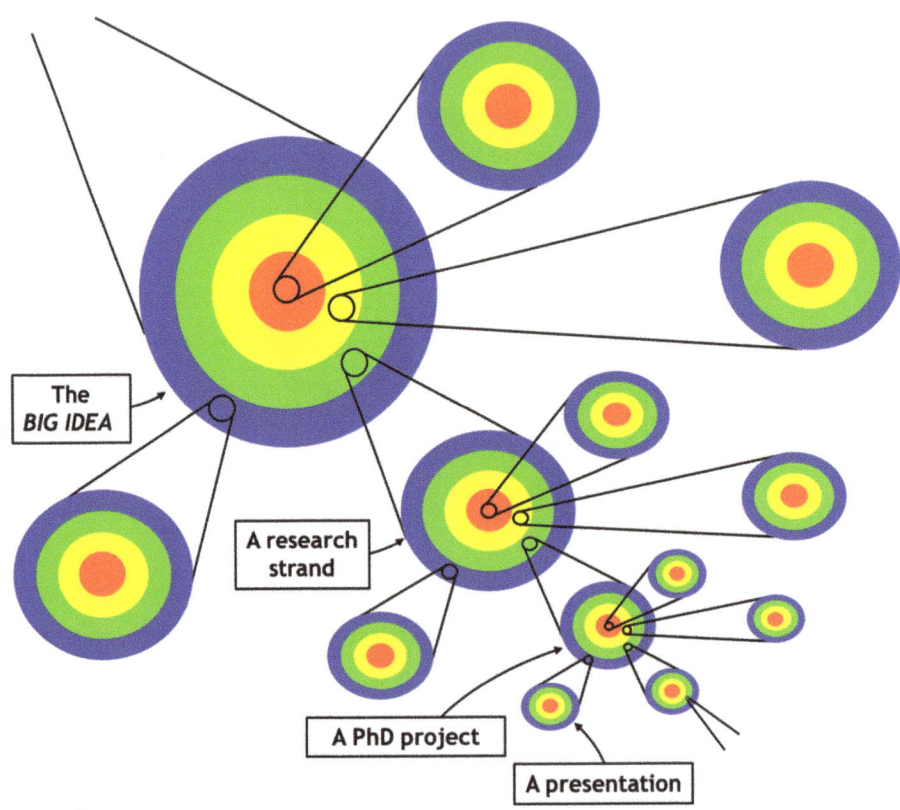

The *InnovAction!* process for deliberate creativity (Section 9.1)

InnovAction!

1. Define the 'focus of attention'.

2. Individually and in silence, write down everything you know about the agreed focus of attention.

3. Share.

4. Then choose one feature, and ask 'How might this be different?'

5. Let it be …

6. …and then, when that discussion runs out of steam, choose another feature and repeat Steps 4 and 5.

My process for wise evaluation
(Section 15.2)

1. **Imagine that the idea has successfully been implemented.** What does that 'future world' look like? And what is *different* as compared to today?

2. **What are the consequences?** Which of those differences make the 'world a better place' as compared to today, and better for whom? And which consequences are not-so-good? And for whom? How are different individuals, communities and constituencies likely to feel?

3. **The journey: what issues had to be addressed, and resolved, to implement the idea successfully?** This includes all issues likely to be encountered, for example, technical difficulties, the availability (or otherwise) of resources, and political and organisational issues such as rivalry, jealousy, both within and beyond the organisation.

4. **How might all the problems best be solved?** The not-so-good consequences, and the issues-to-overcome, are all problems that might be 'blockers'. But not necessarily… What ideas can be generated to solve and overcome these problems? How might the idea be enriched so that these difficulties are significantly reduced, if not eliminated?

5. **The numbers:** Now that the idea, and its implications are well understood, how can the benefits, and all the costs, best be quantified?

6. **The decision:** Is the idea accepted or rejected, or is it worth spending some time and money to reduce some uncertainties, so providing more, and better, insight to inform the yes/no decision?

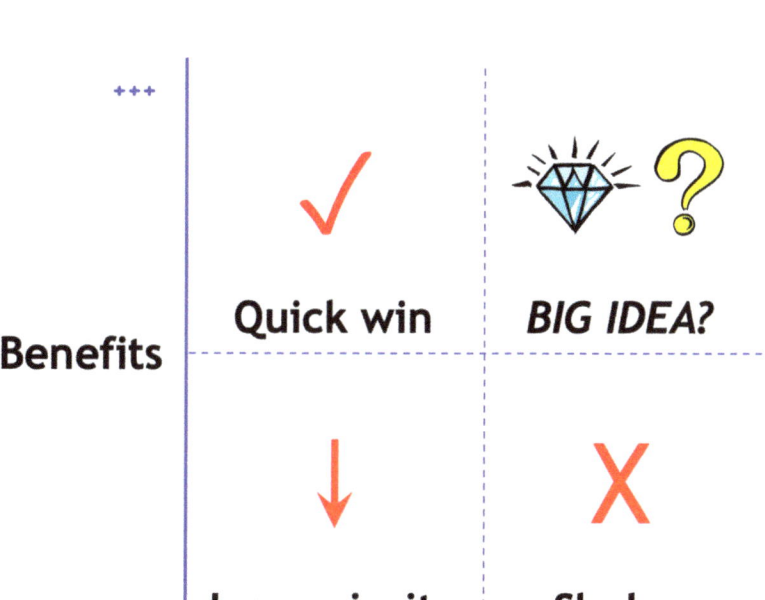

A check-list of cultural 'enablers'
(Chapter 17)

A check-list of cultural 'motivators'
(Chapter 18)

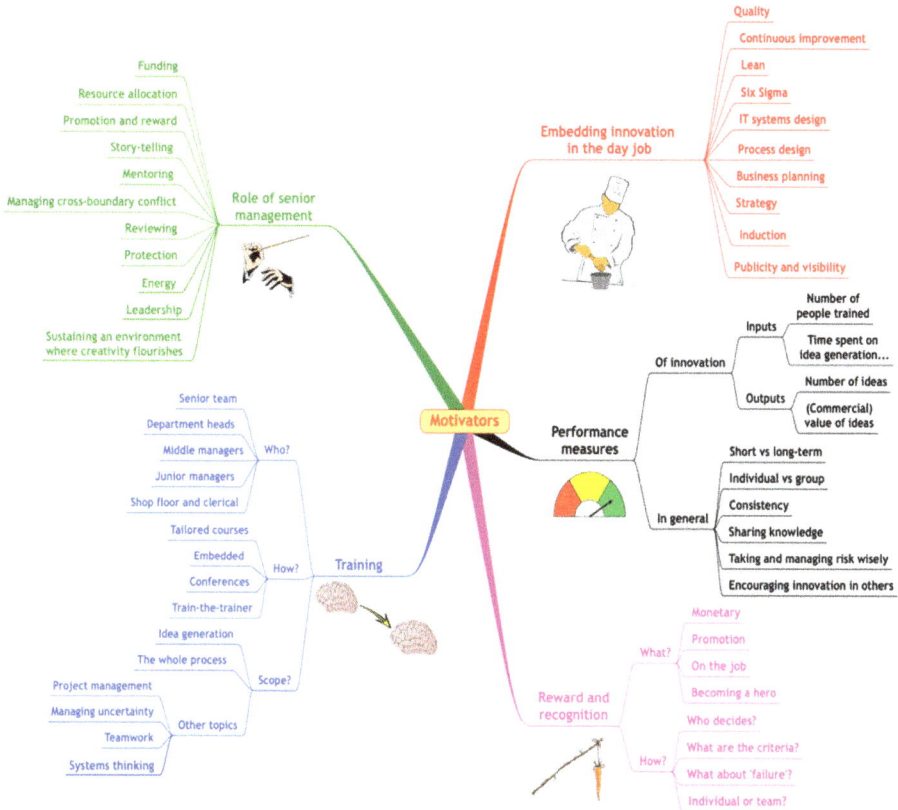

IOP Publishing

Creativity for Scientists and Engineers
A practical guide
Dennis Sherwood

Further reading

Historic sources

Kepler Johannes 1597 *Mysterium Cosmographicum* (Tübingen)
Kepler Johannes 1609 *Astronomia Nova* (modern Engl. transl.) 2015 (Santa Fe, NM: Green Lion Press)

Creativity in the sciences

Harman Peter and Mitton Simon (ed) 2002 *Cambridge Scientific Minds* (Cambridge: Cambridge University Press)
Higham Nicholas and Sherwood Dennis 2022 *How to Be Creative: A Practical Guide for the Mathematical Sciences* (Philadelphia, PA: SIAM – Society for Industrial and Applied Mathematics)
Koestler Arthur 2014 *The Sleepwalkers* (London: Penguin)
Poincaré Henri 1910 Mathematical creation *The Monist* **20** 321–35
Pólya George 1957 *How to Solve It: A New Aspect of Mathematical Method* 2nd edn (New York: Doubleday)
Pycior Helena, Slack Nancy and Abir-Am Pnina (ed) 1996 *Creative Couples in the Sciences* (New Brunswick, NJ: Rutgers University Press)
Simonton Dean 2004 *Creativity in Science: Chance, Logic, Genius and Zeitgeist* (Cambridge: Cambridge University Press)
Sobel Dava 1995 *Longitude: The True Story of the Lone Genius Who Solved the Greatest Scientific Problem of his Time* (London: Fourth Estate)

Creativity in general

de Bono Edward 1973 *PO: Beyond Yes and No* (London: Penguin Books)
de Bono Edward 2015 *Serious Creativity* (London: Vermilion)
de Bono Edward 2016 *Six Thinking Hats* (London: Penguin Life)
de Bono Edward 2015 *The Mechanism of Mind* (London: Vermilion)
de Bono Edward 1974 *The Use of Lateral Thinking* (London: Penguin Books)
Kelley Tom 2016 *The Art of Innovation* (London: Profile Books)

Koestler Arthur 1964 *The Act of Creation* (London: Hutchinson)
Sherwood Dennis 2001 *Smart Things to Know about Innovation and Creativity* (Oxford: Capstone Publishing)
Stover Dawn and others 2017 *The Science of Creativity* (New York: Scientific American eBook)

Also of interest

Boorstin Daniel 1992 *The Creators: A History of Heroes of the Imagination* (New York: Random House)
Fernández-Armesto Felipe 2003 *Ideas that Changed the World* (Kindersley: Dorling)
Chaline Eric 2009 *History's Worst Inventions – and the People Who Made Them* (New York: Fall River Press)
Frugoni Chiara 2003 *Books, Banks, Buttons – and Other Inventions from the Middle Ages* (New York: Columbia University Press)
Johnson Paul 2006 *Creators: From Chaucer and Dürer to Picasso and Disney* (New York: HarperCollins)
Watson Peter 2005 *Ideas: A History from Fire to Freud* (London: Weidenfeld & Nicolson)

Index

2D images, 13.7
3D images, 13.7
'10,000 hours', 8.2

α-helix, 5.14.4, Figure 5.15, 5.14.7, 5.14.9, 5.14.11, 6.5
α-keratin, 6.5
β-keratin, 6.5

A, adenine, 5.14.1, 5.14.8, 5.14.9, 5.14.10, 5.14.11, 5.15
A level school examination, 14.4.6
'academic debate', 16.2
Académie Montmor, 5.8
Acanthostega, 6.6
acoustic leak detection, of water leaks, 10.7
The Act of Creation, 4.1, 4.5, Chapter 4 (end), 11.2.3
Adams, John, 5.6
Adams, John Couch, Prologue
adenine, A, 5.14.1, 5.14.8, 5.14.9, 5.14.10, 5.14.11, 5.15
adhesive, biomimetic, 13.5, Figure 13.9
The Adoration of the Magi, Chapter 6 (end)
Advanced Laser Interferometer Gravitational-wave Observatory, aLIGO, 13.2, Figure 13.2, Figure 13.3, Figure 13.4
advertising slogans, 11.3
Agarwal, Professor Ritesh, 13.13
Age of Enlightenment, 5.6
agency, as the ability to do things, 13.9
Albers, Josef, 7.5, Figure 7.2
Albert Einstein Institute, Hannover, 13.2
Alexander VI, Pope, Chapter 6 (end)
Alice in Wonderland, 5.1
aLIGO, 13.2, Figure 13.2, Figure 13.3, Figure 13.4
Allen, Les, 13.7

Alstom Transport, UK, 13.8
Alzheimer's disease, 8.2
amino acid, 18.4
Ampère, André-Marie, 5.11
amyloid protein, 8.2
analogy, retrofit, 11.4.2
analysis, cost-benefit, 14.4.7
Anderson, Hans Christian, 12.3
angular momentum, orbital, 13.7
Another Essay in Political Arithmetick Concerning the Growth of the City of London, 5.6
antimony – germanium – tellurium alloy, 13.13
antithesis, and thesis and synthesis, 8.6
aphelion, Prologue
appetite for risk, Figure 16.1, 17.6, Figure 17.1
Apple, 5.5
appraisal, investment, 14.4.7
appraisal, personal, 16.4, 18.2
Archimedes screw, 6.2
archive, of ideas, 12.9.2, 15.5, Figure 16.1, 17.4, Figure 17.1
Arcimboldo, Giuseppe, Prologue
arguments against an idea, 14.4.6
arguments for an idea, 14.4.6
Aristarchus, Prologue
Aristotle, Prologue
Armstrong, Sir William, 6.5
Arouet, François-Marie (Voltaire), 5.6
arrogance, 8.5.4, 8.5.5, 16.2, 18.3
Art, 5.2
 and chemistry, 5.4
 and emergence, 7.5
 and Koestler's Law components, 7.1
 and physics, 5.5
 and subjectivity, 7.5, 7.6
artist's brush, 5.4
artist's easel, 5.4
Arup & Partners, 5.12

assumptions, challenge of, springboard, 11.2.1, 11.2.3
Astbury, William, 5.14.3, 5.14.4, 5.14.6, 5.14.9, 5.14.10, 5.14.11, 6.1
   and 'da Vinci problem', 6.3, 6.4, 6.5
Astronomia Nova, Prologue, 2.1
ATLAS project, 13.6
attention, focus of, Step 1 of *InnovAction!* process, 9.1, 9.2, 9.3
attitude to challenge, Figure 16.1, 17.6, Figure 17.1
attitude to 'failure', Figure 16.1, 17.6, Figure 17.1
Austen, Jane, 5.1
authority, 8.5.4
authority bias, 15.1.3
automorphic forms, 2.3
Avery, Oswald, 5.13.1, Chapter 5 (end)
Aviation Maintenance Human Factors, CAP 716, 13.8
axon, 13.4, Figure 13.6

'badness', of an idea, 7.6, 14.2, 14.3, 15.2.2, 15.2.3, 15.2.7
backwards, working, retrofit, 11.4.6
balance, of process for wise evaluation, 15.1.4
ball valve, 13.11
Baird, John Logie, 13.7
bar code, 4.5, 6.7
Bardeen, John, 1.4
Barnes, Professor Bill, 13.10
barriers, to creativity, 8.3, 8.5, 12.6, 16.2
base pair, 5.14.8, 5.14.9, 5.14.10, 5.14.11, 6.3
bases, sequence of in DNA, 18.4
basketball, 4.6
Battló, Casa, 5.8, 5.10
Battló y Casanovas, Josep, 5.10
bearing, 13.5
beauty, of emergent patterns, 7.6, 8.7
Beethoven, Ludwig van, Prologue, 1.4, 2.1, 4.2, 4.4, 4.7, 5.1
Begbroke Science Park, Oxford, 13.13

behaviours, 16.4, Figure 16.1, 17.6, Figure 17.1, 18.1
Beighton, Elwyn, 5.14.3, Figure 5.14, 5.14.6, 5.14.9, 6.1
Belbin, Meredith, 15.2.7
belief, organisational, in importance of creativity and innovation, 17.2
belief, personal, effect on wise evaluation, 15.2.2
Bell, Florence, 5.14.3, 5.14.4, 5.14.6, 5.14.9, 5.14.10, 5.14.11, 6.1
   and 'da Vinci problem', 6.3, 6.4, 6.5
Bell Burnell, Dame Jocelyn, 11.2.1
benefits, of an idea, 15.2.3, 15.2.4, 15.4, 18.3
   intangible, 15.2.6
   measurement and estimation of, 15.2.6
Bergman, Ingmar, Chapter 4 (end)
Berkeley, University of California at, 13.5
Berlin, University of, Chapter 5 (end), 8.6
Bernal, JD, 5.14.4
Bernard of Chartres, 5.7
Bessel function, 5.14.5
Besso, Michele, 5.13
Bhaskaran Professor Harish, 13.13
bias, 14.4.6, 15.1.3, 15.1.4
bias, authority, 15.1.3
'big' idea, 15.1.7, 15.2.1, 15.3, 15.5, Figure 15.2, 17.4
   budget for, 17.1
   funding of, 17.2
   management of development and implementation, 17.3
   time required for development and implementation, 18.3
Bille, Beate, 5.6
Bille, Steen, 5.6
Bille-Brahe family, Prologue (end)
biodiversity hotspot, 13.6
biology, 13.4, 13.12
biomimetic adhesive, 13.5, Figure 13.9
Birkbeck College, University of London, 5.14.4

bisociation, Prologue, 1.8, 4.5, 4.6, 6.6, 11.3, 13.4
black hat, as in *Six Thinking Hats*, 15.4, 15.5, 17.6, 18.3
blackmouth catshark, 13.6
Bletchley Park, 4.7
Bliss, Timothy, 8.2
blockers, to creativity, 8.3, 8.5, 12.6
blue hat, as in *Six Thinking Hats*, 15.4
Blue Plan-it®, 13.14
Bodle Technologies Limited, 13.13
'bombe', 4.7
bond, hydrogen, see *hydrogen bond*
'bottom-up' budgets, 17.1
boundaries, of cost-benefit analysis, 14.4.7, 15.2.6
boundaries, organisational, sharing across, Figure 16.1, 17.6, Figure 17.1
Borgia, Cesare, Chapter 6 (end)
Boyle, Robert, 5.8
Boyle's law, 5.8
Boys, Sir Charles, 13.2
Bragg, Lawrence, 5.14.1, 5.14.5, 5.14.11
Bragg, William, 5.14.1, 5.14.3
Braginsky, Vladimir, 13.2
Brahe, Otto, 5.6
Brahe, Tycho, Prologue, Prologue (end), 2.3, 4.1, 5.6, 5.7, 5.8
brain, 1.2, 2.3, Chapter 8 (end)
  connectivity of, 8.2
  development of, 8.2
  and learning, 8.2
  plasticity, 8.2
Brain Prize, 8.2
brainstorming, 1.5, 3.8, 9.3
Brexit, and wise evaluation, 15.2.2
briefs, for creativity workshops, 12.5
Brin, Sergey, 1.9
British Columbia, University of, 13.9
BritNed interconnector, 13.3
Brodie, John, 4.6
Bronowski, Jacob, 1.3

bronze, sintered, 13.5, Figure 13.7, Figure 13.9
Brown, Dr Jess, 13.14
Brownian motion, 5.13, 13.2
Browning, Robert, 5.1
brush, artist's, 5.4
bubble, soap, colours of, 13.13
bubblegum coral, 13.6
budgets, 16.4, Figure 16.1, 17.1, 17.2, Figure 17.1
bulb, electric light, 5.8, 5.9
Buonarroti, Michelangelo, 5.4
Burdon-Sanderson, John, 2.1
Bürgi, Joost, Prologue
bushing, 13.5
business planning, Figure 16.2, 18.5, Figure 18.1
Bussell, Dame Darcey, 8.2

C, cytosine, 5.14.1, 5.14.8, 5.14.9, 5.14.10, 5.14.11, 5.15
C5 electric car, 1.9
CAP 716, Aviation Maintenance Human Factors, 13.8
CAP 737, Civil Aviation Authority Flight Crew Human Factors Handbook, 13.8
'CCV' paper, 5.14.5, 5.14.6, 5.14.11, 6.3
calculus, dispute between Newton and Leibniz, 5.8
California, University of, Berkeley, 13.5
California Institute of Technology, Caltech, 6.1
Cambridge, University of, 5.14.1, 5.14.5, 5.14.6, 5.14.8, 5.14.9, 5.14.11, 6.1, 18.4
  Sainsbury Laboratory of Plant Science, 17.5
camera, digital, 13.7
camera, single-pixel, 13.7
Cantor, Georg, 1.3
carbon footprint of science, reduction, 13.10

cards, 'meteor', 10.5.4, 12.4.2, 12.8, 12.9.1
Carlyle, Thomas, 5.6
Carollo Engineers, 13.14
Carroll, Lewis, 5.1
Casa Battló, 5.8, 5.10, Figure 5.5
catecholamine, 13.5, Figure 13.8
cathode ray tube, 13.13
catshark, 13.6
causal loop diagram, 9.4.12
Cavendish Laboratory, University of Cambridge, 5.14.5, 5.14.6, 6.1
cavity, optical
CD, 13.12
celestial sphere, Prologue, 1.3, Figure P.1
cell, stem, 13.12
cell biology, 13.12
cell culture, 13.12
centromere, 8.6, Figure 8.4
CERN Large Hadron Collider, 13.2
Cézanne, Paul, 5.2
Chain, Sir Ernst, 2.1
challenge, attitude to, Figure 16.1, 17.6, Figure 17.1
challenge, Koestler, 4.8, 6.1
challenge assumptions, springboard, 11.2.1, 11.2.3
chair, of creativity workshop, 9.4.3
champion, of creativity and innovation, 18.4
Chargaff, Erwin, 5.14.8, 5.14.10, 5.14.11, Chapter 5 (end), 6.1
  and 'da Vinci problem', 6.3
Chargaff's first rule, 5.14.8, 5.14.10
chemical reactions, control of, 13.10
chemistry, 5.3, 13.10
  and art, 5.4
chess, ideas for games based on, example of *InnovAction!* process 10.1, 10.2, 10.3, 16.2
Chevreul, Michel, 7.5
Chibnall, Professor Albert, 18.4
cholesteryl benzoate, 5.5

chromatography, partition, Chapter 5 (end), 7.3
chromosomes, 8.6, Figure 8.4
chronometer, Prologue
Civil Aviation Authority Flight Crew Human Factors Handbook, CAP 737, 13.8
Clack, Jenny, 6.6
Coates, Michael, 6.6
Coca Cola, 11.3
Cochran, William, 5.14.5, 5.14.11
Cochran, Crick and Vand paper, 5.14.5, 5.14.6, 5.14.11, 6.3
Cockburn, Dr Hermione, OBE, FRSE, 13.6
coincidence, 5.8
co-invention, 5.8
Cold Spring Harbor Laboratory, 5.14.11
cold-water coral, 13.6, Figure 13.10
cold-water reef, 13.6, Figure 13.10
collaboration
  across organisational boundaries, 18.3
  example creativity workshop brief, 12.5, 12.6
collaborators, ideas for finding, example of *InnovAction!* process, 10.7
collagen, fibrular, 13.12, Figure 13.19
College, 'Invisible', 5.8
Collingridge, Graham, 8.2
Collins, David, 6.7
Colorado, University of, 13.7
  Boulder, 13.13
colour, 5.2, 5.4, 5.5, 13.4
coloured hats, as in *Six Thinking Hats*, 15.4, 15.5
colouring sheet, magic, 13.6, Figure 13.11
Columbia University, 5.14.8, Chapter 5 (end), 6.1
command-and-control culture, 17.6
Common Sense, 5.6
communication, Figure 16.1, 17.1, Figure 17.1

competitor's *BIG IDEA*, example creativity workshop brief, 12.5
completeness, of process for wise evaluation, 15.1.5
compliance, with organisational rules, 16.3, 17.6
components, and Koestler's Law location, 6.3, 7.1, 8.1
Composition with Red, Yellow, Blue, and Black, 5.2
computing, quantum, 13.7
conferences, ideas for, 9.2
configuration, as different way of being different, 10.6.1
connectivity of brain cells, 8.2, 8.3
'cons', of an idea, 14.4.6
consequences, Step 2 of wise evaluation process, 15.2.3, 15.2.4, 15.2.5, 15.2.6, 15.4
consequences, unintended, 2.6
consistency, of process for wise evaluation, 15.1.2
content innovation, 3.2, 9.2, 12.5
continuous improvement, Figure 16.2, 18.5, Figure 18.1
control, and command, culture, 17.6
control of chemical reactions, 13.10
Copenhagen, University of, Prologue (end), 5.14.6
Copernicus, Nicholas, Prologue, 4.1, 5.7
coral, 13.6
coral, bubblegum, 13.6
coral, cold-water, 13.6, Figure 13.10
Corey, Robert, 5.14.7, 5.14.8, 5.14.11, 6.1
  and 'da Vinci problem', 6.3, 6.4
corona treatment, 13.5
cost-benefit analysis, 14.4.7, 15.2.6
costs
  associated with development and implementation of an idea, 12.6, 14.1, 15.2.6, 17.1, 17.2
  of creativity workshops, 17.2
coupling, strong, 13.10, Figure 13.15
coupling, weak, 13.10
courtesy, within creativity workshop, 9.5
Covid-19
  impact on inter-personal interaction, 17.5
  impact on school examinations, 14.4.6
  impact on train operations, 13.8
  vaccine development, 16.1
cow, in Kansas, 13.11
'creative person', myth of, 4.2, 4.3
Creative Scotland, 13.12
Creative Whack Pack, 11.1.1
creativity, 2.3
  analogy, retrofit, 11.4.2
  and arrogance, 8.5.4, 8.5.5
  barriers to, 8.3, 8.5, 12.6
  belief in importance of, 17.2
  and bisociation, Prologue
  blockers of, 8.3, 8.5, 12.6
  building a culture of, Chapter 16, Chapter 17, Chapter 18, see also *culture, organisational*
  challenge assumptions, springboard, 11.2.1, 11.2.3
  choice of springboard or retrofit, 11.5
  as a communal activity, 5.8, 9.1
  and compliance, 17.6
  and culture, see *culture, organisational*
  and curiosity, Prologue
  decomposing and recombining, springboard, 11.2.2
  and deconstruction of existing patterns, 8.4, 9.5
  definition, 1.1, 4.1, 6.1, 7.6
  dictionary definitions, 1.1
  encouragement in others, 16.4
  enriched definition of, 7.6
  examples of advertising slogans, 11.3
  examples of new games based on chess, 10.1, 10.2, 10.3
  examples in art, 5.2
  examples in chemistry, 5.3

examples in economics, 5.6
examples in engineering, Chapter 13
examples in history, 5.6
examples in literature, 5.1
examples in music, 4.4
examples in philosophy, 5.6
examples in politics, 5.6
examples in science, Chapter 13
and failure, Prologue
and fear, 8.5.3, 8.5.5
as first of four processes for innovation, 2.1, 2.3
and hard work, Prologue
"How might this be different?", 9.5, 9.6, 9.7, 10.5
and imagination, 11.5
and incubation, 2.3
InnovAction! process for, 6.1, Chapter 9, 11.1
'input' measures of, Figure 16.2, 18.2. Figure 18.1
journeying, retrofit, 11.4.4
and knowledge, Prologue
Koestler's definition of, 4.1, 6.1
Koestler's definition of, enriched, 7.6
Koestler's law of, Chapter 4, 4.1, 6.1
lateral thinking, 4.5, 11.2.3
and laziness, 8.5.2, 8.5.5
learning as a barrier, 8.3
and learning from mistakes, Prologue
and 'love', Prologue, 8.5.1, 8.5.5
and luck, Prologue
metaphor retrofit, 11.4.2
need to 'unwire' neural circuits, 8.3
and novelty, 4.4
and observation, Prologue, 2.1
other people's shoes, retrofit, 11.4.3
organisational priority, 16.4
'output' measures of, Figure 16.2, 18.2. Figure 18.1
and pattern formation, Prologue, 4.4, 6.1
and permission, Prologue
PO, springboard, 11.2.3, 11.4.1

and problem-solving, Prologue
as a process of pattern formation, 4.4
processes for, Chapter 9, Chapter 11
performance measures of, Figure 16.2, 18.2. Figure 18.1
random words, retrofit, 11.3, 11.4.1
recognition for, 16.4
recombining, and decomposing, springboard, 11.2.2
retrofit processes for, 11.3, 11.4, 11.5
reward for, 16.4
and risk, Prologue, 2.4, 8.3
role of outsiders, 16.3
simile, retrofit, 11.4.2
six domains of, Chapter 3
skills required, 2.6
as something you're born with, 4.3
sound-bite definition, 1.2, 1.9
springboard processes for, 11.1, 11.2, 11.5
and teams, 2.3, 3.6, 8.7, 12.2
and tenacity, Prologue, 2.1
time required for, 2.3
training, Figure 16.2, 18.3, Figure 18.1
and unlearning, 8.4
and value and utility, 1.9
visioning, retrofit, 11.4.5, 11.4.6
working backwards, retrofit, 11.4.6
creativity workshop, Chapter 12
afterwards, 12.9
agenda, 12.4.2
archive of ideas, 12.9.2, Figure 16.1, 17.4, Figure 17.1
barriers to discussion and exploration, 12.6
briefs, 12.5
chair, role of, 9.4.3
composition, 2.3, 12.3
conditions and mood, 9.5, 10.5
conduct, 9.4, 12.4
constraints, how to deal with, 12.6
cost of, 17.2
costs, how to deal with, 12.6
courtesy, 9.5

duration, 2.3, 12.4, 12.4.1, 12.4.2, 12.4.3
examples of briefs, 12.5
evaluation of ideas at, 12.4.3, 12.5
evaluation of ideas following, 12.9.3
facilitation, 9.4.3
facilities, 12.4.2
funding for, 17.2
group dynamics, 9.4.2, 10.5
group size, 12.3
hierarchy within, 9.5, 9.6, 12.3
idea archive, 12.9.2
importance of not evaluating at, 12.7
importance of observation, 12.1, 12.3
importance of prior training, 12.3
and *InnovAction!* process, Chapter 9
justification of, 17.2
keeping a record, 10.5.4
meteor cards, 12.4.2, 12.8, 12.9.1
organisation, 9.1
and organisational culture, 3.8
participants, 12.3
productivity of, 12.8
recording of, 12.8, 12.9.1
report, 12.9.1
resources, how to deal with, 12.6
respect, 9.5
'safety', 9.5, 9.6
structure, 9.1, 12.4
themes, 12.2
timing, 9.6
venue, 12.4, 12.4.1, 12.4.2
Crick, Francis, 4.4, 5.14.1, 5.14.5, 5.14.7, 5.14.8, 5.14.9, Figure 5.17, 5.14.11, 6.1
and 'da Vinci problem', 6.3
criteria, financial, for evaluation, 14.4.7
cross-boundary collaboration, 18.3
cross-boundary conflict, resolution of, Figure 16.2, 18.4, Figure 18.1
cross-boundary sharing of people, knowledge, equipment, resources..., Figure 16.1, 17.6, Figure 17.1
cross-over, of chromosomes, 8.6, Figure 8.4

cryptography, quantum, 13.7
crystal, liquid, 5.5, 13.13
crystal, photonic, 13.12
crystallography, X-ray, 5.14.1, 5.14.11
cubism, 5.1
culture, cell, 13.12
culture, organisational, Chapter 16, Chapter 17, Chapter 18
and 'alignment' of enablers and motivators, 16.4, 18.6
amorphous nature of, 16.4, 18.6
and appetite for risk, Figure 16.1, 17.6, Figure 17.1
and appraisal, 16.4, 18.2
and attitude to challenge, Figure 16.1, 17.6, Figure 17.1
and attitude to 'failure', Figure 16.1, Figure 16.2, 17.6, Figure 17.1, 18.1, Figure 18.1
and behaviours, 16.1, Figure 16.1, 17.6, Figure 17.1, 18.1
and belief in importance of creativity and innovation, 17.2
and budgets, 16.4, Figure 16.1, 17.1, 17.2, Figure 17.1
command-and-control, 17.6
complexity of, 16.4, 18.6
and compliance with rules, 16.3, 16.7, 17.6
and cross-boundary sharing, Figure 16.1, 17.6, Figure 17.1
and curiosity, 16.3
and day-job, Figure 16.2, 18.5, Figure 18.1
difficulty of changing, 16.4, 18.6
and discipline, 17.6
and empowerment, 17.6
enablers, 16.4, Figure 16.1, Chapter 17, Figure 17.1
and encouragement of creativity in others, 16.4
and experimentation, 17.6
and 'failure', Figure 16.1, Figure 16.2, 17.6, Figure 17.1, 18.1, Figure 18.1

and fear, 16.3, 17.6
flexibility of, 17.6
and flexibility in forming
   teams, 17.6
and funding, Figure 16.1, 17.2,
   Figure 17.2
and hierarchy, Figure 16.1, 17.6,
   Figure 17.1
importance of, 1.2, 3.8
intangibility of, 16.4, 18.6
influence on evaluation, 3.8, 15.1.6
and language, 16.1, 16.2,
   Figure 16.1, 17.6, Figure 17.1
and managing development and
   implementation, 17.3
and managing projects, 17.3
motivators, 16.4, Figure 16.2,
   Chapter 18, Figure 18.1
need for breaking down into
   manageable 'pieces', 16.4
not a 'free-for-all', 16.2
and observation, 16.3
and organisational structure, 17.6
and 'outsiders', 16.3
and performance measures, 16.4,
   Figure 16.2, 18.3, Figure 18.1
and permission, 16.3
and physical environment, 15.4,
   Figure 16.1, 17.5, Figure 17.2
and project management, 17.3
and recognition, 16.4, Figure 16.2,
   18.1, 18.2, Figure 18.1
and remuneration, 18.1, 18.2
and return to 'line' from project
   teams, 17.6
and reward, 16.4, Figure 16.2, 18.1,
   18.2, Figure 18.1
and risk, 16.1, 17.1
and role of senior management, 17.2,
   Figure 16.2, 18.4, Figure 18.1
and rules, 16.2
and sharing knowledge, Figure 16.1,
   17.6, Figure 17.1
and sharing people, Figure 16.1,
   17.6, Figure 17.1

and sharing resources, Figure 16.1,
   17.6, Figure 17.1
starts at the top, 17.2, 18.3
and status, Figure 16.1, 17.6,
   Figure 17.1
and teams, 17.6
and training, 16.4, Figure 16.2, 18.3,
   Figure 18.1
culture change
   difficulty of, 16.4, 18.6
   importance of breaking down into
      manageable 'pieces', 16.4, 18.6
   leadership of, Figure 16.1, 18.4,
      Figure 18.1, 18.6
   'magnetic field' metaphor, 16.4
   must come from top, 18.3
'Cultures of Creativity' exhibition at
   Nobel Prize Museum, 5.14.4
curiosity, Prologue
   importance of as regards creativity,
      1.8, 2.6, 12.1, 16.3
   and *InnovAction!* process, 9.1
Curtis, Professor Adam, 13.12
cytosine, C, 5.14.1, 5.14.8, 5.14.9,
   5.14.10, 5.14.11, 5.15

Dalby, Professor Matthew, 13.12
Dalì, Salvador, 4.2
danger of being in 'love' with one's own
   ideas, Prologue, 8.5.1, 8.5.5, 14.3
Darkness at Noon, Chapter 4 (end)
Darwin, Charles, 1.4, 1.8, 5.8
da Vinci, Leonardo, 1.6, Chapter 6,
   Chapter 6 (end), 11.1.1
   helicopter design, 1.6, 6.1,
      Figure 6.1, 6.2
'da Vinci problem', 4.4, Chapter 6, 8.4,
   8.7, 13.13
   how to address, 6.6, 6.7, 6.8
   importance of interaction, 6.6
   importance of patience, 6.7
Davy, Sir Humphrey, 5.9
day-job, embedding creativity and
   innovation within, Figure 16.2,
   18.5, Figure 18.1

dBX SHIELD™, 13.11, Figure 13.18
DC electric motor, 5.8, 5.11, Figure 5.6
de Beers, 11.3
de Bono, Edward, 4.5, 8.4, 11.2.3, 11.4.1, 15.4
'debate, academic', 16.2
Debra, Professor Dan, 13.2
decision, Step 6 of wise evaluation process, 15.2.7
   importance of human judgement, 15.2.7
decision support system, 13.14
The Decline and Fall of the Roman Empire, 5.6
decomposing and recombining, springboard, 11.2.2
deconstruction, of existing patterns, 8.1, 8.4, 8.6
DeepOcean1, 13.3
del Verrochio, Andrea, Chapter 6 (end)
de Martinville, Édouard-Léon Scott, Chapter 5 (end)
de Moleyns, Frederick, 5.9
Dench, Dame Judy, 5.6
dendrite, 13.4, Figure 13.6
deoxyribose, 5.14.1, 5.15
Descartes, René, 5.6, 5.7, 5.8, 5.13
development, 2.5, 15.6
   budgets for, Figure 16.1, 17.1, Figure 17.1
   costs of, 15.2.6
   funding of, Figure 16.1, 17.2, Figure 17.1
   journey associated with, 15.2.4, 15.2.5
   management of, Figure 16.1, 17.3, Figure 17.1
   resources required for, 15.2.4, 15.2.6
   risk associated with, 15.2.4
   skills required, 2.6
   solutions to problems encountered during, 15.2.5
   as third of four processes for innovation, 2.1, 2.5, 15.6
   time required for, 15.2.6
diagram
   causal loop, 9.4.12

rose, 2.2, Figure 2.3
   Target, see *Target Diagram*
Dickens, Charles, 1.8
Diderot, Denis, 5.6
difference
   as distinct from novelty, 1.7, 10.4
   importance of, Prologue, 1.7, 1.8, 9.2, 9.5, 12.1
different ways of being different, 10.6, Figure 10.1
   sequence, flow and configuration, 10.6.2
   size and scale, 10.6.1
   function and scope. 10.6.3
   roles and responsibilities, 10.6.4
diffraction, 5.14.1
diffraction, X-ray, 5.14.1
diffraction pattern of DNA, 5.14.2, Figure 5.12, 5.14.3, 5.14.2, Figure 5.13, 5.14.3, 5.14.10, 6.1 5.14.10, 6.1
   and 'da Vinci problem', 6.3
diffraction pattern of a helix, 5.14.2, 5.14.10
   and 'da Vinci problem', 6.3
discovery, 1.4
disease, Alzheimer's, 8.2
dish, Petri, 13.12
display, liquid crystal (LCD), 13.13
DNA, structure of, 4.4, 5.8, 5.14, Figure 5.18, 5.15, 6.1
   as carrier of genetic information, 5.14.1, 5.14.8, 5.14.10, 5.15, 8.6
   and chromosomes, 8.6
   and 'da Vinci problem', 6.3, 6.4
   density measurement, 5.14.3, 5.14.6, 5.14.10
   double helix, 5.14.10
   'fingerprinting', 18.4
   'pile of pennies' suggestion, 5.14.3, 5.14.4, 5.14.9, 6.3, 6.4
   sequence of bases, 18.4
   triple helix suggestion, 5.14.7, Figure 5.16, 5.14.8, 5.14.10, 5.14.11, 6.4

X-ray diffraction pattern, 5.14.2, Figure 5.12, 5.14.3, 5.14.2, Figure 5.13, 5.14.3, 5.14.10, 6.1 5.14.10, 6.1
Dodgson, the Reverend Charles (Lewis Carroll), 5.1
domains, of creativity, Chapter 3
  easier and harder, 3.6, 3.7, 9.5, 9.7
Donohue, Jerry, 5.14.9, 5.14.10, 5.14.11
dopamine, 13.5, Figure 13.8
Doppler effect, 5.13
The Double Helix, 5.14.11
double helix structure of DNA, 5.14.1, 5.14.10, 6.1
DRAG© valve, 13.11, Figure 13.16, Figure 13.17
Drexel University, 4.5
Duchesne, Ernest, 2.1
Duckhouse, Rachel, 13.12, Figure 13.20
DVD, 13.12
Dylan, Bob, 4.4, 5.1
Dynamic Earth, Edinburgh, 13.6
dynamics, group, 9.4.2
dynamo, 5.11, Figure 5.7
  'Jumbo', 6.5

E (extroversion), Myers-Briggs Type Indicator, 15.2.7
The Eagle pub, Cambridge, 5.14.9, Figure 5.17
easel, artist's, 5.4
eccentricity, of planetary orbits, Prologue
economics, 5.6
ecosystem, 13.6
Edinburgh, University of, 5.6, 13.6
Edison, Thomas, Prologue, 1.4, 5.9, Chapter 5 (end), 17.6
  and 'da Vinci problem', 6.5
Edison Electric Light Company, Chapter 5 (end)
Edison-Swan company, 5.9
Einstein, Albert, 1.3, 1.4, 4.2, 5.13, 7.3, 13.2
effect, Doppler, 5.13

effect, 'Medusa', 13.9, Figure 13.14
effect, photo-electric, 5.13
EGaIn (eutectic gallium indium), 13.4, Figure 13.6
elastic tunnelling, 13.4
ElecLink interconnector, 13.3
electric light bulb, 5.8, 5.9, 17.6
electric motor, DC, 5.8, 5.11, Figure 5.6
Electrographic Vote Recorder, Chapter 5 (end)
electron, hot, 13.4, Figure 13.6
electron beam lithography, 13.12
elements, periodic table of, 7.3
Elia, transmission system operator in Belgium, 13.3
Ellington, Duke, 4.4
ellipsometry, 13.10, Figure 13.15
Elmer, Mike, 13.3
emergence, Chapter 7, 8.7
  and *InnovAction!* process, 9.1
'emperor's new clothes' questions, 12.3
empowerment, 17.6
enablers, cultural, of innovation, 16.4, Figure 16.1, Chapter 17, Figure 17.1, 18.6
  and behaviours, Figure 16.1, 17.6, Figure 17.1
  and budgets, Figure 16.1, 17.1, Figure 17.1
  and funding, Figure 16.1, 17.2, Figure 17.1
  and idea archive, Figure 16.1, 17.4, Figure 17.1
  and language, Figure 16.1, 17.6, Figure 17.1
  and managing development and implementation, Figure 16.1, 17.3, Figure 17.1
  and physical environment, 15.4, Figure 16.1, 17.5, Figure 17.5
encouragement of creativity in others, 16.4
The Encyclopedia of Chess Variants, 10.4

Encyclopédie, 5.6
endoscope, 13.7
enemies, of innovation, 8.5.5
ENGIE Fabricom, building contractor, 13.3
Enigma decipherment, 4.7
Enlightenment, Age of, 5.6
enol form of thymine and guanine, 5.14.9
entanglement, quantum, 13.7
entrepreneurship, 2.1
environment, physical, to encourage innovation, 15.4, Figure 16.1, 17.5, Figure 17.2
environmental science, 13.6
EQUANS, building contractor, 13.3
equations, Maxwell, 5.13
equipment, sharing of, 13.10
ERC (European Research Council), 13.12
error, human, 13.8
Eryops, 6.6
Essays on the Principles of Morality and Natural Religion, 5.6
etching, 13.12
ETH Zürich, 5.13
'Eureka' moment, Prologue, 4.2
European Research Council (ERC), 13.12
Eusthenopteron, 6.6
eutectic gallium indium (EGaIn), 13.4, Figure 13.6
evaluation, wise, 2.4, 7.6, Chapter 14, Chapter 15
  after idea generation, 12.9.3
  and arguments against, 14.4.6
  and arguments for, 14.4.6
  and balance, 15.1.4
  and bias, 14.4.6, 15.1.3, 15.1.4
  and 'big' ideas, 15.1.7, 15.2.1, 15.3, 15.5, Figure 15.2
  and boundaries, 14.4.7
  and Brexit, 15.2.2
  at creativity workshops, 12.4.3, 12.7
  benefits of idea, 15.2.3, 15.2.4, 18.3
  brief at creativity workshop, 12.5
  and completeness, 15.1.5
  and 'cons', 14.4.6
  consequences, Step 2 of process, 15.2.3, 15.2.4, 15.2.5, 15.2.6, 15.4
  and consistency, 15.1.2
  and cost-benefit analysis, 14.4.7
  and de Bono's *Six Thinking Hats*, 15.4, 15.5
  decision, Step 6 of process, 15.2.7
  and dislikes, 14.4.5
  distinct from idea generation, 10.4
  'evaluation lite', 15.5, Figure 15.2
  and evidence, 14.3
  examples of bad, 14.2, 14.4.6
  and fairness, 15.1.3
  features of a wise evaluation process, 15.1
  and feelings, 15.2.4, 15.4
  financial criteria, 14.4.7
  and financial pay-back, 14.4.7
  and forecasts, 14.4.7, 15.2.6
  half-way house, 15.2.7, 15.3
  and honesty, 15.2.6
  human judgements, 15.2.7
  imagine the future, Step 1 of process, 15.2.2, 15.2.6
  importance of, 2.4, 14.1
  influence of organisational culture on, 3.8
  and internal rate of return, 14.4.7
  and investment appraisal, 14.4.7
  issues to manage, 15.2.3, 15.2.4, 15.2.5, 18.3
  journey of development and implementation, Step 3 of process, 15.2.4, 15.2.5, 15.2.6, 15.4
  and language, 14.4.4
  later, not sooner, 12.7
  and 'likes', 14.4.5
  'lite', 15.5, Figure 15.2, 17.3, 17.4
  and management of risk, 14.1, 15.2.1
  and net present value, 14.4.7
  not at creativity workshops, 12.7
  numbers, Step 5 of process, 15.2.6, 15.4

often-used methods, 14.4
and openness, 15.1.9
and organisational culture, 3.8, 15.1.6
and organisational politics, 14.4.5, 15.2.2, 15.2.4, 15.2.6
outcomes, 15.2.7, 15.3
panel, 15.2.7
and people's feelings, 15.2.4, 15.4
and personal beliefs, 15.2.2
personal characteristics of wise evaluator, 15.2.7
and power, 14.4.3
and pragmatism, 15.1.7
and prejudices, 15.2.2
problems associated with idea, 15.2.3, 15.2.4, 15.2.5
process for, 2.4, Chapter 15, 15.2.1
and projections, 14.4.7
and 'pros', 14.4.6
and resources required for development and implementation, 15.2.4
and risk, 14.1, 15.2.1, 15.2.4
as second of four processes for innovation, 2.1, 2.4
and self-interest, 15.2.2
skills required, 2.6, 15.2.7
and 'small' ideas, 15.1.7
and speed, 15.1.6
and spreadsheets, 14.4.7
solutions to problems, 15.2.5
stage-gate process, 15.3, Figure 15.1
and *status quo*, 15.2.3, 15.2.4, 15.2.5
steps, 15.2, 15.2.2, 15.2.3, 15.2.4, 15.2.5, 15.2.6, 15.2.7
team roles, 15.2.7, 15.4
'the boss knows best', 14.4.2
training, 18.3
and transparency, 15.1.8
evaluation, wise, process for, Chapter 15, 15.2.1
balance within, 15.1.4
benefits of idea, 15.2.3, 15.2.4, 18.3
completeness of, 15.1.5
consequences associated with the idea, Step 2, 15.2.3, 15.2.4, 15.2.5, 15.2.6, 15.4
consistency of, 15.1.2
decision, Step 6, 15.2.7
'evaluation lite', 15.5, Figure 15.2, 17.3
fairness of, 15.1.3
features of, 15.1
and feelings, 15.2.4, 15.4
and forecasts, 15.2.6
half-way house, 15.2.7, 15.3, 15.6
and honesty, 15.2.6
and human judgement, 15.2.7
imagine the future, Step 1, 15.2.2, 15.2.3, 15.2.6
issues to manage, 15.2.3, 15.2.4, 15.2.5, 18.3
journey of development and implementation, Step 3, 15.2.4, 15.2.5, 15.2.6, 15.4
numbers, Step 5, 15.2.6, 15.4
openness of, 15.1.9
and organisational politics, 15.2.2, 15.2.4, 15.2.6
outcomes, 15.2.7, 15.3
and people's feelings, 15.2.4, 15.4
and personal beliefs, 15.2.2
pragmatism of, 15.1.7
and prejudices, 15.2.2
problems associated with idea, 15.2.3, 15.2.4, 15.2.5
process description, 15.2
relationship to de Bono's *Six Thinking Hats*, 15.4, 15.5
and resources required for development and implementation, 15.2.4
and risk, 15.2.4
and self-interest, 15.2.2
solutions to problems, 15.2.5
speed of, 15.1.6
stage-gate process, 15.3, 15.6, Figure 15.1
and *status quo*, 15.2.3, 15.2.4, 15.2.5
team roles, 15.2.7, 15.4
transparency of, 15.1.8

'evaluation lite', 15.5, Figure 15.2, 17.3, 17.4
evaluator, characteristics of, 15.2.7
Everitt, Professor Francis, 13.2
evidence, for wise evaluation
  absence of, 14.3
  of costs and revenues, 14.4.7
evolution by natural selection, 5.8
Exeter, University of, 13.10
expectation of outcomes, and 'failure', 17.6
experience, Chapter 8, 8.4, 8.7
experience, as the ability to feel things, 13.9
experiment, Michelson-Morley, 5.13
experimentation, 17.6
extinction spectrum, 13.10
extroversion (E), as in Myers-Briggs Type Indicator, 15.2.7

F (feeling), Myers-Briggs Type Indicator, 15.2.7
Fabry-Pérot resonance, 13.2, Figure 13.2
facilitation, of creativity workshop, 9.4.3
factors, human, 13.8
failure, Prologue
  and negligence, 17.6
  organisational attitude to, Figure 16.1, 17.6, Figure 17.1
  and outcome different from expectations, 17.6
  and reward and recognition, Figure 16.2, 18.1, Figure 18.1
fairness, of process for wise evaluation, 15.1.3
Faraday, Michael, 5.11, Figure 5.6
Farr, William, 2.2
'fat', in budgets, 17.1
fear, 8.5.3, 8.5.5, 16.3, 17.6
Federal Polytechnic Institute, Zürich, (ETH), 5.13
feeling (F), as in Myers-Briggs Type Indicator, 15.2.7
feelings, associated with an idea, 15.2.4, 15.4

ferrule, 5.4
fibroblast, 13.12
fibrular collagen, 13.12, Figure 13.19
Fichte, Johann, 8.6
Fields Medal, Prologue
filament, of electric light bulb, 5.9, 17.6
film, ultra-thin, 13.13
financial criteria for evaluation, 14.4.7
Finney, Sarah, 6.6
first rule, Chargaff's, 5.14.8
Fitzgerald, George, 5.13
flat screen, 13.13
Fleming, Sir Alexander, 2.1, 11.2.1
flexibility, of creative organisational cultures, 17.6
Florey, Baron Howard, 2.1
flow, turbulent, 13.11
fluctuations, quantum vacuum, 13.10
fluoropolymer, 13.5, Figure 13.9
flow, as different way of being different, 10.6.1
focus of attention, Step 1 of *InnovAction!* process, 9.1, 9.2, 9.3
Fontaine, Hippolyte, 5.11
football net, 4.6
footprint, carbon of science, reduction of, 13.10
Ford, Henry, 9.2
forecasts, and evaluation, 14.4.7, 15.2.6
FORESEEN network, 13.4
Forlanini, Enrico 6.1, Figure 6.2
forms, automorphic, 2.3
Formula 1 racing, 13.8
Fourier transform, 5.14.1
  of a helix, 5.14.5, 5.14.11
fractal, 7.5
fragments, of patterns, Prologue, see also *components, of Koestler's Law*
Francis I, King of France, Chapter 6 (end)
Franklin, Benjamin, 5.6, 5.9
Franklin, Rosalind, 5.14.2, 5.14.6, 5.14.8, 5.14.9, 5.14.10, 5.14.11, 6.1
Fredericks, Vsevolod, 5.5
free flight sales promotion, Hoover, 14.2, 14.3

French revolution, Chapter 6 (end)
friction, 13.5, 13.11
Friend, Peter, 6.6
Fuchsian functions, 2.3
function, as different way of being different, 10.6.1
function, Bessel, 5.14.5
functions, Fuschian, 2.3
funding, of creativity and innovation, 16.1, Figure 16.1, 17.2, Figure 17.2
Furberg, Sven, 5.14.4, 5.14.9, 5.14.10, 5.14.11
future, imagining of, Step 1 of wise evaluation process, 15.2.2, 15.2.3, 15.2.6
G, guanine, see *guanine*

Gadegaard, Professor Nikolaj, FRSE, 13.12
Galilean transformation, 5.13
Galilei, Galileo, Prologue, 4.1, 5.7, 5.8, 5.13
Galle, Johann, Prologue
Galway Statement, 13.6
Gaudí, Antoni, 5.10
Gauss, Carl Friedrich, 5.13
GCSE school examinations, 14.4.6
Geim, André, 1.8
General Electric, Chapter 5 (end)
General Motors, 6.7
general relativity, 7.3
genetic information, DNA as carrier of, 5.14.1, 5.14.8, 5.14.10, 5.15, 8.6
genetics, 8.6
de Gennes, Pierre-Gilles, 5.5
GEO600 interferometer, Hannover, 13.2
germanium – antimony – tellurium alloy, 13.13
Gershwin, George, 4.4
GGB LLC, 13.5
ghost imaging, 13.7
The Ghost in the Machine, 4.1, Chapter 4 (end)

Gibbon, Edward, 5.6
Gilbert, William, Prologue
Gladwell, Malcolm, 8.2
Glasgow, University of, 5.6, 5.14.5, 13.2, 13.7, 13.12
  Institute of Gravitational Research, School of Physics and Astronomy, 13.2
glass, stained, 13.4
gold nanorod, 13.4, Figure 13.6
Golding, William, 1.9
'goodness', of an idea, 7.6, 14.3, 15.2.2, 15.2.3, 15.2.7
Google, 1.9, 17.5
Gosling, Raymond, 5.14.2, 5.14.9, 5.14.10
Gramme, Zénobe, 5.11, Figure 5.7
Grand Prix, Malaysian, 2015, 13.8
graphene, 1.8
gravitation,
  inverse square law of, Prologue, 1.4, 5.7
  dispute over, 5.8
  Kepler's (flawed) inverse law, Prologue
  Newton's law of, 5.7
gravitational waves, 13.2
  Einstein's doubt, 13.2
  first detection of, 13.2, Figure 13.1
  prediction of, 5.13, 13.2
Gravity Probe B satellite, 13.2
Graz, Prologue
Great Barrier Reef, 13.6
green hat, as in *Six Thinking Hats*, 15.4
Greenland, 6.6
Griffith, Frederick, 5.13.1
group composition, in creativity workshops, 12.3
group dynamics, 9.4.2
group size, in creativity workshops, 12.3
guanine, G, 5.14.1, 5.14.8, 5.14.10, 5.14.11, 5.15
  enol and keto forms, 5.14.9
Gustafson, Eric, 13.2

H4 chronometer, Prologue
Habsburg dynasty, Prologue, Prologue (end), Chapter 4 (end)
haemoglobin, 5.14.6
'half-way house', in evaluation process, 15.2.7, 15.3
Hamilton, Lewis, 13.8
Hamlet, Prologue
Hanford interferometer, Washington State 13.2
hard wiring of neural networks, 8.2
Harrison, John, Prologue
Harry Potter, 1.9
Harvard University, Chapter 8 (end)
hats, coloured, as in *Six Thinking Hats*, 15.4, 15.5, 17.6, 18.3
Haydn, Josef, 1.4
Haynes, Colin, 13.8
He, Charlie, 13.14
'head start' programmes, Chapter 8 (end)
healthcare, 13.12
Heatley, Norman, 2.1
Hebb, Donald, 8.2, Chapter 8 (end)
  theory of learning, 8.2, 11.1.1, 13.4
  theory of learning, evidence for, 8.2
Hebb's law, of learning, Chapter 8 (end)
Hebb-Williams maze, Chapter 8 (end)
Hebbian learning, Chapter 8 (end)
Hebrides Terrace Seamount, 13.6
Hegel, Georg, 8.6
Heilmeier, George, 5.5
helical wavefront, 13.7
helicopter
  da Vinci's design, 6.1, Figure 6.1, 6.2
  Forlanini's design, 6.1, Figure 6.2
helix, α, 5.14.4, Figure 5.15, 5.14.7, 5.14.9, 5.14.11, 6.5
helix, double, structure of DNA, 5.14.10, 6.1
helix, Fourier transform of, 5.14.11
helix, triple, suggestion for DNA structure, 5.14.7, Figure 5.16, 5.14.8, 5.14.10, 5.14.11
helix, X-ray diffraction pattern of, 5.14.2, 5.14.4, 5.14.6, 5.14.10
Hendrix, Jimi, 7.5
heredity, and DNA, 5.14.1, 5.14.8, 5.14.10, 5.15
Heriot-Watt University, 13.6
Herodotus, 5.6
Herschel, William, Prologue
hierarchy, within creativity workshop, 9.5, 9.6
  cultural influence on innovation, Figure 16.1, 17.6, Figure 17.1
high-performing team, 3.6, 4.4, 9.3
hinge, 13.5
history, 5.6
Hockney, David, 1.8, 5.5, Figure 5.3
Holmes, Sherlock, 1.8
Home, Henry, Lord Kames, 5.6
homopolar DC electric motor, Figure 5.6
honesty, importance of in wise evaluation, 15.2.6
Hooke, Robert, 5.7, 5.8
Hooke's law, 5.8
Hoover, free flight sales promotion, 14.2, 14.3
Horizon 2020, 13.6
Horrocks, Jeremiah, 5.7
Hospital, St Mary's, London, 2.1
Hosseini, Peiman, 13.13
hot electron, 13.4, Figure 13.6
hotspot, of biodiversity, 13.6
Hough, Professor Sir James, OBE, FRS, 13.2
"How might [this] be different?", 1.8, 6.7, 9.5, 10.5, 16.2
  different ways of being different, 10.6
  Step 4 of ***InnovAction!*** process, 9.1, 9.5
How to Solve It, 11.1.1, 11.2.2, 11.4.6
How to Think Like da Vinci: Seven Steps to Genius Every Day, 11.1.1
Hughes Research Laboratory, California, 6.7
human error, 13.8
human factors, 13.8

human judgment, importance of in wise evaluation process, 15.2.7
Hume, David, 5.6
The Hunting of the Snark, 5.1
Hutcheson, Francis, 5.6
Hven, Prologue
hybrid materials, 13.10
hydrogen bond
  and α-helix, 5.14.4
  and enol and keto forms of thymine and guanine, 5.14.9
hyssop, 2.1

I (introversion), Myers-Briggs Type Indicator, 15.2.7
IBM, 5.5, 6.7
Ichthyostega, 6.6
idea, see also *creativity* and *creativity workshop*
  archive, 12.9.2, 15.5
  bad, 2.4, 7.6, 14.2, 14.3, 15.2.3, 15.2.4, 15.2.7, 15.3
  'being in love with one's own', Prologue, 8.5.1, 8.5.5, 14.3
  benefits of, 15.2.3, 15.2.4, 15.2.6, 15.4, 18.3
  'big', 15.1.7, 15.2.1, 15.5, Figure 15.2, 17.1, 17.2, 17.3, 17.4, 18.3
  budgets for, 17.1
  consequences of, Step 2 of wise evaluation process, 15.2.3, 15.2.4, 15.2.5, 15.2.6, 15.4
  and cost-benefit analysis, 15.2.6
  costs of development and implementation 15.2.6
  danger of being in 'love' with one's own ideas, Prologue, 8.5.1, 8.5.5, 14.3
  enrichment as a result of dialogue, 1.2
  evaluation of, 2.4, 7.6, 12.7, Chapter 14, Chapter 15
  evaluation 'lite', 15.5, Figure 15.2
  feelings associated with, 15.2.4, 15.4
  good, 2.4, 7.6, 14.3, 15.2.3, 15.2.4, 15.2.7, 15.3
  imagining of, Step 1 of wise evaluation process, 15.2.2
  importance of physical environment, 17.5
  and incubation, 2.3
  initial fragility, 1.2, 16.2
  intangible benefits of, 15.2.6
  and investment, 2.4
  issues to manage, 15.2.3, 15.2.4, 15.2.5, 15.4, 18.3
  journey associated with development and implementation, 15.2.4, 15.2.5
  and judgement, 12.7, see also *evaluation*
  as mental activity, 1.2
  not all are good, 14.3
  and organisational politics, 15.2.4
  origination of, 5.8
  'orphan', 17.1
  as outcome, 1.3
  and people's feelings, 15.2.4
  problems associated with, 15.2.3, 15.2.4, 15.2.5, 15.4, 18.3
  quality of, 4.7
  as question, 1.4
  resources required for development and implementation, 15.2.4, 15.2.6
  risk associated with, 17.1
  'small', 15.1.7
  solutions to problems encountered in development and implementation, 15.2.5
  time required for development and implementation, 15.2.6
  uncertainty associated with, 17.1
  and utility, 1.9, 7.6
  and value, 1.9, 7.6
  wise evaluation of, 2.4, 7.6, 12.7, Chapter 14, Chapter 15
idea archive, 12.9.2
idea generation, see *creativity* and *creativity workshop*
ideation, see *creativity* and *creativity workshop*
IDEO, 17.5
IFA2 interconnector, 13.3
The Ideas Book: 60 Ways to Generate Ideas More Effectively, 11.1.1

'Il faut penser à côté', 11.2.3
imagination, 2.6, 11.5
imagine the future, Step 1 of wise evaluation process, 15.2.2, 15.2.3, 15.2.6
imagineer, 2.6
imaging, ghost, 13.7
IMI CCI, 13.11
IMI Critical Engineering, 13.11
implementation, 2.5, 15.6
  costs of, 15.2.6
  definition, 2.1
  as fourth of four processes for innovation, 2.1, 2.5, 15.6
  journey associated with, 15.2.5
  management of, Figure 16.2, 17.3, Figure 17.1
  resources required for, 15.2.4, 15.2.6
  risk associated with, 15.2.4
  skills required, 2.6
  solutions to problems encountered during, 15.2.5
  time required for, 15.2.6
Impression, soleil levant, 5.1
impressionism, 5.1
improvement, continuous, Figure 16.2, 18.5, Figure 18.1
improvement, process, Figure 16.2, 18.5, Figure 18.1
improvement, quality, Figure 16.2, 18.5, Figure 18.1
incubation of ideas, 2.3, 7.6, 8.7
index, refractive, 13.10
induction training, Figure 16.2, 18.5, Figure 18.1
inelastic tunnelling, 13.4
information, genetic, 8.6
Ingres, Jean-Auguste, Chapter 6 (end)
injection moulding, 13.12, Figure 13.19
InnovAction! process for creativity, 6.1, 8.7, Chapter 9, 11.1, Chapter 12
  compared with challenging assumptions, 11.2.1
  compared with decomposing and recombining, 11.2.2
  compared with PO, 11.2.3
  compared with visioning, 11.4.5
  define focus of attention, Step 1, 9.1, 9.2
  describing what you know, 9.3, 10.2
  different ways of being different, 10.6
  example relating to detection of water leaks, 10.7
  example relating to finding collaborators, 10.7
  example relating to games based on chess, 10.1, 10.2, 10.3
  example relating to managing knowledge, 10.7
  example relating to 'Which team?', 10.7
  focus of attention, 9.2
  "How might this be different?", Step 4. 9.1, 9.5, 10.5, 10.6
  idea generation, 9.5, 9.6, 9.7, 10.3, 10.5
  and imagination, 11.5
  importance of difference, 9.2, 9.5, 12.1
  importance of teamwork, 6.1
  let it be, Step 5, 9.1, 9.6
  as put into practice in creativity workshops, Chapter 12
  repeat for another feature, Step 6, 9.1, 9.7
  share, Step 3, 9.1, 9.4, 10.2
  six steps, 9.1
  as a 'springboard', 11.1.2
  write down what you know, Step 2, 9.1, 9.3
  and imagination, 11.5
innovation
  belief in importance of, 17.2
  and content, 3.2
  and culture, 3.8, Chapter 16, Chapter 17, Chapter 18, see also *culture, organisational*
  'input' measures of, Figure 16.2, 18.2. Figure 18.1
  and making enemies, 8.5.5
  organisational priority of, 16.4

'output' measures of, Figure 16.2, 18.2. Figure 18.1
performance measures of, Figure 16.2, 18.2. Figure 18.1
and process, 3.3
and relationships, 3.6
risk associated with, 17.1
as a sequence of four processes, 2.1, 7.6
and strategy, 3.4
and structures, 3.5
uncertainty associated with, 17.1
and You!, 3.7, 9.2, 9.5, 9.7
Innsbruck, University of, 13.7
'input' measures of creativity and innovation, Figure 16.2, 18.2, Figure 18.1
An Inquiry into the Nature and Causes of the Wealth of Nations, 5.6
Insight and Outlook, 4.1, 4.5, Chapter 4 (end)
Installation for 300 speakers, Pianola and vacuum cleaner, 5.5
Institute of Gravitational Research, School of Physics and Astronomy, University of Glasgow, 13.2
Institute of Physics, Ukraine, 13.7
instrumentation, example creativity workshop brief, 12.5
insulin, 18.4
intangible benefit, of an idea, 13.7
intellectual property rights, 5.8
The Interaction of Colour, 7.5, Figure 7.2
interconnected of neural networks, 8.2, 8.3
interferometer, and gravitational waves, 13.2, Figure 13.2
interferometer, Michelson, 13.2, Figure 13.2
internal rate of return, financial criterion for evaluation, 14.4.7
INTP team role, 15.2.7
introversion (I), as in Myers-Briggs Type Indicator, 15.2.7
intuition (N), as in Myers-Briggs Type Indicator, 15.2.7
invariance, 5.13
invention, 1.4
inverse law of gravitation, Prologue
inverse square law
    of gravitation, Prologue, 1.4, 5.7
    dispute over, 5.8
    of light, Prologue
investment, and ideas, 24
investment appraisal, 14.4.7
'Invisible College', 5.8
iPad, 5.5
iPhone, 1.7
issues to manage, of an idea, 15.2.3, 15.2.4, 15.2.5, 15.4, 18.3
    solutions to, 15.2.5
IT system development, Figure 16.2, 18.5, Figure 18.1

J (judging), Myers-Briggs Type Indicator, 15.2.7
J-Power Systems, 13.3
Jarvik, Erik, 6.6
Jefferson, Thomas, 5.6
Jenkins, Professor Rob, 13.9
journey of development and implementation, Step 3 of wise evaluation process, 15.2.4, 15.2.5, 15.2.6, 15.4
journeying, retrofit, 11.4.4
judgement, of ideas, see *evaluation*
judging (J), as in Myers-Briggs Type Indicator, 15.2.7
'Jumbo' dynamo, 6.5

KAGRA interferometer, Japan, 13.2
Kahneman, Daniel, 15.2.7
Kames, Lord, Henry Home, 5.6
Kendrew, John, 5.14.6
Kepler, Johannes, Prologue, Prologue (end), 1.3, 1.4, 2.1, 2.3, 4.1, 4.2, 4.4, 4.5, 4.7, Chapter 4 (end), 5.7, 5.8, 5.14.2, 5.14.11, Chapter 6 (end)
and Platonic solids, Prologue, Figure P.1, 1.3, 5.8

keratin, 5.14.3, 5.14.4, 5.14.11, 6.5
α-keratin, 6.5
β-keratin, 6.5
keto form of thymine and guanine, 5.14.9
King's College London, 5.14.2, 6.1 13.4
   Department of Physics, 13.4
Kingstone, Professor Alan, 13.9
Kinnersley, Ebenezer, 5.9
knowledge, Prologue, Chapter 8, 8.4, 8.7
   and creativity, Prologue
   and organisational power, 18.2, 18.3
   paradox, Prologue
   sharing, across organisational boundaries, Figure 16.1, 17.6, Figure 17.1
knowledge management, ideas for, example of *InnovAction!* process, 10.7
knowledge transfer, Figure 16.1, 17.1, Figure 17.1
Koestler, Arthur, Prologue, 4.1, 4.5, 8.5.5
   definition of creativity, 4.1, 6.1
   Insight and Outlook, 4.1, Chapter 4 (end)
   The Act of Creation, 4.1, Chapter 4 (end)
   The Ghost in the Machine, 4.1, Chapter 4 (end)
   The Sleepwalkers, Prologue, 2.3, 4.1, Chapter 4 (end), 8.5.5
Koestler challenge, 4.8, 6.1
Koestler's Law of creativity, Chapter 4, 4.1, 4.5, 4.6, 6.1, 8.7, 10.4, Chapter 12
   and archive of ideas, 12.9.2, 17.4
   and 'da Vinci' problem, 6.3, 8.4
   and deconstruction of existing patterns, 8.1, 8.7
   DNA case study, 5.14.11
   and emergence, 7.1
   examples of, Chapter 5, Chapter 13

   and formation of new patterns, 8.1, 8.7
   and genetics, 8.6
   and "how might this be different?", 9.5
   and idea quality, 7.1
   and *InnovAction!* process, 9.1
   and lateral thinking, 11.2.3
   location of components, 8.1
   limitations of, 4.7, 7.1
   need for all components, 6.3
   problem of missing components, 6.3
   and random word process, 11.3
   and unlearning, 8.4
   and wise evaluation, 14.3
Kresge Auditorium, MIT, 5.12, Figure 5.11
Kroc, Ray, 2.1

'lab-on-a-chip', 13.4
Laguerre-Gaussian mode, 13.7
landscape metaphor for learning, 8.2, 8.3, 9.3, 9.4.1, 11.1.1
   and unlearning, 8.4, 9.3, 11.1.1
language, importance of, 12.7, 16.1, 16.2, Figure 16.1, 17.6, Figure 17.1
   emotional nuances, 15.4
   and wise evaluation, 14.4.4, 15.4
Large Hadron Collider, CERN, 13.2
Larmor, Joseph, 5.13
laser, 6.7, 13.7
Laser Interferometer Gravitational-wave Observatory, LIGO, 13.2
lateral thinking, 4.5, 11.2.3, 15.4
von Laue, Max, 5.14.1
law, Boyle's, 5.8
law, Hebb's, of learning, Chapter 8 (end)
law, Hooke's, 5.8
law, inverse of gravitation, Prologue
law, inverse square of gravitation, 1.4, 5.7,
   dispute over, 5.8

law, inverse square of light, Prologue, 5.7
Law, Koestler's of creativity, see *Koestler's Law of creativity*
laws of motion, Newton's, 5.7
   invariance of, 5.13
laziness, 8.5.2, 8.5.5, 17.6
LCD (liquid crystal display), 13.13
LCD screen, 13.13
leadership, Figure 16.2, 18.4, Figure 18.1
leaks, from water pipes, 10.7
lean production, Figure 16.2, 18.5, Figure 18.1
learning, Chapter 8, 8.7
   as barrier to creativity, 8.3
   and creativity, 2.6
   from mistakes, Prologue
   Hebb's law, Chapter 8 (end)
   Hebb's theory of, 8.2, 11.1.1, 13.4
   Hebbian, Chapter 8 (end)
   landscape metaphor, 8.2, 8.3, 8.4, 9.3, 9.4.1, 11.1.1
   neurophysiological basis of, 8.2
   and strengthening of neural networks, 8.2
   and unlearning, 8.4
   'valleys' of, 8.2, 8.3
   value of, 8.3
learning trap, 8.3, 12.3
LED, 5.5, 13.7
Leeds, University of, 5.14.1, 5.14.3, 5.14.6
Lego, 4.4, 4.7
Leibniz, Gottfried, 1.4, 5.8
Leiden, University of, 13.7
Lehmann, Otto, 5.5
Lennon, John, 5.1
let it be, Step 5 of *InnovAction!* process, 9.1, 9.6
Levi, Primo, 5.3
Lévi-Strauss, Claude, 1.3
light, 13.7
light, speed of, 5.13

light bulb, 5.8, 5.9, 6.5
LIGO, 13.2
Lincoln, Abraham, 1.3, 11.4.3
liquid crystal display (LCD), 13.13
'listeners', for detecting water leaks, 10.7
listening stick, for detecting water leaks, 10.7
liquid crystal, 5.5, 13.13
liquid crystal screen, 5.5
Lister, Joseph, 2.1
'lite', evaluation, 15.5, Figure 15.2, 17.3, 17.4
literature, 5.1
lithography, electron beam, 13.12
Livingstone interferometer, Louisiana, 13.2
lobster, squat, 13.6
Locke, John, 5.6
London, University of
   Birkbeck College, 5.14.4
   King's College, 5.14.2, 6.1, 13.4
London Centre for Nanotechnology, 13.4
longitude, measurement of, Prologue
Lophelia pertusa, 13.6, Figure 13.10
Lord of the Flies, 1.9
Lorentz, Hendrik, 5.13
Lorentz transformation, 5.13, 13.2
LoVe Observatory, 13.6
love, for your own ideas, danger of, Prologue, 8.5.1, 8.5.5, 14.3
'low priority', as in 'evaluation lite', 15.5, Figure 15.2
lubrication, 13.5
luck, Prologue

Mach, Ernst, 5.13
Machiavelli, Niccolò, 8.5.5
magic colouring sheet, 13.6, Figure 13.11
De Magnete, Prologue
Maiman, Theodore, 6.7
Malaysian Grand Prix 2015, 13.8
management, senior, role of, 17.2

management of development and implementation, Figure 16.1, 17.3, Figure 17.1
managing projects, 17.3, Figure 16.2, 18.3, Figure 18.1
Manchester, University of, 5.6
Marconi, Guglielmo, 1.4
Marine Simulation, 13.6
Mars, orbit of, Prologue, 1.3, 2.1, 2.3, 4.4
Martin, Archer, Chapter 5 (end), 7.3
maser, 6.7
materials, hybrid, 13.10
materials science, 13.10
Mathematical Creation, 2.3
Mathematical Principles of Natural Philosophy (*Principia Mathematica*), 5.7, 5.8
Massachusetts Institute of Technology, 5.12
mauveine, Prologue, 5.4
Maxwell, James Clerk, 5.13, 13.7
Maxwell equations, 5.13
maze, Hebb-Williams, Chapter 8 (end)
MC (merocyanine), Figure 13.15
McCartney, Paul, 5.1
McDonalds, 2.1
McGehee, Professor Mike, 13.13
McGill University, Chapter 8 (end)
The Mechanism of Mind, 8.4, 11.2.3, 15.4
Medical Research Council (MRC), 18.4
Medical Research Council Biophysics Unit, King's College London, 5.14.2
Medici family, Chapter 6 (end)
Medusa, 13.9
'Medusa' effect, 13.9, Figure 13.14
measures, of performance, 16.4. Figure 16.2, 18.2. Figure 18.1
meiosis, 8.6, Figure 8.4
memory, 13.4
memristor, 13.4
Mendeleev, Dmitri, 7.3
Menlo Park laboratory, Chapter 5 (end)
mental 'landscape', 8.2
Merck, 2.1
merocyanine (MC), Figure 13.15
Messersmith, Professor Phillip, 13.5
de Mestral, George, 4.5
metamaterial, 13.4, Figure 13.6
metaphor, retrofit, 11.4.2
'meteor cards', 10.5.4, 12.4.2, 12.8, 12.9.1
Michaelangelo, 5.4
Michelson, Albert, 5.13, 13.2
Michelson-Morley experiment, 5.13, 13.2
microlithograhy, 13.12
microscope, 13.7
Middleton, Adam, 13.3
mind-map, 9.3
Mingulay, 13.6
missing component, example creativity workshop brief, 12.5
mistakes
  and learning, Prologue
  learning from, Prologue
  and negligence, Prologue
MIT, 5.12, 13.2
mobile phone, 13.6, 13.13
modelling, optical transfer matrix, 13.13
modulus, Youngs, see *Young's modulus*
mollusc, 13.5
momentum, orbital angular, 13.7
Mona Lisa, Chapter 6 (end)
Mondrian, Piet, 5.2, Figure 5.1
Monet, Claude, Prologue, 5.2, 5.4, Figure 5.2, 5.10
Monitor-Evaluator team role, 15.2.7
Monte Carlo simulation, 13.14
Montessori methods of schooling, Chapter 8 (end)
Morley, Edward, 5.13, 13.2
Morris, Richard, 8.2
Moscow, University of, 13.2
motion, Brownian, 5.13, 13.2

motion, Newton's laws of, 5.7, 7.3
  invariance of, 5.13
motivators, cultural, of innovation,
    16.4, Figure 16.2, Chapter 18,
    Figure 18.1
  and the day job, Figure 16.2, 18.5,
    Figure 18.1
  and performance measures,
    Figure 16.2, 18.2, Figure 18.1
  and recognition, Figure 16.2, 18.1,
    Figure 18.1
  and reward, Figure 16.2, 18.1,
    Figure 18.1
  and role of senior management,
    Figure 16.2, 18.4, Figure 18.1
  and training, 15.4, 17.5, Figure 16.2,
    18.3, Figure 18.1
motor, DC electric, 5.8, 5.11,
    Figure 5.6
mould, healing properties of, 2.1
moulding, injection, 13.12,
    Figure 13.19
mountains and valleys metaphor for
    learning, 8.2, 8.3, 9.3, 9.4.1, 11.1.1
Mozart, Wolfgang Amadeus, 1.4, 8.2
MRC (Medical Research Council),
    5.14.2, 18.4
Murphy, building contractor, 13.3
music, 4.4, 6.1
  and emergent patterns, 7.3
  and Koestler's law components, 8.1
  and subjectivity, 7.5, 7.6
mussel, 13.5
multi-tasking, dangers of, 13.8
Myers-Briggs (personality) Type
    Indicator, 15.2.7
myoglobin, 5.14.6
Mysterium Cosmographicum, Prologue

N (intuition), Myers-Briggs Type
    Indicator, 15.2.7
Naismith, James, 4.6
nanolithography, 13.12
nanopattern, 13.12
nanorod, 13.4, Figure 13.6

nanostructures, 13.12
nanotechnology, 13.4
Napoleon, 5.10
National Grid plc, 13.3
Nature, 5.14.3, 5.14.10
Nature Materials, 13.12
negligence, 17.6
Nemo high voltage power connection
    link, 13.3
net, football, 4.6
net present value (NPV), financial cri-
    terion for evaluation, 14.4.7
network, neural, 8.2, 13.4
neural network, 8.2, 13.4
  hard wiring of, 8.2
  interconnectedness of, 8.2, 8.3
  need to 'unwire' to enable creativity,
    8.3
neuron, 8.2, 8.3
'neurons that fire together wire
    together', Chapter 8 (end)
neurophysiology, 8.2, 13.4
neuroplasticity, 8.2
neuroscience, 13.4
neurotransmitter, 8.2, 13.4, Figure 13.6
new, 1.6
  as distinct from different, 1.7, 1.8
  difficulties associated with, 1.6
  missing from Koestler's Law, 4.4
new product development, 18.5
Newton, Isaac, Prologue, 1.4, 2.1, 4.1
  funeral of, and influence on Voltaire,
    5.6
  inverse square law of gravitation,
    1.4, 5.7
  laws of motion, 5.7, 7.3
  invariance of, 5.13
  standing on giants' shoulders, 2.1
  visible spectrum, 13.7
Newton's rings, 13.13
Newtonian transformation, 5.13
Nicholas II, Tsar of Russia, Prologue
Nicholson, John, 6.6
Nightingale, Florence, 2.1
nine dots puzzle, 1.8, Figure 1.1, 9.8

Nobel Peace Prize, 5.14.1, 5.14.11
Nobel Prize, Prologue, Chapter 5 (end), 8.2, 15.2.2
  in Chemistry, 5.14.4, 5.14.6, 7.3, 18.4
  in Economics, 15.2.7
  in Literature, 1.9, 4.1, 4.4
  in Physics, 1.8, 5.5, 5.13, 5.14.1
  in Physiology or Medicine, 2.1, 5.14.1, 8.2
Nobel Prize Museum, 'Cultures of Creativity', 5.14.4
noise, reduction of, 13.11
Nokia, 1.7
NOPO, 11.2.3
Norse mythology, 9.2
North Sea Link interconnector, 13.3
Northwestern University, 13.5
novelty, see *new*
NPV (net present value), financial criterion for evaluation, 14.4.7
nucleotide base, 5.13.1, 5.15
  sequence of in DNA, 18.4
numbers, Step 5 of wise evaluation process, 15.2.6, 15.4

observation, Prologue
  and discovery of penicillin, 2.1
  importance of as regards creativity, 1.8, 2.1, 9.3, 12.1, 12.3, 16.3
  and *InnovAction!* process, 9.1, 9.3
octopus, 13.6
Odin, 9.2
Office of Qualifications and Examinations Regulation (Ofqual), 14.4.6
Ofqual (Office of Qualifications and Examinations Regulation), 14.4.6
  example of unwise evaluation, 14.4.6
oil click, colours of, 13.13
Olosuga, Professor David, 5.6
openness, of process for wise evaluation, 15.1.9
Opera House, Sydney, 5.8, 5.12, Figure 5.8, Figure 5.9, Figure 5.10

operation, provocative, see *PO*
operations improvement, example creativity workshop brief, 12.5
optical cavity, 13.10
optical transfer matrix modelling, optics, 13.7
orbital angular momentum, 13.7
The Organisation of Behaviour, 8.2, Chapter 8 (end)
organisational boundaries, sharing across, Figure 16.1, 17.6, Figure 17.1
organisational culture, see *culture, organisational*
organisational politics, influencing wise evaluation, 14.4.5, 15.2.2, 15.2.4, 15.2.6
organisational rules, 16.2
organisational structure, and creative culture, 17.6
L'Oréal, 11.3
original, 5.8
origination of ideas, 5.8
'orphan' idea, 17.1
other people's shoes, retrofit, 11.4.3
Otho, Valentinus, Prologue
outcome, expectation of, and 'failure', 17.6
Outliers, 8.2
'output' measures of creativity and innovation, Figure 16.2, 18.2, Figure 18.1
outreach, 13.6
'outsiders', contribution to creativity, 16.3
Oxford, Begbroke Science Park, 13.13
Oxford, University of, 5.1, 5.14.4, 13.13

P (perceiving), Myers-Briggs Type Indicator, 15.2.7
Padgett, Professor Miles, OBE, FRS, FRSE, 13.7
Page, Larry, 1.9
PageRank algorithm, 1.9

Paine, Tom, 5.6
paint, 5.4
paint brush, 5.4
paint tube, 5.4, 6.5
Palo Alto, 13.2
participants, at creativity workshops, 12.3
partition chromatography, 7.3
parts, and wholes, 7.5
Pasteur, Louis, 2.1
Pasteur Institute, Chapter 5 (end)
patent
   for bar code, 6.7
   for electric light bulb, 5.9, Figure 5.4, 6.5
   for football net, 4.6
   and Thomas Edison, Chapter 5 (end)
patterns
   deconstruction of existing, 8.1, 8.6, 8.7
   and emergence, 7.1, 7.2, 7.3
   importance of as regards creativity, Prologue, 4.4, 6.1
   quality of, 4.7, 7.1
   recombination to form new, 8.6, 8.7
   and subjectivity, 7.5, 7.6
patterns, stress in materials, 13.7
Patterson, Emma, 13.6
Pauling, Linus, 5.14.4, Figure 5.15, 5.14.7, 5.14.8, 5.14.9, 5.14.10, 5.14.11, 6.1, 6.5
   and 'da Vinci problem', 6.3, 6.4
pay-back, financial criterion for evaluation, 14.4.7
Pearl Street power station, New York, 6.5
people, sharing across organisational boundaries, Figure 16.1, 17.6, Figure 17.1
penicillin, 2.1, 11.2.1
Penicillium notatum, 2.1
pennies, pile of, suggested structure for DNA, 5.14.3, Figure 5.13, 5.14.4, 5.14.9, 6.3, 6.4, 6.5
Pennsylvania, University of, 13.13
Pennsylvania Railroad, 6.7

people's feelings, associated with an idea, 15.2.4, 15.4
perceiving (P), as in Myers-Briggs Type Indicator, 15.2.7
performance measures,
   of creativity and innovation, Figure 16.2, 18.2. Figure 18.1
   organisational, 16.4, Figure 16.2, 18.2. Figure 18.1
perihelion, Prologue
The Periodic Table, 5.3
periodic table of the elements, 7.3
Perkin, William, Prologue, 5.4
permission, Prologue, 3.8, 12.1, 16.3
   and *InnovAction!* process, 9.1
Perseus, 13.9
personal innovation, 3.7
Peruggia, Vincenzo, Chapter 6 (end)
Perutz, Max, 5.14.6
Petri dish, 13.12
Petty, Sir William, 5.6, 5.8
Pfizer, 2.1
phase transition, topological, 13.10
philosophy, 5.6
phonautograph, Chapter 5 (end)
phone, mobile, 13.6, 13.13
phone, smart, 13.7, 13.13
phonograph, Chapter 5 (end)
'Photo 51' X-ray diffraction pattern of DNA, 5.14.2, Figure 5.12, 5.14.3, 5.14.6, 5.14.7, 5.14.8, 5.14.9, 6.1
photo-electric effect, 5.12
photon, 13.7, 13.10
photonic crystal, 13.12
photonics, 13.4, 13.7
physical environment, to encourage innovation, 15.4, Figure 16.1, 17.5, Figure 17.2
physics and art, 5.5
Piano, Renzo, Chapter 4 (end)
Picasso, Pablo, 1.4, 5.1
pigments, 5.2, 5.4, 5.5, 13.4
'pile of pennies', suggested structure for DNA, 5.14.3, Figure 5.13, 5.14.4, 5.14.9

and 'da Vinci problem', 6.3, 6.4, 6.5
Pissarro, Camille, 5.4
pixel, 7.5, 13.7, 13.13
planetary sphere, Prologue, 1.3, Figure P.1
planning, business, Figure 16.2, 18.5, Figure 18.1
plant science, Sainsbury Laboratory, 17.5
plasma treatment, 13.5
plasmon, 13.4, Figure 13.6
plasmonics, 13.4, 13.7
plasticity, brain, 8.2
Plato, Prologue
Platonic solids, Prologue, Figure P.1, 1.3, 1.9, 2.3, 5.8
Playfair, William, 2.2
PLH (poly-L-histidiine), 13.4, Figure 13.6
PO, springboard, 11.2.3, 11.4.1
  PO-1, 11.2.3
  PO-2, 11.2.3, 11.4.1
  PO-3, 11.2.3
PO: A Device for Successful Thinking, 11.2.3
PO: Beyond Yes and No, 11.2.3, 15.4
Poincaré, Henri, 2.3, 5.13, 13.2
pointillism, 7.5
polariser, 13.13
polariton, 13.10
politics, 5.6
politics, organisational, influencing wise evaluation, 14.4.5, 15.2.2, 15.2.4, 15.2.6
Pollock, Jackson, 7.5
poly-L-histidine (PLH), 13.4, Figure 13.6
Pólya, George, 11.1.1, 11.2.2, 11.4.6
Pomuk, Orhan, Chapter 4 (end)
Popper, Karl, Chapter 4 (end)
power, organisational, 8.5.4, 18.2, 18.3
pragmatism, of process for wise evaluation, 15.1.7
Prague, Prologue

prejudice, effect on wise evaluation, 15.2.2
Pride and Prejudice, 5.1
The Prince, 8.5.5
Principia Mathematica, 5.7, 5.8
Principles of Philosophy, 5.7
Prize, Nobel, see *Nobel Prize*
Prize, Sterling, for architecture, 17.5
problem, absence of, Prologue, 9.2
problem, da Vinci, see '*da Vinci problem*'
problem solving, 9.2
problem statement, 9.2
problems, associated with an idea, 15.2.3, 15.2.4, 15.2.5, 15.4, 18.3
  solutions to, 15.2.5
process
  for idea generation, Chapter 9, 9.1
  for wise evaluation, Chapter 15, 15.2
process improvement, Figure 16.2, 18.5, Figure 18.1
process innovation, 3.3, 9.2, 12.5, 16.1
  example creativity workshop brief, 12.5
product development, new, 18.5
product improvement, example creativity workshop brief, 12.5
production, lean, Figure 16.2, 18.5, Figure 18.1
project management, 17.3, Figure 16.2, 18.3, Figure 18.1
project team
  disbanding of, when project ends, 17.6
  formation, 17.6
projections, and evaluation, 14.4.7
promotion, Hoover's free flight sales, 14.2, 14.3
promotion, within organisation, Figure 16.2, 18.1, 18.2, Figure 18.1
'pros' of an idea, 14.4.6
protein, 13.4, 13.5, 13.12, 18.4
protein, amyloid, 8.2
prosthaphaeresis, Prologue

provocative operation, PO, springboard, see *PO*
PTFE, 13.5, Figure 13.7, Figure 13.9
Ptolemy, 13.7
pulsar, 11.2.1
purine, 5.14.1
puzzle, nine dots, 1.8, Figure 1.1, 9.8
puzzle, simplest overall shape, 8.4, Figure 8.1, Figure 8.2, Figure 8.3, 8.5
puzzle, to illustrate difficulty of, 8.4, Figure 8.1, Figure 8.2, Figure 8.3, 8.5
pyrimidine, 5.14.1

quality of idea, 4.7, 7.1
  and emergence, 7.1
quality improvement, Figure 16.2, 18.5, Figure 18.1
quantum computing, 13.7
quantum cryptography, 13.7
quantum entanglement, 13.7
quantum teleportation, 13.7
quantum vacuum fluctuations, 13.10
'quick win', as in 'evaluation lite', 15.5, Figure 15.2, 17.1, 17.4
quinine, Prologue, 5.4
QuiverVision, 13.6

Räikkönnen, Kimi, 13.8
railway, safety on, 13.8
Rand, John, 5.4, 6.5
random words, retrofit, 11.3, 11.4.1
Raphael, 5.4
rate of return, internal, criterion for evaluation, 14.4.7
ray, Richardson's, 13.6, Figure 13.11
RCA laboratories, 5.5
reactions, chemical, control of, 13.10
Reason, Professor James, 13.8
receptor, of neurotransmitter, 13.4, Figure 13.6
recognition, 16.4, Figure 16.2, 18.1, 18.2, Figure 18.1

recombination of chromosomes, 8.6
re-combination, of elements to form new patterns, 8.1, 8.6
recombining, and decomposing, springboard, 11.2.2
red hat, as in *Six Thinking Hats*, 15.4, 15.5, 17.6
redfish, 13.6
reduction of noise, 13.11
reef, coral, 13.6
reef, cold-water, 13.6, Figure 13.10
Reef, Great Barrier, 13.6
Regensburg, Prologue
Reinitzer, Friedrich, 5.5
reflection, 13.7
reflection spectrum, 13.10
refraction, 13.7
refractive index, 13.10, 13.13
relationship innovation, 3.6, 9.2, 9.5, 9.7, 12.5
relativity, general, 7.3
relativity, special, 5.8, 5.13, 7.3, 13.2
Rembrandt van Rijn, 1.4
remote operating vehicle (ROV), 13.6
remuneration, 18.1, 18.2
Renoir, Pierre-Auguste, 5.4
repeat for another feature, Step 6 of ***InnovAction!*** process, 9.1, 9.7
reproduction, sexual, 8.6
resonator, tuneable, 13.13
resonance, Fabry-Pérot, 13.2, Figure 13.2
resources
  allocated to development and implementation, 17.3
  required for the development and implementation of an idea, 12.6, 14.1, 15.2.1, 15.2.4, 15.2.6
  sharing of, 13.10, Figure 16.1, 17.6, Figure 17.1
respect, within creativity workshop, 9.5
responsibility, as different way of being different, 10.6.1

retrofit processes for idea generation, 11.1.1, 11.3, 11.4, 11.5
    choice as regards springboards, 11.5
    journeying, 11.4.4
    other people's shoes, 11.4.3
    PO-2, 11.4.1
    random words, 11.3
    simile, metaphor and analogy, 11.4.2
    visioning, 11.4.5
    working backwards, 11.4.6
return, internal rate of, financial criterion for evaluation, 14.4.7
revolution, French, Chapter 6 (end)
reward, 16.4, Figure 16.2, 18.1, 18.2, Figure 18.1
Rheticus, Georg, Prologue
Richardson's ray, 13.6, Figure 13.11
Riede, Professor Moritz, 13.13
rights, intellectual property, 5.8
rings, Newton's, 13.13
risk, Prologue, 2.4, 8.3, 16.1, 16.3
    appetite, for, Figure 16.1, 17.6, Figure 17.1
    as managed by wise evaluation, 14.1, 15.2.1, 15.2.4
    associated with innovation, 17.1
Roberts, Professor J Murray, 13.6
Rockall Bank, 13.6, Figure 13.10
role, as different way of being different, 10.6.1
roles, in teams, 15.2.7
Romans, 13.11
rose diagram, 2.2
ROV (remote operating vehicle), 13.6
Rowan, Professor Sheila, CBE FRS, 13.2
Rowlandson, Thomas, 4.6, Figure 4.1
Rowling, JK, 1.9
Royal Ballet, 8.2
Royal Institute of British Architects, 17.5
Royal Institution, 5.9
Royal Society, 5.8
ruby laser, 6.7
Rudolf II, Holy Roman Emperor, Prologue, Prologue (end)
Rudolphine Tables, Prologue
rule, Chargaff's first, 5.14.8, 5.14.10
rules
    compliance with, 16.3, 16.7, 17.6
    for evaluation of ideas, 2.4
    importance of keeping to, Prologue
    organisational, 16.2, 16.3
    within a creative culture, 17.6
Russell, Bertrand, Chapter 4 (end)

S (sensing), Myers-Briggs Type Indicator, 15.2.7
Saarinen, Eero, 5.12, Figure 5.11
safety, on railways, 13.8
'safety', within creativity workshop, 9.5
Sainsbury Laboratory of Plant Science, Cambridge, 17.5
St Andrews, University of, 13.7
St Mary's Hospital, London, 2.1
sales promotion, Hoover's free flight, 14.2, 14.3
Sanger, Fred, 18.4
scale, as different way of being different, 10.6.1
Scandal in Bohemia, 1.8
Schweitzer, Albert, Chapter 4 (end)
Science, 13.9
science, carbon footprint, reduction, 13.10
science, environmental, 13.6
science, materials, 13.10
science centre, 13.6
The Science of Mechanics, 5.13
scope, as different way of being different, 10.6.1
screen, flat, 13.13
screen, liquid crystal, 5.5, 13.13
screw, Archimedes, 6.2
La Seine à la Grande Jatte, 7.5, Figure 7.1
Self, Dick, 13.11
self-facilitation, of creativity workshop, 9.4.3
self-interest, effect on wise evaluation, 15.2.2

Semens-Flanagan, Mike, 13.11
semiconductor technology, 13.12
senior management
   and creativity training, Figure 16.2, 18.3, Figure 18.1
   role of, 17.2, Figure 16.2, 18.3, 18.4, Figure 18.1
sensing (S), as in Myers-Briggs Type Indicator, 15.2.7
sequence, as different way of being different, 10.6.1
sequence
   of amino acids in insulin, 18.4
   of bases in DNA, 18.4
Serious Creativity, 11.2.3, 15.4
Seurat, Georges, 7.5, Figure 7.1
sexual reproduction, 8.6
Sforza family, Chapter 6 (end)
Shakespeare, William, Prologue, 1.4, 5.1
share, Step 3 of ***InnovAction!*** process, 9.1, 9.4, 10.1, 10.2
sharing
   across organisational boundaries, Figure 16.1, 17.6, Figure 17.1
   of equipment and resources, 13.10, Figure 16.1, 17.6, Figure 17.1
   of knowledge, Figure 16.1, 17.6, Figure 17.1
Sheffield, Lucius, 5.4
Sheffield, Dr Washington, 5.4
'shelve', as in 'evaluation lite', 15.5, Figure 15.2, 17.4
shoes, other people's, retrofit, 11.4.3
Siemens Energy, 13.3
silica suspension, use in detecting gravitational waves, 13.2, Figure 13.3
Silver, Bernard, 4.5, 6.7
simile, retrofit, 11.4.2
simplest overall shape puzzle, 8.4, 8.5
simulation, Monte Carlo, 13.14
Sinclair, Sir Clive, 1.9
single-pixel camera, 13.7

sintered bronze, 13.5, Figure 13.7, Figure 13.9
Sisley, Alfred, 5.4
Six Sigma, 18.5
Six Thinking Hats, 15.4
size, as different way of being different, 10.6.1
skills, required for creativity and innovation, 2.6
The Sleepwalkers, Prologue, Chapter 4 (end), 8.5.5
slogans, advertising, retrofit, 11.3
'small' ideas, 15.1.7
smart phone, 13.7, 13.13
Smith, Adam, 5.6
soap bubble, colours, 13.13
solutions, Step 4 of wise evaluation process, 15.2.3, 15.4
solutions to problems associated with an idea, 15.2.5
solving problems, 9.2
Sonning Prize, Chapter 4 (end)
Souriau, Paul, 4.5, 11.2.3
Sousa, John Philip, 4.4
Spark Creativity: 50 Ways to Ignite Bright Ideas, 11.1.1
special relativity, 5.8, 5.13, 7.3, 13.2
Spectacular! app, 13.6
spectrum, extinction, 13.10
spectrum, reflection, 13.10
speed, of process for wise evaluation, 15.1.6
speed of light, 5.13
sphere, celestial, Prologue, 1.3, Figure P.1
sphere, planetary, Prologue, 1.3, Figure P.1
SPI (spiropyran), Figure 13.15
spiropyran (SPI), Figure 13.15
spreadsheet, potential use (and misuse) in evaluation, 14.4.7, 15.2.6, 17.2
Sprengel, Hermann, 5.9
springboard processes for idea generation, 11.1, 11.2, 11.6
   challenge assumptions, 11.2.1

choice as regards retrofits, 11.6
decomposing and recombining, 11.2.2
InnovAction!, 11.1.2
PO, 11.2.3
squat lobster, 13.6
stage gate process, 15.3, 15.6, Figure 15.2
stained glass, 13.4
Stanford University, 13.2, 13.13
Starr, John, 5.9
Statement, Galway, 13.6
statement, of problem to be solved, 9.2
status, cultural influence on innovation, Figure 16.1, 17.6, Figure 17.1
status quo, as 'base line' for wise evaluation, 15.2.3, 15.2.4, 15.2.5
stem cell, 13.12
Sterling Prize for architecture, 17.5
Stjerneborg, Prologue
strategy, example creativity workshop brief, 12.5
strategy innovation, 3.4, 9.2, 12.5
stress patterns in materials, 13.7
strong coupling, 13.10, Figure 13.15
structure, organisational
   example creativity workshop brief, 12.5
   and creative culture, 17.6
structure innovation, 3.5, 12.5
subjectivity, as regards interpretation of emergent patterns, 7.5, 7.6
Subterranean Homesick Blues, 4.4, 5.1
Sumitomo, 13.3
SUMPAC, 1.6
Sutherland, Dr Justin, 13.14
Swan, Sir Joseph, 5.9, 6.5
   patent for light bulb, 5.9, Figure 5.4
Swan Lake, 8.2
Sydney Opera House, 5.8, 5.12, Figure 5.8, Figure 5.9, Figure 5.10
synapse, 8.2, 8.3
synapse, synthetic, 13.4
Synge, Richard, Chapter 5 (end), 7.3
synthesis, and thesis and antithesis, 8.6

synthetic synapse, 13.4
system, decision support, 13.14
system, IT, development of, Figure 16.2, 18.5, Figure 18.1
systems thinking, 9.4.12, Figure 16.2, 18.3, Figure 18.1

T (thinking), Myers-Briggs Type Indicator, 15.2.7
T, thymine, see *thymine*
Target Diagram, 2.1, 3.1, Figure 2.1, Figure 2.2, Figure 2.4, Figure 3.1, Figure 3.2, 7.6
   in context, Figure 2.4, Figure 3.2
   and domains of creativity, Chapter 3, Figure 3.1, 9.5, 9.7
   easier and harder domains, 3.6, 3.7, 9.5, 9.7
   and focus of attention, 9.2
   as a fractal, 2.2, Figure 2.2
   and organisational culture, 3.8, Figure 3.2
   richer picture, 2.2, Figure 2.2
   and skills, 2.5
task switching, 13.8
team roles, 15.2.7, 15.4
team, high-performing, 3.6, 4.4, 9.3
team, project
   disbanding of, when project ends, 17.6
   formation, 17.6
teamwork, 1.7, 3.6, 9.3
   example creativity workshop brief, 12.5, 12.6
   ideas for, examples, 9.3
   and *InnovAction!* process, 9.1, 10.4, 12.2
   'Which team?', example of *InnovAction!* process, 10.7
   'Which team?', example creativity workshop brief, 12.5
technology, semiconductor, 13.12
technology, valve, 13.11
teleportation, quantum, 13.7
television, 13.7, 13.13

tellurium – antimony – germanium alloy, 13.13
tenacity, Prologue, 2.1
tensor, 5.13
themes, for creativity workshops, 12.2
Theory of Modern Sentiments, 5.6
theory of learning, Hebb's, 8.2, 11.1.1
thesis, antithesis, synthesis, 8.6
thinking (T), as in Myers-Briggs Type Indicator, 15.2.7
'thinking aside', 4.5, 11.2.3
Thinking, Fast and Slow, 15.2.7
thinking, lateral, 4.5, 11.2.3, 15.4
thinking, systems, 9.4.12, Figure 16.2, 18.3, Figure 18.1
thinking, vertical, 11.2.3
Thirty Years War, Prologue, Prologue (end)
Through the Looking Glass, 5.1
Thucydides, 5.6
thymine, T, 5.14.1, 5.14.8, 5.14.10, 5.14.11, 5.15
  enol and keto forms, 5.14.9
time required for development and implementation of an idea, 15.2.6
Titus Andronicus, 5.1
toothpaste tube, 5.4
'top-down' budgets, 17.1
'top-down' organisational culture, 17.6
topological phase transition, 13.10
Townes, Charles, 6.7
training
  in creativity tools and techniques, 12.3, 16.4, Figure 16.2, 17.1, Figure 17.1, 18.3, Figure 18.1
  induction, Figure 16.2, 18.5, Figure 18.1
  of senior management too, Figure 16.2, 18.3, Figure 18.1
transfer, of knowledge, Figure 16.1, 17.1, Figure 17.1
transform, Fourier, 5.14.1
transformation, Galilean, 5.13
transformation, Lorentz, 5.13
transformation, Newtonian, 5.13
transformation, Voigt, 5.13
transistor, 13.12
transit of Venus, 5.7
transition, topological phase, 13.10
transparency, of process for wise evaluation, 15.1.8
trap, learning, 8.3, 12.3
tribological agent, 13.5
triple helix hypothesis for structure of DNA, 5.14.7, Figure 5.16, 5.14.8, 5.14.10, 5.14.11
  and 'da Vinci problem', 6.3, 6.4
tube, cathode ray, 13.13
tube, paint, 5.4, 6.5
tube, toothpaste, 5.4
Tübingen, University of, Prologue (end)
tuneable resonator, 13.13
tunnelling, 13.4, Figure 13.6
turbulent flow, 13.11
Turing, Alan, 4.7
Turner, JMW, 5.2
TWA flight centre, New York, 5.12, Figure 5.11
Two Treatises on Government, 5.6

U, uracil, 5.15
Ukraine, Institute of Physics, 13.7
ultra-thin film, 13.13
uncertainty, if innovation, 17.1
'unintended consequences', 2.6
University of Berlin, Chapter 5 (end), 8.6
University of British Columbia, 13.9
University of California, Berkeley, 13.5
University of Cambridge, 5.14.1, 5.14.5, 5.14.6, 5.14.8, 5.14.9, 5.14.11, 6.1, 18.4
  Sainsbury Laboratory of Plant Science, 17.5
University of Chicago, Chapter 8 (end)
University of Copenhagen, Prologue (end), 5.14.6
University of Colorado, 13.7
  Boulder, 13.13

University of Edinburgh, 5.6, 13.6
University of Exeter, 13.10
University of Glasgow, 5.6, 5.14.5,
    13.2, 13.7, 13.12
  Institute of Gravitational Research,
    School of Physics and Astronomy,
    13.2
University of Innsbruck, 13.7
University of Leeds, 5.14.1, 5.14.3,
    5.14.6
University of Leiden, 13.7
University of London
  King's College, 5.14.2, 6.1, 13.4
  Birkbeck College, 5.14.4
University of Manchester, 5.6
University of Oxford, 5.1, 5.14.4, 13.13
University of Moscow, 13.2
University of Pennsylvania, 13.13
University of St Andrews, 13.7
University of Tübingen, Prologue (end)
University of Vienna, Chapter 5 (end)
University of York, 13.9
University of Zürich, 5.13
University, Drexel, 4.5
University, Harvard, Chapter 8 (end)
University, Heriot-Watt, 13.6
University, McGill, Chapter 8 (end)
University, Northwestern, 13.5
University, Stanford, 13.2, 13.13
University, Yale, Chapter 5 (end)
unlearning, 8.4, 8.7, 10.4
  difficulty of, 8.5.5, 9.6
  and *InnovAction!* process, 9.1, 9.3
  key principle underpinning, 8.5
  puzzle to illustrate difficulty of, 8.4,
    Figure 8.1, Figure 8.2, Figure 8.3,
    8.5
'unwiring' of neural circuits, 8.3
Up at a Villa – Down in the City, 5.1
uracil, U, 5.15
Uranienborg, Prologue (end)
US Declaration of Independence, 5.6
US Department of Agriculture, 2.1
The Use of Lateral Thinking, 11.2.3,
    15.4

utility, of an idea 1.9, 7.6
Utzon, Jørn, 5.12, Figure 5.9

vaccine, Covid-19 development, 16.1
vacuum fluctuations, quantum, 13.10
'valleys', of learning, 8.2, 8.3, 9.3,
    9.4.1, 11.1.1
  and unlearning, 8.4, 9.3, 11.1.1
value, of an idea, 1.9, 7.6, 17.2
value, net present (NPV), financial
    criterion for evaluation, 14.4.7
valve technology, 13.11
valve, ball, 13.11
valve, dBX SHIELD^{TM}, Figure 13.18
valve, DRAG$^{©}$, 13.11, Figure 13.16,
    Figure 13.17
van Eyck, Jan, 5.4, Figure 5.2
Van Gogh, Vincent, 1.8, 5.4
Vand, Vladimir, 5.14.5, 5.14.11
Vasari, George, Chapter 6 (end)
velcro, 4.5
Venus
  orbit of, Prologue
  transit of, 5.7
le Verrier, Urbain, Prologue
vertical thinking, 11.2.3
Vienna, University of,
    Chapter 5 (end)
Viking Link interconnector, 13.3
Virgo interferometer, Pisa, 13.2
visioning, retrofit, 11.4.5, 11.4.6
Voigt, Woldemar, 5.13
Voigt transformation, 5.13
Voltaire (François-Marie Arouet), 5.6
von Oech, Roger, 11.1.1
Vörlander, Daniel, 5.5

Wallace, Alfred, 5.8
War, Thirty Years, Prologue, Prologue
    (end)
Water ARC$^{®}$, 13.14
water leaks, ideas for detection,
    example of *InnovAction!*
    process, 10.7
Watson, Dr John, 1.8

Watson, James, 4.4, 5.14.2, 5.14.6,
    5.14.7, 5.14.8, 5.14.9, Figure 5.17,
    5.14.10, 5.14.11, 6.1
  and 'da Vinci problem', 6.3
wavefront, helical, 13.7
waves, gravitational, see *gravitational waves*
weak coupling, 13.10
Weiss, Rai, 13.2
Wellcome Sanger Institute, 18.4
West Orange laboratory, Chapter 5 (end)
white hat, as in *Six Thinking Hats*, 15.4
whole greater than sum of parts, 7.5, 13.6
  example creativity workshop brief, 12.5
Sir William Dunn School of Pathology, University of Oxford, 5.1
Wilkins, Maurice, 5.14.1, 5.14.2, 5.14.6, 5.14.8, 5.14.10, 5.14.11, 6.1
Wilkinson, Professor Chris, 13.12
Williams, Charles, 13.5
Williams, Kenneth, Chapter 8 (end)
win, quick, 15.5, Figure 15.2, 17.1, 17.4
wisdom, 2.6, 15.2.6
wise evaluation, see *evaluation*
Woodland, Joe, 4.5, 6.7
wool, 5.14.3, 6.5
words, random, retrofit, 11.3, 11.4.1
work, hard, Prologue
working backwards, retrofit, 11.4.6
workshop, creativity, see *creativity workshop*
Wright, Orville, 1.6

write down what you know, Step 2 of ***InnovAction!*** process, 9.1, 9.3, 10.1
Wynne, John, 5.5

X-ray crystallography, 5.14.1, 5.14.11
X-ray diffraction, 5.14.1
X-ray diffraction pattern of DNA, 5.14.2, Figure 5.12, Figure 5.14, 5.14.4, 5.14.6, 6.1
  first, 5.14.3, Figure 5.13
X-ray diffraction pattern of a helix, 5.14.2, 5.14.4, 5.14.6

Yahoo, 1.9
Yale University, Chapter 5 (end)
yellow hat, as in *Six Thinking Hats*, 15.4, 15.5, 18.3
Young's modulus
  electrically induced changes, 13.13
  of silica fibre, 13.2, 13.13
York, University of, 13.9
You!, 3.7, 9.2, 9.5, 9.7
Yousafzai, Malala, 5.14.1

Zayats, Professor Anatoly, 13.4
Zeilinger, Anton, 13.7
zeitgeist, 5.8, 5.14.4
Zola, Émile, 1.4
Zürich, ETH (Federal Polytechnic Institute), 5.13
Zürich, University of, 5.13

Ørsted, Hans Christian, 5.11